Proceedings of GeoShanghai 2018 International Conference: Ground Improvement and Geosynthetics

Lin Li · Bora Cetin · Xiaoming Yang

Editors

Proceedings of GeoShanghai 2018 International Conference: Ground Improvement and Geosynthetics

 Springer

Editors
Lin Li
Department of Civil
   and Environmental Engineering
Jackson State University
Jackson, MS
USA

Xiaoming Yang
Oklahoma State University
Oklahoma, OK
USA

Bora Cetin
Iowa State University
Ames, IA
USA

ISBN 978-981-13-0121-6   ISBN 978-981-13-0122-3   (eBook)
https://doi.org/10.1007/978-981-13-0122-3

Library of Congress Control Number: 2018939621

Printed on acid-free paper

This Springer imprint is published by the registered company Springer Nature Singapore Pte Ltd. part of Springer Nature
The registered company address is: 152 Beach Road, #21-01/04 Gateway East, Singapore 189721, Singapore

Preface

The 4th GeoShanghai International Conference was held on May 27–30, 2018, in Shanghai, China. GeoShanghai is a series of international conferences on geotechnical engineering held in Shanghai every four years. The conference was inaugurated in 2006 and was successfully held in 2010 and 2014, with more than 1200 participants in total. The conference offers a platform of sharing recent developments of the state-of-the-art and state-of-the-practice in geotechnical and geoenvironmental engineering. It has been organized by Tongji University in cooperation with the ASCE Geo-Institute, Transportation Research Board, and other cooperating organizations.

The proceedings of the 4th GeoShanghai International Conference include eight volumes of over 560 papers; all were peer-reviewed by at least two reviewers. The proceedings include Volumes 1: Fundamentals of Soil Behavior edited by Dr. Annan Zhou, Dr. Junliang Tao, Dr. Xiaoqiang Gu, and Dr. Liangbo Hu; Volume 2: Multi-physics Processes in Soil Mechanics and Advances in Geotechnical Testing edited by Dr. Liangbo Hu, Dr. Xiaoqiang Gu, Dr. Junliang Tao, and Dr. Annan Zhou; Volume 3: Rock Mechanics and Rock Engineering edited by Dr. Lianyang Zhang, Dr. Bruno Goncalves da Silva, and Dr. Cheng Zhao; Volume 4: Transportation Geotechnics and Pavement Engineering edited by Dr. Xianming Shi, Dr. Zhen Liu, and Dr. Jenny Liu; Volume 5: Tunneling and Underground Construction edited by Dr. Dongmei Zhang and Dr. Xin Huang; Volume 6: Advances in Soil Dynamics and Foundation Engineering edited by Dr. Tong Qiu, Dr. Binod Tiwari, and Dr. Zhen Zhang; Volume 7: Geoenvironment and Geohazards edited by Dr. Arvin Farid and Dr. Hongxin Chen; and Volume 8: Ground Improvement and Geosynthetics edited by Dr. Lin Li, Dr. Bora Cetin, and Dr. Xiaoming Yang. The proceedings also include six keynote papers presented at the conference, including "Tensile Strains in Geomembrane Landfill Liners" by Prof. Kerry Rowe, "Constitutive Modeling of the Cyclic Loading Response of Low Plasticity Fine-Grained Soils" by Prof. Ross Boulanger, "Induced Seismicity and Permeability Evolution in Gas Shales, CO_2 Storage and Deep Geothermal Energy" by Prof. Derek Elsworth, "Effects of Tunneling on Underground Infrastructures" by Prof. Maosong Huang, "Geotechnical Data Visualization and Modeling of Civil

Infrastructure Projects" by Prof. Anand Puppala, and "Probabilistic Assessment and Mapping of Liquefaction Hazard: from Site-specific Analysis to Regional Mapping" by Prof. Hsein Juang. The Technical Committee Chairs, Prof. Wenqi Ding and Prof. Xiong Zhang, the Conference General Secretary, Dr. Xiaoqiang Gu, the 20 editors of the 8 volumes and 422 reviewers, and all the authors contributed to the value and quality of the publications.

The Conference Organizing Committee thanks the members of the host organizations, Tongji University, Chinese Institution of Soil Mechanics and Geotechnical Engineering, and Shanghai Society of Civil Engineering, for their hard work and the members of International Advisory Committee, Conference Steering Committee, Technical Committee, Organizing Committee, and Local Organizing Committee for their strong support. We hope the proceedings will be valuable references to the geotechnical engineering community.

Shijin Feng
Conference Chair
Ming Xiao
Conference Co-chair

Organization

International Advisory Committee

Herve di Benedetto	University of Lyon, France
Antonio Bobet	Purdue University, USA
Jean-Louis Briaud	Texas A&M University, USA
Patrick Fox	Penn State University, USA
Edward Kavazanjian	Arizona State University, USA
Dov Leshchinsky	University of Illinois, USA
Wenhao Liang	China Railway Construction Corporation Limited, China
Robert L. Lytton	Texas A&M University, USA
Louay Mohammad	Louisiana State University, USA
Manfred Partle	KTH Royal Institute of Technology, Switzerland
Anand Puppala	University of Texas at Arlington, USA
Mark Randolph	University of Western Australia, Australia
Kenneth H. Stokoe	University of Texas at Austin, USA
Gioacchino (Cino) Viggiani	Université Joseph Fourier, France
Dennis T. Bergado	Asian Institute of Technology, Thailand
Malcolm Bolton	Cambridge University, UK
Yunmin Chen	Zhejiang University, China
Zuyu Chen	Tsinghua University, China
Jincai Gu	PLA, China
Yaoru Lu	Tongji University, China
Herbert Mang	Vienna University of Technology, Austria
Paul Mayne	Georgia Institute of Technology, USA
Stan Pietruszczak	McMaster University, Canada
Tom Papagiannakis	Washington State University, USA
Jun Sun	Tongji University, China

Scott Sloan   University of Newcastle, Australia
Hywel R. Thomas   Cardiff University, UK
Atsashi Yashima   Gifu University, Japan

Conference Steering Committee

Jie Han   University of Kansas, USA
Baoshan Huang   University of Tennessee, USA
Maosong Huang   Tongji University, China
Yongsheng Li   Tongji University, China
Linbin Wang   Virginia Tech, USA
Lianyang Zhang   University of Arizona, USA
Hehua Zhu   Tongji University, China

Technical Committee

Wenqi Ding (Chair)   Tongji University, China
Charles Aubeny   Texas A&M University, USA
Rifat Bulut   Oklahoma State University, USA
Geoff Chao   Asian Institute of Technology, Thailand
Jian Chu   Nanyang Technological University, Singapore
Eric Drumm   University of Tennessee, USA
Wen Deng   Missouri University of Science and Technology,
                                       USA
Arvin Farid   Boise State University, Idaho, USA
Xiaoming Huang   Southeast University, China
Woody Ju   University of California, Los Angeles, USA
Ben Leshchinsky   Oregon State University, Oregon, USA
Robert Liang   University of Dayton, Ohio, USA
Hoe I. Ling   Columbia University, USA
Guowei Ma   Hebei University of Technology, China
Roger W. Meier   University of Memphis, USA
Catherine O'Sullivan   Imperial College London, UK
Massimo Losa   University of Pisa, Italy
Angel Palomino   University of Tennessee, USA
Krishna Reddy   University of Illinois at Chicago, USA
Zhenyu Yin   Tongji University, China
ZhongqiYue   University of Hong Kong, China
Jianfu Shao   Université des Sciences et Technologies
                                       de Lille 1, France
Jonathan Stewart   University of California, Los Angeles, USA

Wei Wu	University of Natural Resources and Life Sciences, Austria
Jianhua Yin	The Hong Kong Polytechnic University, China
Guoping Zhang	University of Massachusetts, USA
Jianmin Zhang	Tsinghua University, China
Xiong Zhang (Co-chair)	Missouri University of Science and Technology, USA
Yun Bai	Tongji University, China
Jinchun Chai	Saga University, Japan
Cheng Chen	San Francisco State University, USA
Shengli Chen	Louisiana State University, USA
Yujun Cui	École Nationale des Ponts et Chaussees (ENPC), France
Mohammed Gabr	North Carolina State University, USA
Haiying Huang	Georgia Institute of Technology, USA
Laureano R. Hoyos	University of Texas at Arlington, USA
Liangbo Hu	University of Toledo, USA
Yang Hong	University of Oklahoma, USA
Minjing Jiang	Tongji University, China
Richard Kim	North Carolina State University, USA
Juanyu Liu	University of Alaska Fairbanks, USA
Matthew Mauldon	Virginia Tech., USA
Jianming Ling	Tongji University, China
Jorge Prozzi	University of Texas at Austin, USA
Daichao Sheng	University of Newcastle, Australia
Joseph Wartman	University of Washington, USA
Zhong Wu	Louisiana State University, USA
Dimitrios Zekkos	University of Michigan, USA
Feng Zhang	Nagoya Institute of Technology, Japan
Limin Zhang	Hong Kong University of Science and Technology, China
Zhongjie Zhang	Louisiana State University, USA
Annan Zhou	RMIT University, Australia
Fengshou Zhang	Tongji University, China

Organizing Committee

Shijin Feng (Chair)	Tongji University, China
Xiaojqiang Gu (Secretary General)	Tongji University, China
Wenqi Ding	Tongji University, China
Xiongyao Xie	Tongji University, China

Yujun Cui	École Nationale des Ponts et Chaussees (ENPC), France
Daichao Sheng	University of Newcastle, Australia
Kenichi Soga	University of California, Berkeley, USA
Weidong Wang	Shanghai Xian Dai Architectural Design (Group) Co., Ltd., China
Feng Zhang	Nagoya Institute of Technology, Japan
Yong Yuan	Tongji University, China
Weimin Ye	Tongji University, China
Ming Xiao (Co-chair)	Penn State University, USA
Yu Huang	Tongji University, China
Xiaojun Li	Tongji University, China
Xiong Zhang	Missouri University of Science and Technology, USA
Guenther Meschke	Ruhr-Universität Bochum, Germany
Erol Tutumluer	University of Illinois, Urbana—Champaign, USA
Jianming Zhang	Tsinghua University, China
Jianming Ling	Tongji University, China
Guowei Ma	Hebei University of Technology, Australia
Hongwei Huang	Tongji University, China

Local Organizing Committee

Shijin Feng (Chair)	Tongji University, China
Zixin Zhang	Tongji University, China
Jiangu Qian	Tongji University, China
Jianfeng Chen	Tongji University, China
Bao Chen	Tongji University, China
Yongchang Cai	Tongji University, China
Qianwei Xu	Tongji University, China
Qingzhao Zhang	Tongji University, China
Zhongyin Guo	Tongji University, China
Xin Huang	Tongji University, China
Fang Liu	Tongji University, China
Xiaoying Zhuang	Tongji University, China
Zhenming Shi	Tongji University, China
Zhiguo Yan	Tongji University, China
Dongming Zhang	Tongji University, China
Jie Zhang	Tongji University, China
Zhiyan Zhou	Tongji University, China
Xiaoqiang Gu (Secretary)	Tongji University, China
Lin Cong	Tongji University, China
Hongduo Zhao	Tongji University, China

Fayun Liang   Tongji University, China
Bin Ye   Tongji University, China
Zhen Zhang   Tongji University, China
Yong Tan   Tongji University, China
Liping Xu   Tongji University, China
Mengxi Zhang   Tongji University, China
Haitao Yu   Tongji University, China
Xian Liu   Tongji University, China
Shuilong Shen   Tongji University, China
Dongmei Zhang   Tongji University, China
Cheng Zhao   Tongji University, China
Hongxin Chen   Tongji University, China
Xilin Lu   Tongji University, China
Jie Zhou   Tongji University, China

Contents

About the Editors

Lin Li received his PhD from University of Wisconsin-Madison, with a focus on Geotechnical Engineering. He has been Geotechnical Laboratory Director in the Department of Civil and Environmental Engineering, Jackson State University since 2005. He is currently a Fellow of American Society of Civil Engineers.

Bora Cetin completed his PhD in Geotechnical Engineering at the University of Maryland, College Park. He is currently an Assistant Professor in the Geotechnical and Geoenvironmental Engineering Area in the Department of Civil, Construction, and Environmental Engineering at Iowa State University.

Xiaoming Yang completed his PhD in Geotechnical Engineering at the University of Kansas. He is currently an Assistant Professor at Oklahoma State University. He is a Member of American Society of Civil Engineers.

Tensile Strains in Geomembrane Landfill Liners

R. Kerry Rowe[(⊠)] and Yan Yu

Department of Civil Engineering, GeoEngineering Centre at Queen's-RMC,
Queen's University, Kingston, ON K7L 3N6, Canada
{kerry.rowe,yan.yu}@queensu.ca

Abstract. A geomembrane (GMB) liner is a key component of the barrier system in many modern engineered landfills. In combination with a clay liner, the GMB minimizes contaminant migration to groundwater and surface water. GMBs in landfill applications are mostly made from high-density polyethylene (HDPE). When in contact with landfill leachate, the HDPE GMBs experiences significant aging and loss of mechanical properties with time. In particular, a loss in stress crack resistance combined with excessive tensile stress/strain can result in GMB cracking and ultimately failure. Thus, to ensure good long-term performance, the maximum tensile strain sustained by an HDPE GMB should be limited to an acceptable level. Both the local GMB indentations induced by gravel in an overlying drainage layer or underlying clay liner and the down-drag load in the GMB on side slopes with settlement of the waste can cause significant tensile strains in HDPE GMBs. This paper reviews key research examining tensile strains developed in GMBs from both sources.

Keywords: Geomembranes · Strains · Landfills · Indentations
Side slopes

1 Introduction

Modern engineered landfills generally require a barrier system below the waste to minimize contaminant escape to the groundwater and surface water, and therefore to reduce the potential impacts on the human health and surrounding environment [31]. A barrier system consists of a high permeable leachate collection system (LCS) and a low permeability liner system. As part of a landfill composite liner, high-density polyethylene (HDPE) geomembranes (GMBs) are excellent barriers for harmful inorganic substances (e.g., heavy metals) typically found in landfills, and when combined with an underlying geosynthetic clay liner or compacted clay liner can perform their intended functions extremely well [29]. However, with time, HDPE GMBs will experience a loss of their mechanical properties [18, 27, 32–34]. When a GMB degradation is such that it can no longer resist the tensile strains/stresses, fully penetrating cracks [1, 12] allow the escape of leachate. Once this escape exceeds allowable design values, the GMB is considered to have reached the end of its service-life [31].

Two key sources that have the potential to cause significant tensile strains in GMBs are: (a) local indentations in the GMB induced by the overlying drainage materials

© Springer Nature Singapore Pte Ltd. 2018
L. Li et al. (Eds.): GSIC 2018, *Proceedings of GeoShanghai 2018 International Conference: Ground Improvement and Geosynthetics*, pp. 1–10, 2018.
https://doi.org/10.1007/978-981-13-0122-3_1

[4, 5, 30] and/or gravel in the underlying clay liner [6], and (b) down-drag load for GMBs on side slopes generated by waste settlement [2, 14, 15, 19, 21, 22, 37, 38, 40, 42, 44–46].

Short-term punctures can generally be minimized by providing sufficient protection to the GMB liner [9, 17, 23, 24, 28, 39], the magnitude of tensile strains that a GMB can sustain without compromising their intended long-term performance reported in the literature varies. To avoid premature GMB failure due to stress cracking, Seeger and Müller [36] indicated that the GMB strain should be less than 3%. Based on the GMBs examined under the simulated field conditions in the geosynthetic liner longevity simulator (GLLS) cells [1, 12, 30], it can be inferred that sustained tensions that induce tensile strains greater than 4–5% should be avoided by the use of a suitable protection layer [1, 12, 30], eliminating potentially problematic gravel from the upper layer of a clay liner or other subgrade, and designing to limit strains from other sources (such as down-drag by waste placement and subsequent settlement/degradation).

Giroud et al. [16] reported that strain concentrations in the vicinity of seams can give rise to failure adjacent to seams. Laboratory testing reported by Kavazanjian et al. [20] indicated that the stain magnification induced by a seam was even greater than estimated by Giroud et al. [16].

The objective of this paper is to summarize the research related to the generation of tensile strains in GMBs used in landfill liners associated with the local GMB indentations induced by granular materials and the down-drag load for GMBs on side slopes due to waste settlement.

2 GMB Tensile Strains from Indentations

2.1 GMB Indentations and Laboratory Testing Methods

A protection layer is required between the GMB and the drainage layer to prevent the short-term puncture of the GMB [24, 28] and to minimize the long-term tensile strains in the GMB [1, 12, 30]. Laboratory testing to establish the short-term (typically in a 10- to 100-h sustained pressure test) GMB tensile strains developed with a proposed protection layer over the GMB is currently the most feasible way to qualify the efficiency of protection layers in term of reducing/limiting the GMB tensile strains [7, 35].

Standard laboratory test methods [3, 10] can be used to examine the short-term GMB tensile strains with and without a protection layer. Large-scale laboratory test apparatus has also been developed to examine the influence of different protection layers on the GMB strains [4, 5, 9, 35]. A very thin lead sheet is used beneath the GMB to record the GMB deformations in these laboratory tests.

2.2 Strain Calculation Methods

The magnitude of the calculated GMB strains based on the indentations recorded in the lead sheet is highly dependent on the method used to calculate the strain [3, 8, 11, 25, 39]. The local membrane strain is calculated by fitting a circular segment to the indentation in the lead sheet used in the BAM [8] and ASTM D5514 [3] approaches (noting that there is an error in the equation given in ASTM D5514 [3]). The LEF-2 [25] strain calculation method estimates the incremental strains for each 3-mm segment of the measurement axes of the indentation. Recognizing that a 1.5 mm-thick (or greater) GMB has bending as well as membrane strains, an improved alternative approach to calculate the incremental strains was proposed in Tognon et al. [39] using the vertical deformed GMB profile recorded in the lead sheet to assess both membrane and bending incremental strains. All these methods err in underestimating the strains since they neglect the horizontal displacements; the Tognon et al. [39] method being the best of the methods based on vertical displacement. Eldesouky and Brachman [11] have recently proposed an alternative method to calculate the GMB incremental strains considering both the vertical and radial displacements of the GMB under the axisymmetric conditions. However, the method presented by Eldesouky and Brachman [11] is only suitable for the axisymmetric conditions and is not yet suitable to be used for GMBs under the gravel drainage layer for landfill applications.

2.3 Influence of Gravel Size on Maximum GMB Tensile Strain

Laboratory experiments reported by Brachman and Gudina [4] for two poorly graded and angular gravels (nominal grain sizes of 25 mm and 50 mm) directly on a 1.5 mm-thick GMB (i.e., no protection layer) over a compacted clay liner at an applied pressure of 250 kPa for 10 h (21 ± 2 °C). The average spacing between the gravel contacts was reported to be 37 mm with a maximum GMB tensile strains of 16% for the 25-mm gravel and 55 mm with maximum GMB tensile strains of 32% for the 50-mm gravel. These strains are unacceptable in landfill applications, necessitating the inclusion of a protection layer between the GMB and overlying gravel layer to limit the GMB tensile strains as discussed below.

2.4 Influence of Geotextile Protection on Maximum GMB Tensile Strain

The physical response of a 1.5-mm thick, HDPE GMB beneath the 50-mm gravel with and without a geotextile protection layer was reported by Brachman and Gudina [5], where the GMB was overlying a needle-punched geosynthetic clay liner (GCL) on a firm foundation layer and the gravel was subjected to an applied pressure of 250 kPa at 21 ± 1 °C.

Based on the physical testing results, the maximum GMB strain was 17% without a geotextile protection layer. Thus, increasing the stiffness of the foundation (a compacted clay liner [4] versus a GCL on the firm foundation [5]) reduced the maximum GMB strain from 32% [4] to 17% [5] when the other conditions were identical.

The use of a geotextile protection layer between the GMB and gravel resulted in smaller GMB strain. For similar needle punched geotextiles, the larger the mass per unit area of the geotextile, the smaller the GMB strain. When using a geotextile with the mass per unit area of 2200 g/m^2 (the highest among three geotextiles tested [5]), the GMB strains were just below 6% compared to 17% without a geotextile protection layer. However, even with a geotextile protection layer, the GMB strain is still considered too large because these strains are for the short-term loading conditions. For the long-term field conditions with elevated temperatures and chemical exposure in a landfill, GMB strains greater than 6% are expected [1, 12].

2.5 Influence of Alternative Protection Layers on Maximum GMB Tensile Strain

A 150-mm thick layer of sand as the protection layer on the GMB was also examined [4] under the same testing conditions as the geotextile protection layers. The use of a 150-mm-thick sand layer limited the maximum GMB tensile strain to less than 0.2%. Thus, the sand protection layer (150 mm thick) was very effective in terms of providing protection to the GMB under 50-mm gravel at the vertical pressure of 250 kPa. Laboratory experiments on sand, geocomposites, geonet, and rubber tire shreds as alternative protection layers for a composite GMB-GCL landfill barrier beneath 50-mm gravel [9] also confirmed that of all the protection layers examined, sand was the most effective at limiting strains, although a geocomposite was more effective than the traditional nonwoven.

2.6 Influence of Time and Temperature on Maximum GMB Tensile Strain

The laboratory experiments [4, 5, 9] were based on the applied pressure of 250 kPa held for a duration of 10 h at room temperature (21–22 °C). However, under the field conditions in the landfill, the GMB is expected to be loaded for much longer and subjected to higher temperatures.

Small scale laboratory experiments [35] were performed with testing time up to 10,000 h and temperatures up to 85 °C for a 1.5-mm thick HDPE GMB overlying a compressible compacted clay liner, where the GMB was loaded by an overlying machined probe, simulating a gravel particle, with a sustained vertical force corresponding to that induced by an average applied stress of 250 kPa. The laboratory results indicated that the machined probe was able to closely reproduce the average strains from real 50-mm gravel. For a GMB without a protection layer, the GMB tensile strains increased from 14.9 to 18.0% at a temperature of 55 °C when time was increased from 10 to 10,000 h. Increasing the temperature from 22 to 85 °C increased the GMB tensile strain observed after 1000 h from 13.8% (22 °C) to 20.5% (85 °C).

3   GMB Tensile Strains on Side Slope

3.1   GMB Tears on Side Slope

The use of a proper protection layer between the GMB and gravel drainage layer can be very effective in minimizing the GMB punctures and limiting the GMB strains. However, the GMB can still fail on side slopes due to down-drag load from waste settlement. The field exhumation of a large landfill in South East Asia [14] revealed a failure of the GMB at the crest of the side slope near the bench. A well-documented slope failure of the waste at the Kettleman Hills Landfill [26] also showed tears in the

GMB liner on the side slope associated with the failure developed by sliding along the interfaces between the underlying liner system beneath the waste fill. These field observations of the GMB failures on side slopes highlight the importance of protecting the GMBs not just from the indentations caused by gravel particles but also from the down-drag forces acting on the GMB due to waste settlement.

3.2 Numerical Modelling

The failures of the GMB liners observed in the field are very valuable in terms of recognizing the limitations of the design practice for the geosynthetic liner systems. Field observations improve the understanding of failure mechanisms associated with geosynthetic liner systems [41, 46]. However, it is generally not feasible to conduct the field-scale tests because of the practical difficulties and associated costs of performing these tests. Thus, there is a paucity of field measurements associated with GMB liner strains due to waste settlement [43]. Numerical models are currently the only practical tools for engineers to explore the different design scenarios and to gain confidence when designing the geosynthetic liner systems. Both the finite element method (FEM; [13, 41]) and finite difference method (FDM; [2, 14, 15, 19, 22, 37, 38, 42, 44–46] have been used to numerically model the performance of geosynthetic liner systems. All these numerical models have assumed that the slopes had planar surfaces and the GMBs, according to good practice, were not welded across the side slopes.

3.3 Centrifuge Testing

Centrifuge modeling has also been used to examine the performance of GMB liners under waste settlement [21, 40]. These centrifuge tests used the scaled models and increased the body stresses by centrifugal acceleration. A FDM model [44] was used to model the GMB strains/loads developed in the large-scale centrifuge test of the geomembrane-lined landfill with benches on side slopes similar to those encountered in a canyon landfill subject to waste settlement [21]. The results showed that the calculated GMB strains on benches and waste surface settlement at the landfill centre were generally in good agreement with the measured data [44].

The numerical analyses [44] indicated that the GMB with an axial tensile stiffness $J = 2000$ kN/m yielded a maximum prototype tensile load equal to the tensile strength (i.e., $T_y = 120$ kN/m). If the GMB had a higher stiffness (e.g., $J = 4000$ kN/m) and strength (e.g., $T_y = 240$ kN/m), the GMB maximum tensile load was 205 kN/m (i.e., less than the yield strength 240 kN/m) and the maximum tensile strain was 5.1% (< yield strain $\varepsilon_y = 6.0\%$; [44]). Thus a GMB with axial tensile stiffness $J = 4000$ kN/m could prevent the geomembrane from yielding and reduce the maximum strain to about 5% other things being the equal. However, this would imply the need for an unrealistically thick GMB liner to control the maximum tensile strain and an alternative approach is needed.

3.4   Influence of Slope Inclination on GMB Tensile Strains

Yu and Rowe [45] numerically examined a full-scale landfill profile with a slope inclination of 1H:1V and two 4-m wide intermediate benches below the ground surface. The foundation was competent rock and the GMB was 1.5 mm-thick [45]. The numerical results showed that the calculated maximum GMB tensile strain was 8.6% for the short-term waste settlement (Case 1) and increased to 19.8% for the long-term waste settlement (Case 2). Changing the slope inclination from 1H:1V to 2H:1V decreased the maximum GMB tensile strain from 8.6 to 4.4% for Case 1 and from 19.8 to 10.7% for Case 2. A further reduction in slope inclination to 3H:1V resulted in the maximum tensile strains of 2.0 and 2.1% for Case 1 and Case 2, respectively. Thus reducing the slope inclination had a very positive effect in terms of reducing maximum GMB tensile strains for both short-term and long-term waste settlement, and the use of a slope inclination of 3H:1V limited the maximum GMB tensile strains to acceptable design level for the conditions examined.

4   Conclusions

To ensure a long service-life of a high-density polyethylene geomembrane (GMB) exposed to leachate in a landfill, it is necessary to limit the tensile strains/stresses in the GMB to an acceptably low level. Two potential sources of strain have been considered herein; namely, (i) strains due to local GMB indentations induced by overlying coarse gravel in a leachate collection system or by gravel in an underlying compacted clay liner, and (ii) the down-drag load due to waste settlement for GMBs on side slopes. The key research related to limiting both sources of GMB strains was discussed. The findings associated with local GMB indentations induced by the gravel used in a modern leachate collection system under a 250 kPa vertical pressure (strains would be somewhat larger at higher pressures) are summarized below:

- Without a protection layer between the GMB and overlying gravel, GMBs over a compacted clay liner may experience short-term tensile strains of 16% for the 25 mm gravel and 32% for the 50 mm gravel under the short-term (10-h) physical loading conditions. These GMB strains are too large to be acceptable for landfill applications.
- Without protection and with 50 mm gravel, the maximum tensile strain of 32% for a GMB over a compacted clay liner was almost twice as high as the 17% for a GMB over a hydrated (water content 128%) geosynthetic clay liner (GCL) resting on a firm foundation.
- For GMBs over a needle-punched GCL with a geotextile protection layer, none of the geotextiles with the mass per unit area up to 2200 g/m^2 were able to reduce the short-term GMB strains to acceptable level for 50 mm gravel. The 2200 g/m^2 geotextile protection layer between the GMB and 50 mm gravel particles reduced the short-term GMB tensile strains to just below 6%.
- A multilayered geotextile with needle punched nonwoven core between two thinner and stiffer outer layers which enhanced tensile stiffness and a mass per unit area of about 1100 g/m^2 was effective in reduced the short-term GMB strains generated by

50 mm gravel to 3.9% and to 2.6% when the mass per unit area was increased to 3000 g/m^2.

- The use of a geonet as a protection layer was not acceptable since it was unable to limit the GMB strains to acceptable level, with maximum GMB strains of 13–15% being measured.
- A 150-mm thick layer of tire shreds limited the maximum GMB strains to 5.2–6.3%, and even lower to 2.3–2.8% when using a single layer of geotextile (with a mass per unit area of 570 g/m^2) was placed between the tire shreds and the GMB.
- A 150-mm-thick sand layer protection layer limited the maximum GMB tensile strain to less than 0.2%.
- The observed magnitude of GMB tensile strains was dependent on the length of sustained loading and the temperature. For a GMB without a protection layer overlain by 50-mm gravel, increasing the time of loading from 10 to 10,000 h at a temperature 55 °C increased the GMB tensile strain from 14.9 to 18.0%. An increase in temperature from 22 to 85 °C after 1000 h increased the GMB tensile strains from 13.8 to 20.5%.

For the cases and conditions examined, the key findings associated with the down-drag load for GMBs on side slopes are:

- Decreasing the slope inclination from 1H:1V to 3H:1V reduced the maximum GMB tensile strains for both short-term (e.g., immediately after landfill closure) and long-term waste settlement. For GMBs on side slopes without an axially stiff geotextile reinforcing layer above the GMB, the maximum GMB tensile strain was 8.6% for 1H:1V and decreased to 4.4% for 2H:1V and 2.0% for 3H:1V immediately after landfill closure. After long-term waste settlement, the maximum GMB tensile strains were 19.8, 10.7, and 2.1% for 1H:1V, 2H:1V, and 3H:1V, respectively. Thus, a slope inclination steeper than 3H:1V resulted in maximum GMB strains that were not acceptable for landfill applications without special measures being taken to limit the strains.
- The use of a high stiffness geotextile reinforcing layer above the GMB reduced the maximum GMB tensile strains to less than 2%. However, the geotextile itself became an engineering concern for a slope inclination 1H:1V. When using a geotextile with an axial tensile stiffness J_{gt} = 4200 kN/m, the maximum geotextile tensile strains were 5.0% for immediately after landfill closure and 9.1% after long-term waste settlement. An increase of the geotextile stiffness to J_{gt} = 8000 and 10000 kN/m resulted in a decrease of the maximum geotextile strain to 3.7 and 3.3% after landfill closure, respectively and to 6.1 and 5.3% after long-term waste settlement, respectively. Thus the maximum geotextile strains are likely too large to be acceptable when using a slope inclination of 1H:1V.
- Decreasing the slope inclination from 1H:1V to 2H:1V (with a geotextile J_{gt} = 8000 kN/m) reduced the maximum geotextile tensile strain from 3.7 to 2.5% immediately after landfill closure and from 6.1 to 3.7% after long-term waste settlement. Thus, a proper selection of the slope inclination and geotextile tensile stiffness can reduce both the GMB and geotextile strains to acceptable levels.

Acknowledgements. The work reported in this paper was supported by a grant (A1007) from the Natural Sciences and Engineering Research Council of Canada (NSERC).

References

1. Abdelaal, F.B., Rowe, R.K., Brachman, R.W.I.: Brittle rupture of an aged HDPE geomembrane at local gravel indentations under simulated field conditions. Geosynth. Int. **21**(1), 1–23 (2014)
2. Arab, M.G.: The integrity of geosynthetic elements of waste containment barrier systems subject to seismic Loading. Ph.D. thesis, School of Sustainable Engineering and the Built Environment, Arizona State University, Tempe, Arizona, USA (2011)
3. ASTM D5514: Standard test method for large scale hydrostatic puncture testing of geosynthetics. ASTM International, West Conshohocken, PA, USA (2014)
4. Brachman, R.W.I., Gudina, S.: Gravel contacts and geomembrane strains for a GM/CCL composite liner. Geotext. Geomembr. **26**(6), 448–459 (2008)
5. Brachman, R.W.I., Gudina, S.: Geomembrane strains and wrinkle deformations in a GM/GCL composite liner. Geotext. Geomembr. **26**(6), 488–497 (2008)
6. Brachman, R.W.I., Sabir, A.: Geomembrane puncture and strains from stones in an underlying clay layer. Geot. Geomembr. **28**(4), 335–343 (2010)
7. Brachman, R.W.I., Sabir, A.: Long-term assessment of a layered-geotextile protection layer for geomembranes. J. Geotech. Geoenviron. Eng. **139**(5), 752–764 (2013)
8. Bundesanstalt für Materialforschung und -prüfung (BAM): Guidelines for the Certification of Protection Layers for Geomembranes in Landfill Sealing Systems. BAM, Berlin, Germany (2015)
9. Dickinson, S., Brachman, R.W.I.: Assessment of alternative protection layers for a GM/GCL composite liner. Can. Geotech. J. **45**(11), 1594–1610 (2008)
10. EN 13719: Geosynthetics - determination of the long term protection efficiency of geosynthetics in contact with geosynthetic barriers. European committee for standardization (CEN), Brussels (2016)
11. Eldesouky, H.M.G., Brachman, R.W.I.: Calculating local geomembrane strains from a single gravel particle with thin plate theory. Geotext. Geomembr. **46**(1), 101–110 (2018)
12. Ewais, A.M.R., Rowe, R.K., Brachman, R.W.I., Arnepalli, D.N.: Service-life of a HDPE GMB under simulated landfill conditions at 85 °C. J. Geotext. Geoenviron. Eng. **140**(11), 04014060 (2014)
13. Filz, G.M., Esterhuizen, J.J.B., Duncan, J.M.: Progressive failure of lined waste impoundments. J. Geotext. Geoenviron. Eng. **127**(10), 841–848 (2001)
14. Fowmes, G.J.: Analysis of steep sided landfill lining systems. Ph.D. thesis, Department of Civil and Building Engineering, Loughborough University, Loughborough, UK (2007)
15. Fowmes, G.J., Dixon, N., Jones, D.R.V.: Validation of a numerical modelling technique for multilayered geosynthetic landfill lining systems. Geotext. Geomembr. **26**(2), 109–121 (2008)
16. Giroud, J.P., Tisseau, B., Soderman, K.L., Beech, J.F.: Analysis of strain concentration next to geomembrane seams. Geosynth. Int. **2**(6), 1049–1097 (1995)
17. Gudina, S., Brachman, R.W.I.: Physical response of geomembrane wrinkles overlying compacted clay. J. Geotext. Geoenviron. Eng. **132**(10), 1346–1353 (2006)
18. Hsuan, Y.G., Koerner, R.M.: Antioxidant depletion lifetime in high density polyethylene geomembranes. J. Geotext. Geoenviron. Eng. **124**(6), 532–541 (1998)

19. Jones, D.R.V., Dixon, N.: Landfill lining stability and integrity: the role of waste settlement. Geotext. Geomembr. **23**(1), 27–53 (2005)
20. Kavazanjian, E., Andresen, J., Gutierrez, A.: Experimental evaluation of HDPE geomembrane seam strain concentrations. Geosynth. Int. **24**(4), 333–342 (2017)
21. Kavazanjian, E., Gutierrez, A.: Large scale centrifuge test of a geomembrane-lined landfill subject to waste settlement and seismic loading. Waste Manage **68**, 252–262 (2017)
22. Kavazanjian, E., Wu, X., Arab, M., Matasovic, N.: Development of a numerical model for performance-based design of geosynthetic liner systems. Geotext. Geomembr. **46**(2), 166–182 (2018)
23. Koerner, R.M., Hsuan, Y.G., Koerner, G.R., Gryger, D.: Ten year creep puncture study of HDPE geomembranes protected by needle-punched nonwoven geotextiles. Geotext. Geomembr. **28**(6), 503–513 (2010)
24. Koerner, R.M., Wilson-Fahmy, R.R., Narejo, D.: Puncture protection of geomembranes, part III: examples. Geosynth. Int. **3**(5), 655–675 (1996)
25. LFE-2: Cylinder testing geomembranes and their protective materials. A methodology for testing protector geotextiles for their performance in specific site conditions. Environment Agency, UK (2014)
26. Mitchell, J.K., Seed, R.B., Seed, H.B.: Kettleman Hills waste landfill slope failure, I: liner system properties. J. Geotech. Eng. **116**(4), 647–668 (1990)
27. Mueller, W., Jacob, I.: Oxidative resistance of high-density polyethylene geomembranes. Polym. Degrad. Stab. **79**(1), 161–172 (2003)
28. Narejo, D., Koerner, R.M., Wilson-Fahmy, R.F.: Puncture protection of geomembranes, part II: experimental. Geosynth. Int. **3**(5), 629–653 (1996)
29. Rowe, R.K.: Short and long-term leakage through composite liners. The 7th Arthur Casagrande Lecture. Can. Geotech. J. **49**(2), 141–169 (2012)
30. Rowe, R.K., Abdelaal, F.B., Brachman, R.W.I.: Antioxidant depletion from an HDPE geomembrane with a sand protection layer. Geosynth. Int. **20**(2), 73–89 (2013)
31. Rowe, R.K., Quigley, R.M., Brachman, R.W.I., Booker, J.R.: Barrier Systems for Waste Disposal Facilities. Taylor & Francis Books Ltd. (E & FN Spon), London (2004)
32. Rowe, R.K., Islam, M.Z., Hsuan, Y.G.: Leachate chemical composition effects on OIT depletion in an HDPE geomembrane. Geosynth. Int. **15**(2), 136–151 (2008)
33. Rowe, R.K., Islam, M.Z., Hsuan, Y.G.: Effects of thickness on the aging of HDPE geomembranes. J. Geotech. Geoenvironmen. Eng. **136**(2), 299–309 (2010)
34. Rowe, R.K., Rimal, S., Sangam, H.: Ageing of HDPE geomembrane exposed to air, water and leachate at different temperatures. Geotext. Geomembr. **27**(2), 137–151 (2009)
35. Sabir, A., Brachman, R.W.I.: Time and temperature effects on geomembrane strain from a gravel particle subjected to sustained vertical force. Can. Geotech. J. **49**(3), 249–263 (2012)
36. Seeger, S., Müller, W.: Theoretical approach to designing protection: selecting a geomembrane strain criterion. In: Dixon, N., Smith, D.M., Greenwood, J.H., Jones, D.R. V. (eds.) Geosynthetics: Protecting the Environment, pp. 137–152. Thomas Telford, London (2003)
37. Sia, A.H.I., Dixon, N.: Numerical modelling of landfill lining system waste interaction: implications of parameter variability. Geosynth. Int. **19**(5), 393–408 (2012)
38. Thiel, R., Kavazanjian, E., Wu, X.: Design considerations for slip interfaces on steep-wall liner systems. In: Proceedings of the Tenth International Conference on Geosynthetics, Deutsche Gesellschaft für Geotechnik and International Geosynthetics Society German Chapter, Berlin, Germany, p. 6 (2014)
39. Tognon, A.R., Rowe, R.K., Moore, I.D.: Geomembrane strain observed in large-scale testing of protection layers. J. Geotech. Geoenviron. Eng. **126**(12), 1194–1208 (2000)

40. Thusyanthan, N.I., Madabhushi, S.P.G., Singh, S.: Tension in geomembranes on landfill slopes under static and earthquake loading - Centrifuge study. Geotext. Geomembr. **25**(2), 78–95 (2007)

41. Villard, P., Gourc, J.P., Feki, N.: Analysis of geosynthetic lining system (GLS) undergoing large deformations. Geotext. Geomembr. **17**(1), 17–32 (1999)

42. Wu, X.: Effect of waste settlement and seismic loading on the integrity of geosynthetic barrier systems. Masters' thesis, School of Sustainable Engineering and the Built Environment, Arizona State University, Tempe, Arizona, USA (2013)

43. Yazdani, R., Campbell, J.L., Koerner, G.R.: Long-term in situ strain measurements of a high density polyethylene geomembrane in a municipal solid waste landfill. In: Proceedings of the Geosynthetics 1995, Industrial Fabrics Association International, Roseville, MN, pp. 893–906 (1995)

44. Yu, Y., Rowe, R.K.: Modelling deformation and strains induced by waste settlement in a centrifuge test. Can. Geotech. J. (2018). https://doi.org/10.1139/cgj-2017-0558

45. Yu, Y., Rowe, R.K.: Development of geomembrane strains in waste containment facility liners with waste settlement. Geotext. Geomembr. **46**(2), 226–242 (2018)

46. Zamara, K.A., Dixon, N., Fowmes, G., Jones, D.R.V., Zhang, B.: Landfill side slope lining system performance: A comparison of field measurements and numerical modelling analyses. Geotext. Geomembr. **42**(3), 224–235 (2014)

Ground Improvement

Application of Image Analysis on Two-Dimensional Experiment of Ground Displacement Under Strip Footing

Guanxi Yan[✉], Youwei Xu, Vignesh Murgana,
and Alexander Scheuermanna

School of Civil Engineering, Geotechnical Engineering Centre,
University of Queensland, St Lucia, Australia
g.yan@uq.edu.au

Abstract. The two-dimensional experiment of sandboxes for simulating ground failure has become a popular teaching tool in the current geomechanics laboratory. Most of them need cameras with high resolution and well-textured soil for digital image correlation (DIC). Those techniques do not only require a higher budget for the camera but also expensive PIV software for image analysis. Although it is more accurate and robust, it is not often fully available in every geomechanics laboratory. To provide a cheaper substitution, a simple image analyzing algorithm was developed to consistently detect the trajectory of each marker pre-embedded in a 2D sandbox. Through conducting image processing in the sequence of binary images, noise filtering and correlating the markers by smallest Euclidean distance in the sequence of images, all markers can be accurately monitored and tracked. The soil displacement of white sand under the strip footing can be consequentially measured with an acceptable accuracy. Compared to the previous DIC methods, this 2D experiment coupled with this image analyzing algorithm can measure the soil deformation for poor-textured soil. In this paper, a 2D strip footing model test is presented with newly developed methods for the measurement of ground displacement. It is a useful and straightforward experiment for investigating 2D plain-strain foundation failure.

Keywords: Shallow foundation · Strip footing · Image analysis
Ground displacement

1 Introduction

In geotechnical engineering, soil deformation measurement is highly concerned about the safety design and analysis of foundations, pile, and slope. The conventional measuring techniques merely provide the visions of soil settlement on the ground surface. With the development of the digital camera, geotechnical engineering researchers can collect more information about soil deformation from a 2D laboratory experiment [1, 2]. Thus, particle image velocimetry (PIV) has been successfully applied to analyze the characteristics of soil deformation [3–5]. Many advanced PIV techniques with corresponding free toolboxes (coded in Matlab®) are available for 2D Soil deformation measurement in the literature [6–8].

© Springer Nature Singapore Pte Ltd. 2018
L. Li et al. (Eds.): GSIC 2018, *Proceedings of GeoShanghai 2018 International Conference: Ground Improvement and Geosynthetics*, pp. 13–21, 2018.
https://doi.org/10.1007/978-981-13-0122-3_2

Most of their deformation measurement methods are achieved by tracking the highest correlated patches in the sequence of frames. Their digital image correlation (DIC) methods can measure the field of view (FOV) in high resolution (fine mesh), even include detection of rotation and shear deformation of patches [6]. However, if the material has poor texture, these approaches cannot accurately capture each patch, further leading to miscorrelating irrelevant subsets. In this condition, some markers have to be pre-located into FOV so that soil displacement can be locally detected. With the ground moving, those markers might sometimes smear. This leads to the failure of consistently tracking displacement for certain locations of interests.

To provide a straightforward solution for investigating 2D plain-strain shallow foundation failure, a quasi-2D sandbox coupled with a simpler algorithm of image analysis is developed. This 2D experimental setup only requires a general digital camera to detect the ground displacement failure resulted from loading on a strip footing. In the following content, the experimental setup and image analysis are described in detail, followed by the result in regards to the accuracy of image analysis and comparison between conventional DIC and new methods.

2   Experimental Method

2.1   Experimental Setup of 2D Ground Failure Under Strip Footing

The experimental setup is illustrated in Fig. 1. As can be seen from the Fig. 1c, the experimental setup consists of a quasi-2D drainage and seepage tank shown in green in Fig. 1a and b, a loading frame highlighted in black in Fig. 1a and b, and a camera support in front of the tank. The dimension of the experimental view is 122 cm × 45 cm. There was a gravel layer of 10 cm, located at the bottom. The testing soil layer was set above the gravel layer in 35 cm. A steel plate in 5 cm × 2 cm was used to simulate the strip footing in Fig. 1c. The superstructure load was simulated by controlling the number of weights on the pendulous frame. According to Fig. 1a, the weight causes a moment rotating around the loading frame in clockwise and the other moment in the same magnitude but counter-clockwise can yield a higher load on the strip footing. Thus, the multistep loading on the footing can be achieved. As the tank was originally designed to simulate seepage problem in the soil, there are flow input and output under the tank. The groundwater table can also be controlled to investigate both saturated and unsaturated soil. As this study merely aims to demonstrate a cheaper solution for detecting ground movement in 2D, commercial sand (30/60 sand, SAN745-3LA) was adopted, but not for an actual investigation of ground failure for certain soil. The specification of sand is listed in Table 1. The porosity of sand tank was compacted to 39%. Direct shear tests for this sample present cohesion of 3 kPa and friction angle of 31°.

As shown in Fig. 1c, compared to natural sand collected from the beach, this sand has poor texture and is light brown which leads to a low contrast between the background and measuring targets. To be able to be captured by a camera having low resolution, several dark markers were located during the sand filling process, as the arrangement of black points under the strip footing shown in the tank in Fig. 1d. These

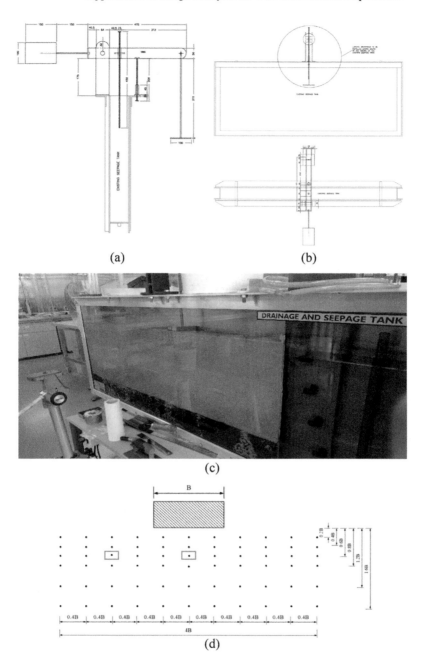

Fig. 1. The experimental setup, (a) the side view, (b) the front view and top view, (c) one capture of initial conditions before loading start, (d) the arrangement of dark markers to enhance the contrast between background and target.

Table 1. The specification of selected sand

Soil sample	30/60 sand
USCS soil classification	Poorly graded sand (SP)
D50 (mm)	0.43
C_u	1.2
C_c	1
G_s (ρ_s/ρ_w)	2.655
Mineralogy	SiO_2
Sand (%)	99.99
Fine (%)	0.01
$\rho_{dry,max}$	1.73
$\rho_{dry,min}$	1.58

makers generate a good contrast between measuring targets and sand background so each particle can be captured using image analysis. Therefore, through analyzing a sequence of frames recording ground movement, the location and displacement of each maker can be precisely determined. The camera used in this study is a standard camera, RICOH WG-4 GPS. The single lens camera is unnecessary for image analysis but can contribute to a better quality of pictures.

By controlling the loading frame, 5.5 kPa was applied on strip footing every minute from 0 kPa at 0 min initially up to 44.5 kPa at eight mins finally. Linear Variable Displacement Transducer (LVDT) was set above the strip footing to record the vertical displacement of footing consistently. With an aim to only measure the ground displacement under the footing, this loading process is only for generating a ground failure but nothing else for specific analysis of foundation failure referring to any discussion on maximum bearing capacity calculation.

2.2   Image Analysis for Measurement of Ground Displacement

After completed the experiment, the images capturing particle movement were loaded into Matlab for image analysis. The algorithm of image analysis can be summarized in following steps:

(1) Loading each image from one folder using Matlab command "imread";
(2) Binary the RGB image using the command "im2bw" with a threshold of gray value, shown in Fig. 2a;
(3) Labeling each dark marker and applying command "regionprops" to determine the perimeter (P) and area (A) of each dark marker;
(4) Filtering out the largest and smallest area which violate the reasonable size for the black marker;
(5) Filtering out the captured points which violate the circularity larger than 1.8 and less than 0.7;
(6) Filtering out the noises in the area occupied by footing;
(7) After filtering all noise and irrelevant targets, saving all the centroids for each marking point into a series of matrices;

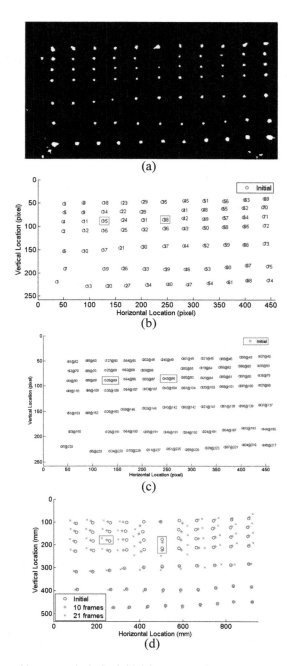

Fig. 2. The output of image analysis for initial frame, (a) Binary image generated after step (7), (b) Index labeled for captured markers after step (11), (c) The centroids labeled for step (11), (d) The sequence of captured markers moving with time in frame numbers.

(8)   Reading the location of each marker from each frame as shown in Fig. 2c, and then searching the closest marker from the next frame.
(9)   Saving the centroids of closest markers from each adjacent pair of frames into a new matrix;
(10) Filtering out the centroids which perfectly unchanged from first to last frames;
(11) Visualizing the captured markers on a 2D image for the first frame and labeling the matrix index of the position of each marker on the image as shown in Fig. 2b;
(12) According to this image, manually selecting points of interest for displacement calculation and visualizing the deformation for each marker in sequence of frames as shown in Fig. 2d.

The calibration between physical length to image pixel is 0.47 mm/pixel with a standard deviation of ±0.02 mm/pixel. Compared to DIC method which analyzing the deformation of each patch as a subset of an image, this method only measures the maker displacement. Without detecting the initial size of each patch (Representative Elementary Volume in 2D), there is no chance to calculate the elastic-plastic strain and shear strain for each subset, and even plot strain distribution in contour for entire modeling domain. However, it is still an effective solution for capturing poor-textured ground displacement using a normal camera in low resolution.

3   Result and Discussion

3.1   Comparison Between DIC and the Image Analyzing Method

Due to the poor quality of image using the camera in low resolution, a comparison is conducted between the DIC method and the newly developed image analysis algorithm. Figure 3a shows the captured location of each marker using conventional DIC method, and Fig. 3b shows the new algorithm can accurately capture the measuring targets for displacement calculation. Apparently, the conventional method cannot even capture the texture in soil precisely for such a poor textured soil. As the strip footings and marking lines are presented in the image domain, the conventional method, usually available in many none open-sourced toolboxes of DIC, cannot be flexibly adjusted to eliminate the noises. These noises left in final result leads to certain amount of useless data. It causes waste of labor on manually eliminating errors resulting from the inability of the method. Also, some of the makers cannot be captured, so some points of interest might not be able to be analyzed, such as the several makers missed in the triangular elastic zone under strip footing shown in Fig. 3a. Compared to the conventional DIC, which cannot capture the targets and further miscorrelate patches in the sequence of frames, the simple particle tracking method cannot only captured the marker precisely but also can correlate the marker between each adjacent pair of frames using smallest Euclidean distance. It improves the accuracy of identification of each soil texture but also simplifies the correlating method.

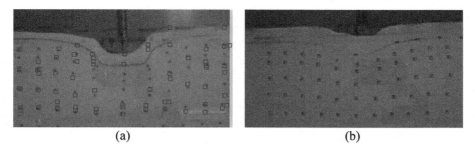

(a)   (b)

Fig. 3. The comparison between DIC method using ImageJ and new image analysis method for particle tracking: (a) Marker captured by DIC (b) Marker captured by the new method.

3.2   Displacement Measurement of the Highlighted Markers

After image processing, based on the step (12), the displacements for marker No. 15 and 38 highlighted in Figs. 1d and 2, are calculated. The Fig. 4a shows the horizontal displacement-time, and the Fig. 4b presents the vertical displacement-time. As can be seen from the Fig. 2d, the marker No. 38 shows only a vertical settlement with slight horizontal movement. The horizontal displacement of No. 38 just slightly increases from 0 to 1.4 mm, while its vertical displacement gradually increases from 0 to 4 mm at frame No. 15 and later steeply rises to 11 mm for a new equilibrium state. The trend of this data manifests that the ground initiates the deformation slowly in an elastic stage. Later, when a high stress applied is over the maximum bear capacity of this loose compacted sand, a large ground settlement occurs as a manifestation of plastic deformation. As for marker No. 15, in contrast to No. 33, there is little heaving happened from 0 to 2.8 mm, as the negative vertical displacement in Fig. 4b, but a

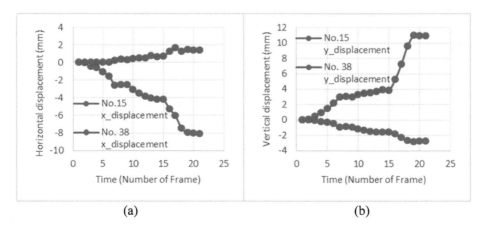

(a)   (b)

Fig. 4. The displacement measured for marker No. 15 and 38: (a), the horizontal displacement against time (moving right is positive); (b), the vertical displacement against time in a number of frames (moving downward is positive).

significant horizontal displacement towards left-hand side from 0 to 8 mm, as the negative horizontal displacement in Fig. 4a. In similar to the suddenly increasing vertical displacement for marker No. 38 in Fig. 4b, the horizontal displacement for marker No. 15 also shows a sharp increment after frame No. 15. Also, according to the displacement logging in Fig. 4, the ground approach a new equilibrium state after frame No. 19, as there is no more progress of vertical displacement for No. 38 and horizontal displacement for No. 15. This comparison somehow demonstrates the consistency of measurement in regards to equilibrium states of the sand ground. It therefore also demonstrates the functions and accuracy of the experimental setup with the new-developed image analysis.

4   Concluding Summary

This study presents a quasi-2D experimental setup for investigating ground movement for loading on a strip footing. Due to the poor-textured sand filled into 2D soil tank and normal camera in low resolution applied for capturing deforming ground, for measuring ground displacement, a new image analysis method is developed to track the movement of black markers buried in the sand. The image analysis algorithm is coded in Matlab using commands available in Image Toolbox. By involving multi-filtering process, all blurry points and irrelevant targets are removed from result analysis. The marker movement is derived by tracking each marker in a sequence of frames using smallest distance. Displacement can be determined based on the distance between centroids of the markers in the current frame and the initial frame. Each marker can be accurately captured and consistently measured. The displacement logging can manifest the ground movement from one to a new equilibrium state. The images and results, generated by this image analysis, complete the last puzzle for implementing a 2D bearing capacity experiment. It is a useful and straightforward solution for researching and educating 2D plain-strain problem in the geotechnical laboratory.

References

1. Iskander, M.: Modelling with Transparent Soils: Visualizing Soil Structure Interaction and Multi Phase Flow Non-intrusively. Springer, Heidelberg (2010). https://doi.org/10.1007/978-3-642-02501-3
2. Iskander, M., Liu, J.: Spatial deformation measurement using transparent soil. Geotech. Test. J. 33(4), 314–321 (2010)
3. Ahmadi, H., Hajialilue-Bonab, M.: Experimental and analytical investigations on bearing capacity of strip footing in reinforced sand backfills and flexible retaining wall. Acta Geotech. 7(4), 357–373 (2012)
4. Boldyrev, G., Melnikov, A., Barvashov, V.: Particle image velocimetry and numeric analysis of sand deformations under a test plate. In: The 5th European Geosynthetics Congress (2012)
5. Chen, Y., She, Y., Cai, J., Zai, J.: Displacement field research of soil beneath shallow foundation based on digital image correlation method. In: International Conference on Experimental Mechnics 2008 and Seventh Asian Conference on Experimental Mechanics. International Society for Optics and Photonics (2008)

6. Stanier, S.A., Blaber, J., Take, W.A., White, D.: Improved image-based deformation measurement for geotechnical applications. Can. Geotech. J. **53**(5), 727–739 (2015)
7. Sveen, J.K.: An introduction to MatPIV v. 1.6. 1. Preprint series. Mechanics and Applied Mathematics (2004). http://urn.nb.no/URN:NBN:no-23418
8. White, D., Take, W., Bolton, M.: Soil deformation measurement using particle image velocimetry (PIV) and photogrammetry. Geotechnique **53**(7), 619–632 (2003)

Calculation Method for Settlement of Stiffened Deep Mixed Column-Supported Embankment over Soft Clay

Guan-Bao Ye, Feng-Rui Rao, Zhen Zhang$^{(\boxtimes)}$, and Meng Wang

Department of Geotechnical Engineering,
Tongji University, Shanghai 200092, China
zhenzhang@tongji.edu.cn

Abstract. Stiffened Deep Mixed (SDM) column is a new ground improvement technique which can be used to significantly increase bearing capacity and reduce settlement of soft soil. In the region consisting of deep thick saturated soft soil, SDM columns have been successfully used to support highways and railway embankments, tanks, and buildings. However, there still has been no feasible method in design of SDM column-reinforced subsoil so far. This paper proposed a method to calculate settlement of SDM columns-supported embankment over soft soil. The total settlement of SDM column-reinforced soft soil is a sum of the compression of the soil within length of stiffened core piles, the compression of the soil from core pile tip to SDM column base and the compression of the soil below SDM column base. Punching effect was considered in developing the method to consider the punching deformation of core pile upward and downward. A full scale test was introduced to verify the feasibility of the proposed method and it yielded a good prediction with the field data.

Keywords: Stiffened deep mixing column · Soft clay · Settlement
Embankment · Theoretical analysis

1 Introduction

Deep saturated soft soil in the east coastal area of China poses many challenges to geotechnical engineers, such as low bearing capacity, excessive settlement, and slope instability. Recently, a new technology called Stiffened Deep Mixed (SDM) column was proposed (Ling et al. 2001). SDM column is formed by inserting a precast concrete core pile into the center of DM column immediately after construction of DM column. Core pile is installed to increase strength of column and reduce ground settlement. Meanwhile, DM column is used to increase skin friction along the core pile shaft. Moreover, bearing capacity provided by the SDM column was similar to the cast-in-place pile with the same diameter and length while the SDM column can save cost nearly by 30% (Qian et al. 2013).

Various studies have been conducted to investigate the behavior of SDM column-reinforced soft soil (Tanchaisawat et al. 2009; Zhao et al. 2010; Ye et al. 2016). However, there was still no feasible design method to calculate the settlement of ground reinforced by SDM columns. This paper developed an analytical method to

© Springer Nature Singapore Pte Ltd. 2018
L. Li et al. (Eds.): GSIC 2018, *Proceedings of GeoShanghai 2018 International Conference: Ground Improvement and Geosynthetics*, pp. 22–29, 2018.
https://doi.org/10.1007/978-981-13-0122-3_3

calculate the settlement of SDM column-supported embankment over soft clay. Punching effect of concrete core pile was considered in the analysis. Finally, the proposed method was applied to predict the settlement of a case history.

2   Calculation Model

Figure 1 shows the typical cross section of SDM column-reinforced soil. Based on the geometry characteristic in cross section, the profile could be divided into three regions: Region I (the soil within the length of concrete core pile), Region II (the soil between the core pile base and the DM column base), and Region III (the soil below the DM column base). Thus, the total settlement of SDM column-reinforced subsoil is a sum of the compressions of the three regions,

$$S_{\text{total}} = S_{\text{I}} + S_{\text{II}} + S_{\text{III}} \tag{1}$$

in which, S_{total} is the total settlement, S_{I} is the compression of Region I, S_{II} is the compression of Region II, and S_{III} is the compression of Region III.

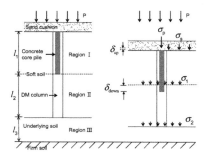

Fig. 1. Cross section of subsoil with SDM column

The main assumptions made in the analysis are summarized as: (a) concrete core pile, DM column and subsoil behave as isotropic linear-elastic materials, and radial deformation is ignored; (b) DM column and soil below the core pile base deform under an equal strain condition; (c) skin friction is distributed linearly along the core pile shaft.

3   Derivation of the Theoretical Method

3.1   Compression of Soil in Region I

Considering the punching effect of the core pile (see Fig. 1), S_{I} can be expressed as,

$$S_{\text{I}} = \delta_{\text{up}} + \delta_{\text{down}} + \delta_{\text{core}} \tag{2}$$

in which, δ_{up} is the upward punching deformation of the core pile, δ_{down} is the downward punching deformation of the core pile, and δ_{core} is the compression of the core pile. Since the surrounding soil settles more than the core pile, negative frictional force occurs on the core pile shaft. Taking the equal settlement plane as a datum plane, Eq. (2) can be rewritten as,

$$S_I = S_{su} + S_{sd} = \left(\delta_{up} + \delta_1\right) + \left(\delta_{down} + \delta_2\right) \tag{3}$$

in which S_{su} and δ_1 are the compressions of the surrounding soil and the core pile above the equal settlement plane, and S_{sd} and δ_2 are those below the equal settlement plane. Ye et al. (2016) illustrated that the distribution of the skin friction along the core pile was curved as shown in Fig. 2(a). For simplicity, the skin friction along the core pile is assumed to have a linear distribution as illustrated in Fig. 2(b).

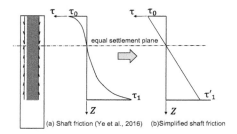

Fig. 2. Simplification of concrete core pile skin friction

The Bjerrum method (1969) is used to calculate the maximum negative friction,

$$\tau_0 = k\sigma_s \tan \varphi_i \tag{4}$$

in which $k = \tan^2(45° + \varphi/2)$, φ is the frictional angle of DM column, φ_i is the frictional angle of the interface between core pile and DM column, and σ_s is the vertical stress on the top of surrounding soil. Thus, the concrete core skin friction can be expressed as, $\tau(z) = \tau_0(1 - z/l_0)$, in which l_0 is the depth of the equal settlement plane, positive value of $\tau(z)$ means negative friction, vice versa.

Based on the equilibrium of the forces in a vertical direction, an equation can be established as $P = m\alpha\sigma_p + (1 - m\alpha)\sigma_s$, in which P is the average loading on the ground, m is the replacement ratio of SDM column, α is the area ratio of core pile, and σ_p and σ_s are the vertical stresses on the top of core pile and surrounding soil. Setting the stress concentration ratio $n = \sigma_p/\sigma_s$, one can obtain $\sigma_s = P/((n - 1)\alpha m + 1)$, and $\sigma_p = nP/((n - 1)\alpha m + 1)$. The equivalent compression modulus of the soil and the DM column in Region I (E_I^{eq}) is considered as an area-weighted average value of DM column and subsoil, i.e., $E_I^{eq} = m(1 - \alpha)E_{DM} + (1 - m)E_{sI}$, in which E_{DM} is the compression modulus of DM column, and E_{sI} is the compression modulus of the soil in

Region I. Figure 3 shows a slice of the unit cell with a thickness of dz. The differential equation can be obtained based on the equilibrium of the forces in a vertical direction,

$$d\sigma_{sz}/dz + \lambda\tau(z) = 0 \tag{5}$$

in which, σ_{sz} is the vertical stress of surrounding soil at depth z; A_s is the area of surrounding soil; c_p is the perimeter of core pile; and $\lambda = c_p/A_s$. Using the boundary condition $\sigma_{sz} = \sigma_s$ at $z = 0$, σ_{sz} can be solved as,

$$\sigma_{sz} = (\lambda\tau_0 z^2)/(2l_0) - \lambda\tau_0 z + \sigma_s \tag{6}$$

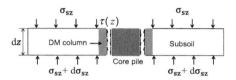

Fig. 3. Slice of unit cell in Region I

At depth z, the vertical stress in the core pile is,

$$\sigma_{pz} = \sigma_p + A_p^{-1}\int_0^z \tau(z)c_p dz = \sigma_p + \gamma(\tau_0 z - \tau_0 z^2/(2l_0)) \tag{7}$$

in which, A_p is the area of core pile, $\gamma = c_p/A_p$. Thus, some terms on the right side of Eq. (3) can be solved as, $S_{su} = \int_0^{l_0}\left(\sigma_{sz}/E_I^{eq}\right)dz$, $S_{sd} = \int_{l_0}^{l_1}\left(\sigma_{sz}/E_I^{eq}\right)dz$, $\delta_1 = \int_0^{l_0}\left(\sigma_{pz}/E_p\right)dz$, and $\delta_2 = \int_{l_0}^{l_1}\left(\sigma_{pz}/E_p\right)dz$.

The upward punching deformation of the core pile can be considered as,

$$\delta_{up} = p_c(\sigma_p - \sigma_s) \tag{8}$$

where p_c is the upward punching under an uniform force. It can be considered as $p_c = L_c/E_c$, where L_c is the thickness of cushion, E_c is the compression modulus of cushion, and E_p is the compression modulus of core pile.

The downward punching deformation of the core pile can be considered as,

$$\delta_{down} = p_s(\sigma_{pl_1} - \sigma_{sl_1}) \tag{9}$$

in which, σ_{pl_1} is the vertical stress at the bottom of core pile; σ_{sl_1} is the vertical stress at the same depth of subsoil; and p_s is the downward punching under an uniform force. Chen (2005) proposed a method to calculate p_s,

$$p_s = (1 - \mu_0^2)\omega\sqrt{A_p}/E_0 \tag{10}$$

$$E_0 = (1 - 2\mu_0^2/(1 - \mu_0))E_{II}^{eq} \tag{11}$$

in which μ_0 is the Poisson's ratio of DM column, ω is a parameter decided by the shape of loading plane. E_{II}^{eq} is the compression modulus of the soil in Region II.

Thus, combining Eq. (8), S_{su}, δ_1 and Eq. (3), it can be written as,

$$(\sigma_s l_0 - \lambda\tau_0 l_0^2/3)/E_I^{eq} = p_c(\sigma_p - \sigma_s) + (\sigma_p l_0 + \gamma\tau_0 l_0^2/3)/E_p \tag{12}$$

Similarly, combining Eq. (9), S_{sd}, δ_2 and Eq. (3), it can be written as,

$$
\begin{aligned}
&\frac{1}{E_I^{eq}}\left(\frac{\lambda\tau_0}{6l_0}l_1^3 - \frac{\lambda\tau_0}{2}l_1^2 + \sigma_s l_1 - \sigma_s l_0 + \frac{\lambda\tau_0}{3}l_0^2\right) \\
&= \frac{1}{E_p}\left(-\frac{\gamma\tau_0}{6l_0}l_1^3 + \frac{\gamma\tau_0}{2}l_1^2 + \sigma_p l_1 - \sigma_p l_0 - \frac{\gamma\tau_0}{3}l_0^2\right) + p_s\left[\sigma_p - \sigma_s + (\gamma+\lambda)\tau_0 l_1 - \frac{(\gamma+\lambda)\tau_0 l_1^2}{2l_0}\right]
\end{aligned}
\tag{13}
$$

By substituting Eq. (4) into Eqs. (12) and (13), it can be written as,

$$n = \frac{\sigma_p}{\sigma_s} = \frac{\theta_1}{\theta_2} = \frac{\theta_3}{\theta_4} \tag{14}$$

in which, n is the stress concentration ratio on the top of core pile, $\theta_1 = \frac{l_0}{E_I^{eq}} + p_c - \frac{\lambda k \tan\varphi_i}{3E_I^{eq}}l_0^2 - \frac{\gamma k \tan\varphi_i}{3E_p}l_0^2$, $\theta_2 = p_c + \frac{l_0}{E_p}$, $\theta_4 = p_s + \frac{(l_1-l_0)}{E_p}$

$$
\begin{aligned}
\theta_3 &= \frac{1}{E_I^{eq}}\left(\frac{\lambda k l_1^3 \tan\varphi_i}{6l_0} - \frac{\lambda k l_1^2 \tan\varphi_i}{2} + l_1 - l_0 + \frac{\lambda k l_0^2 \tan\varphi_i}{3}\right) \\
&\quad - \frac{1}{E_p}\left(-\frac{\gamma k l_1^3 \tan\varphi_i}{6l_0} + \frac{\gamma k l_1^2 \tan\varphi_i}{2} - \frac{\gamma k l_0^2 \tan\varphi_i}{3}\right) - p_s\left(-1 + (\gamma+\lambda)k l_1 \tan\varphi_i - \frac{(\gamma+\lambda)k l_1^2 \tan\varphi_i}{2l_0}\right)
\end{aligned}
$$

In Eq. (14), l_0 is the only unknown variation. By solving Eq. (14), l_0 and n can be obtained, then σ_p and σ_s can be solved. Thus, the compression of Region I can be calculated as,

$$S_I = S_{su} + S_{sd} = (\lambda\tau_0 l_1^3/(6l_0) - \lambda\tau_0 l_1^2/2 + \sigma_s l_1)/E_I^{eq} \tag{15}$$

3.2   Compression of Soil in Region II

Since the difference in the moduli of DM column and surrounding soil is not so great, the compression of the soil in Region II is considered to settle under an equal strain condition. The equivalent modulus of Region II is computed using an area-weighted average value of the compression moduli of the DM column and the surrounding soft soil. The additional stress in Region II is computed using Jones's solution (1962), which is a plane stress solution for the vertical stress caused by embankment load in two-layer or three-layer systems. The compression of the soil in Region II (S_{II}) can be expressed as,

$$S_{II} = [(\sigma_1 + \sigma_2)l_2]/(2E_{II}^{eq}) \tag{16}$$

in which, σ_1 is the vertical stress on the bottom of Region I, σ_2 is the vertical stress on the bottom of Region II, l_2 is the thickness of Region II, and E_{II}^{eq} is the compression modulus of soil in Region II, i.e., $E_{II}^{eq} = mE_{DM} + (1 - m)E_s$.

3.3   Compression of Soil in Region III

The compression of the soil in Region III is computed using Boussinesq's solution:

$$S_{III} = \eta\sigma_2 l_3/E_{III}^{eq} \tag{17}$$

in which η is the coefficient of average superimposed stress, E_{III}^{eq} is the compression modulus of soil in Region III, l_3 is the thickness of Region III.

Based on the above derivations, the solution to calculate the compressions of the three regions were developed and the total settlement of the soil with SDM column can be obtained.

4   Application of the Proposed Method

A test embankment was constructed at the northern part of the Asian Institute of Technology (AIT) Campus, Thailand (Vootttipruex et al. 2011). Figure 4 shows the configuration of the test embankment. The height of the embankment was 6 m consisting of a 1 m thick sand cushion on the pile head. SDM columns were installed in a square pattern at a spacing of 2.0 m with a diameter of 0.6 m and a length of 7 m. The concrete core pile had a square cross section of 0.22 m × 0.22 m and a length of 6 m. Table 1 tabulates the main properties of the soil layers and the SDM columns used in the field test.

According to the Asaoka method (1978), the final settlement can be predicted based on the field data. Table 2 presents the comparison of the settlement calculated using the proposed method with that using the Asaoka method and a good agreement was obtained with each other. The settlements calculated by the technical specification for strength composite piles in China (JGJ/T 327-2014 2014) are also listed in Table 2. It can be seen that this code significantly underestimates the settlement of SDM column-reinforced soil, especially in Region I. It might be due to the reason that this code did not consider the punching effect of core pile. The above analysis is demonstrated that the proposed method is feasible for the calculation of SDM column-supported embankment over soft soil.

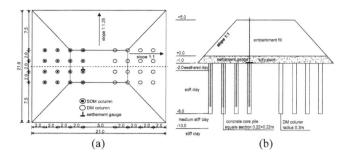

Fig. 4. Test embankment (unit: m): (a) Plan view; (b) Cross section

Table 1. Table captions should be placed above the tables.

Materials	d (m)	E (MPa)	γ (kN/m³)	μ	φ (°)	c (kPa)
Embankment fill		30	16	0.30	20	10
Sand cushion	1	30	17	0.30	25	0
Weathered crust	1	2.5	17	0.40	23	10
Soft clay	6	2.5	15	0.42	23	2
Medium stiff clay	2	5	18	0.38	25	10
Stiff clay	15	9	19	0.33	26	30
DM pile	7	45	15	0.30	30	
Concrete core pile	6	28000	24	0.25		

Note: d = depth; E = compression modulus; γ = unit weight;
μ = poisson's ratio; φ = friction angle; c = cohesion.

Table 2. Comparison of predictions (mm)

	Region I	Region II	Region III	Settlement
Proposed method	42.4	13.1	118.2	173.7
Asaoka method	N/A	N/A	N/A	161.3
(JGJ/T 327-2014)	0.04	13.9	92.9	106.8

5   Conclusions

An analytical method was proposed to calculate the settlement of SDM column-supported embankment over soft clay. The total settlement of the SDM column-reinforced subsoil is a sum of the compression of three regions: Region I (the soil within the length of concrete core pile), Region II (the soil between core pile base and DM column base), and Region III (the soil below DM column base). The punching effect of core pile upward to the sand cushion and downward to the DM column is considered. The solutions for calculating the compressions of the soils in the three regions were developed. Finally the proposed method was used to predict the settlement of a case history. The feasibility of the proposed method was verified by a comparison between the analytical results and the field measurements.

Acknowledgments. The authors appreciate the financial support provided by the Natural Science Foundation of China (NSFC) (Grant No. 51508408 & No. 51078271) and the Pujiang Talents Scheme (No. 15PJ1408800) for this research.

References

Asaoka, A.: Observational procedure of settlement prediction. Soils Found. **18**(4), 87–101 (1978)

Bjerrum, L., Johannesson, I.J., Eide, O.: Reduction of skin friction on steel piles to rock. In: Proceedings of the 7th International Conference on Soil Mechanics and Foundations Engineering, Montreal (1969)

Chen, X.F.: The Theory and Building Cases of Settlement Computation. Science Publication, Beijing (2005)

Jones, A.: Tables of stressed in three-layer elastic systems. Highw. Res. Board Bull. (342), 176–214 (1962)

Ling, G.R., An, H.Y., Xie, D.Z.: Experimental study on concrete core mixing pile. J. Build. Struct. **22**(2), 92–96 (2001). (in Chinese)

Qian, Y.J., Xu, Z.W., Deng, Y.G., Sun, G.M.: Engineering application and test analysis of strength composite piles. Chin. J. Geotech. Eng. **35**(2), 998–1001 (2013)

Tanchaisawat, T., Suriyavanagul, P., Jamsawang, P.: Stiffened deep cement mixing (SDCM) pile: laboratory investigation. In: Excellence in Concrete Construction Through Innovation-Proceedings of the International Conference on Concrete Construction, pp. 39–48. CRC Press, Netherlands (2009)

Technical specification for strength composite piles (JGJ/T327-2014). China Architecture and Building Press, Beijing (2014)

Voottipruex, P., Bergado, D.T., Suksawat, T., Jamsawang, P.: Behavior and simulation of deep cement mixing (DCM) and stiffened deep cement mixing (SDCM) piles under full scale loading. Soils Found. **51**(2), 307–320 (2011)

Ye, G.B., Cai, Y.S., Zhang, Z.: Numerical study on load transfer effect of stiffened deep mixed column-supported embankment over soft soil. KSCE J. Civil Eng. **21**, 703–714 (2016)

Zhao, X., Wu, M., Chen, S.W., Kong, D.D.: Study on bearing behaviors of single axially loaded SDCM pile. In: Deep Foundations and Geotechnical in Situ Testing, Proceedings of the 2010 GeoShanghai International Conference, Shanghai, pp. 277–284 (2010)

Consolidation Analysis of Soft Soil by Vacuum Preloading Considering Groundwater Table Change

Yan Xu[1,2], Guan-Bao Ye[1(✉)], and Zhen Zhang[1]

[1] Department of Geotechnical Engineering, Tongji University,
Shanghai, China
ygbl030@126.com
[2] Key Laboratory of Land Subsidence Monitoring and Prevention,
Ministry of Land and Resources of China,
Shanghai Institute of Geological Survey, Shanghai, China

Abstract. Vacuum preloading is a commonly-used method to improve a large area consisting of soft soil. However, most theoretical and experimental studies have not considered the water table charge during vacuum preloading. This paper analyzed the changes of excess pore water pressure and groundwater table during the process of vacuum preloading, and confirmed that the groundwater table varied during vacuum preloading and had influence on the consolidation process. Based on the field observation, an analytical solution to calculate the consolidation degree of soil by vacuum preloading considering the groundwater table change was developed. Then a parametric study was conducted to investigate the influence factors of the drain spacing ratio, depth of ground water level, soil permeability coefficient ratio of undisturbed area and disturbed area on the consolidation. The developed solution is valuable for the prediction of consolidation degree under vacuum preloading in practice.

Keywords: Vacuum preloading · Partially saturated soil
Consolidation degree · Analytical solution · Ground water level

1 Introduction

In the past three decades, the demands for land resources have increased significantly with the rapid development of China, especially in the east coast area. However, thick layers of soft clays are widely distributed in this area. Such unfavorable geotechnical condition poses many challenges to the constructions of infrastructures and buildings, such as low bearing capacity, excessive settlement and slope instability. Among the various ground improvement techniques for large area soft soil, vacuum preloading method is one of the most cost-efficient method in practice.

Various studies have focused on the consolidation of soil by vacuum preloading (Dong 1992; Xie 1995; Gong and Cen 2002; Xu et al. 2012). However, few of them considered the groundwater table change during vacuum preloading. Dong (2001) established a calculation formulas for the soil foundation's ground water level under vacuum preloading; Ming and Zhao (2005) analyzed groundwater level under different

© Springer Nature Singapore Pte Ltd. 2018
L. Li et al. (Eds.): GSIC 2018, *Proceedings of GeoShanghai 2018 International Conference: Ground Improvement and Geosynthetics*, pp. 30–40, 2018.
https://doi.org/10.1007/978-981-13-0122-3_4

water current boundary; Zhu et al. (2004) and Zhou et al. (2009) found that the groundwater level descended rapidly at the beginning of vacuum preloading and then became relatively stable; Qiu et al. (2006) put forward that vacuum suction could induce unsaturated zone in the soil.

In this paper, the changes of excess pore water pressure and groundwater table during the process of vacuum preloading were analyzed. Based on the field observation, an analytical solution to calculate the consolidation degree of soil by vacuum preloading considering the groundwater table change was developed using the simplified calculation model of moisture mixed fluid. Then a parametric study was conducted to investigate the influence factors of the drain spacing ratio, depth of ground water level, soil permeability coefficient ratio of undisturbed area and disturbed area on the consolidation.

2   Analysis on Field Observation

The selected project is located in the eastern suburb of Shanghai, China. Table 1 shows the subsoil profiles and properties from the ground surface. The groundwater table was at a depth of 0.3 m–2.4 m from the ground surface. The project for vacuum preloading had a dimension of 150 m × 200 m. The PVDs with a cross section of 100 mm×4 mm and a length of 20.5 m were installed in a triangular pattern at a spacing of 1.1 m. A 0.5 m sand blanket was placed on the ground surface acting as a platform for placing the horizontal perforated pipes. A layer of impermeable membrane was used to cover the test area. The pore-water pressure transducers were installed every 2 m along depths to measure the pore-water pressures in soil. The total periods for loading and consolidation were 90 days.

In vacuum preloading, the vacuum pressure under membrane was applied to −85 kPa in 10 days and then kept constant of −85 kPa for 90 days for soil consolidation. Figure 1 illustrates the excess pore water pressure distribution along depth in

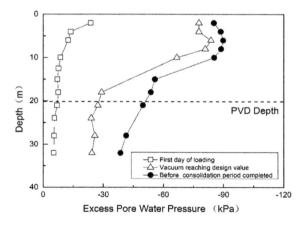

Fig. 1. Distribution of excess pore water pressure along depth

the reinforced area center at different time, i.e., on the first day of loading, on the day when the vacuum pressure reached −85 kPa under the membrane, and on the day before the consolidation period was completed. The excess pore water pressure did not respond promptly during the loading period. It can be explained by that since the permeability of soft soil was small, the vacuum pressure had to overcome the resistance of soil to transfer the negative excess pore water pressure to the deeper soil. It can also be seen that basically the negative excess pore pressure decreased with the depth increased from the ground surface. However, it must be noticed that the negative excess pore pressure appeared to increase within the depth between 2 m and 6 m. The increase of excess pore pressure ranged from 2 m to 6 m can be explained by the decrease of groundwater table.

Table 1. Soil properties on selected project site

Soil layer		D (m)	E (MPa)	γ (kN/m^3)	e_0	c' (kPa)	φ' (°)	k_v (cm/s)	υ
①$_1$	Plain fill	0.80	2.00	17.4	1.21	20	16.0	8.58e−08	0.30
②	Silty clay	1.60	4.23	18.4	0.93	20	19.0	1.08e−07	0.35
③	Soft silty clay	3.00	3.00	17.6	1.12	12	17.5	1.08e−07	0.43
③t	Clayey silt	1.90	8.56	18.7	0.82	5	33.0	8.40e−05	0.35
③	Soft silty clay	2.70	3.00	17.6	1.12	12	17.5	1.08e−07	0.45
④	Very soft clay	9.30	2.03	16.8	1.39	10	12.5	5.06e−08	0.45
⑤$_1$	Clay	7.70	2.84	17.4	1.21	13	13.5	1.27e−07	0.35
⑤$_3$	Silty clay	10.00	4.53	18.0	1.21	18	20.5	1.08e−07	0.35

Note: D = layer thickness; E = elastic modulus; γ = unit weight; e_0 = initial void ratio; c' = effective cohesion; φ' = effective friction angle; k_v = vertical permeability; υ = Poison's ratio.

Figure 2 shows the variation of excess pore water pressure with time at the depth of 6 m during the process of vacuum preloading. Theoretically, the value of excess pore water pressure in the top soli layer equals to the applied vacuum pressure (−85 kPa). Considering the decrease of groundwater table due to the applied vacuum loading, the difference resulting in the excess pore water pressure was approximately 5 kPa. In that case, considering the change of ground water level, the theoretical excess pore pressure in top layers can be −90 kPa, which was agreed with the alternative excess pore pressure from the field. Based on the analysis on the field data, it can be concluded that the groundwater table was decreased during vacuum preloading, resulting in an unsaturated zone of soil with high saturation degree would appear above the groundwater table.

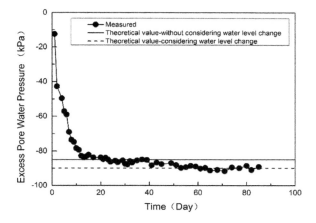

Fig. 2. Distribution curve of excess pore water pressure with time

3   Calculation Model

3.1   Basic Assumptions

Regarding the partially saturated soil with high saturation degree ($S_r > 80\%$) in the process of vacuum preloading, the water flow is mainly followed by the gas phase because the macro-pores were dominated by water phase. For simplicity, the pore gas and water mixture is equivalent to a compressible mixture fluid, and then the partially saturated soil can be reduced to two phase body consisting of the soil skeleton and the mixture fluid.

It is assumed that the permeability coefficient of soil, compression coefficient, effective stress coefficient, and the total stress are constant during consolidation, the consolidation is under one-way drainage condition. The effective stress principle equation can be expressed as (Bishop and Blight 1963)

$$\Delta\sigma' = -\chi\Delta u_w - (1 - \chi)\Delta u_a \qquad (1)$$

in which, u_w is the excess pore water pressure; u_a is the excess pore air pressure; $\Delta\sigma'$ is the increment of effective stress; χ is the Bishop effective stress coefficient; $\chi = S_r/(0.4S_r + 0.6)$; S_r is saturation degree.

Equation (1) indicates that the skeleton stress increment includes two parts: the effective stress caused by hydraulic dissipation $-\chi\Delta u_w$ and effective stress caused by pressure dissipation $-(1 - \chi)\Delta u_a$. The volume compression Δe_w is caused by soil skeleton stress increment $-\chi\Delta u_w$ corresponding to hydraulic dissipation, which is not affected by air pressure. The moisture mixed fluid equation can be expressed as,

$$u_m = (1 - \chi)u_a + \chi u_w \qquad (2)$$

Thus, the Eq. (1) can be rewritten as $\sigma = \sigma' + u_m$, which has a consistent form with the effective stress principle equation of saturated soil.

3.2 Governing Equation

Based on the above assumptions and refer to the simplified consolidation calculation model for partially saturated soil proposed by Cao and Yin (2009), the continuity equation of moisture mixed fluid can be expressed as

$$\frac{\partial \varepsilon_v}{\partial t} = -\frac{k_m}{\gamma_m}\left(\frac{\partial^2 u_m}{\partial r^2} + \frac{1}{r}\frac{\partial u_m}{\partial r}\right) + \frac{\partial \varepsilon_{v1}}{\partial t} \tag{3}$$

$$\frac{\partial \varepsilon_{v1}}{\partial t} = \frac{1}{B_m}\frac{\partial u_m}{\partial t} \tag{4}$$

in which, u_m is the excess moisture mixed fluid pressure; k_m is the permeability coefficient of moisture mixed fluid (assumed to be constant) and can be expressed as $k_m = \frac{2k_{wu}}{S_r}$; k_{wu} is the permeability coefficient of soil and can be expressed as $k_{wu} = k_w\frac{1+e_0}{1+e}\left(\frac{S_\varepsilon}{e_0}\right)^3$; k_w is the permeability coefficient of saturated soil; γ_m is the unit weight of moisture mixed fluid and can be expressed as $\gamma_m = S_r\gamma_w$; $\partial\varepsilon_{v1}/\partial t$ is the compression of residual pore fluid; B_m is the bulk compression modulus of moisture mixed fluid.

According to the variation of water and moisture mixed fluid flow in soil, the continuity equation of water can be established as

$$k_w\left(\frac{\partial^2 u_w}{\partial r^2} + \frac{1}{r}\frac{\partial u_w}{\partial r}\right) = k_m\left(\frac{\partial^2 u_m}{\partial r^2} + \frac{1}{r}\frac{\partial u_m}{\partial r}\right) \tag{5}$$

According to the above basic equation and the equilibrium differential equation:

$$\frac{\partial \varepsilon_v}{\partial t} = -\frac{k_m}{\gamma_m}\left(\frac{\partial^2 u_m}{\partial r^2} + \frac{1}{r}\frac{\partial u_m}{\partial r}\right) + \frac{1}{B_m}\frac{\partial u_m}{\partial t}, \quad r_s < r \le r_e \tag{6}$$

$$\frac{\partial \varepsilon_v}{\partial t} = -\frac{k_{ms}}{\gamma_m}\left(\frac{\partial^2 u_m}{\partial r^2} + \frac{1}{r}\frac{\partial u_m}{\partial r}\right) + \frac{1}{B_m}\frac{\partial u_m}{\partial t}, \quad r_w \le r \le r_s \tag{7}$$

$$\frac{\partial \bar{u}_m}{\partial r} - \frac{1}{a_v}\frac{\partial \varepsilon_v}{\partial r} = -P_0 \tag{8}$$

Childs and Collis-George (1950) experimentally demonstrated that the water flow rate of partially saturated soil was linear proportional to the hydraulic gradient, and Darcy's law was also applicable, which was also discovered by Fredlund and Raharjdo (1993). Therefore, the moisture mixed fluid flow in unit time in the drain can be expressed as

$$\partial q = \pi r_w^2 \partial V = \pi r_w^2 k_p \partial I = \pi r_w^2 \frac{k_p}{\gamma_m}\frac{\partial^2 u_p}{\partial z^2}dz \tag{9}$$

Thus, the water outflowed from the soil equals to the increment of seepage flow upward in the sand drain:

$$\frac{\partial^2 u_p}{\partial z^2} = -\frac{2k_{ms}}{\gamma_m k_p} \frac{\partial u_m}{\partial r}\Big|_{r=r_w} \tag{10}$$

4 Analytical Solution

The origin point of the axisymmetric analytical model was the equivalent sand drain (plastic drainage plate in vacuum preloading) center, and the coordinate system $r - z$ was established. The boundary condition in partially saturated soil can be expressed as, ① $\frac{\partial u_m}{\partial r} = 0$, where $r = r_e$; ② $u_m = u_p$, at $r = r_w$; ③ $u_p = -P_0$, at $z = 0$; ④ $u_p = u_p'$, $u_m = u_w$, $k_{ms}\frac{\partial u_m}{\partial z} = k_{ws}\frac{\partial u_w}{\partial z}$, $k_m\frac{\partial u_m}{\partial z} = k_w\frac{\partial u_w}{\partial z}$, at $z = h$; ⑤ $\bar{u}_m = 0$, at $t = 0$, in which, u_w is the excess pore water pressure in saturated soil; u_p' is the excess pore water pressure in sand drain located in saturated soil; h is the depth of saturated soil; H is sand drain length; k_{ms} is the permeability coefficient of moisture mixed fluid in smear zone; k_w is the permeability coefficient of saturated soil in smear zone; r_e, r_s and r_w were the radius of sand well influence area, smear zone and equivalent sand well, respectively; $-P_0$ is the applied vacuum pressure.

Using boundary conditions ① and ②, the integral of Eqs. (6) and (7) can be expressed as

$$u_m = \frac{T_s}{2} \cdot \left(r_s^2 \ln\frac{r}{r_w} - \frac{r^2 - r_w^2}{2} \right) + \frac{T_m}{2} \cdot R_s \ln\frac{r}{r_w} + u_p, \quad r_w \leq r \leq r_s \tag{11}$$

$$u_m = \frac{T_m}{2} \cdot \left[r_e^2 \ln\frac{r}{r_w} - \frac{r^2 - r_s^2}{2} - L_s \right] + \frac{T_s}{2} \cdot \left(L_s - \frac{R_{sw}}{2} \right) + u_p, \quad r_s < r \leq r_e \tag{12}$$

in which, $T_m = \frac{\gamma_m}{k_m}\left(\frac{\partial \varepsilon_v}{\partial t} - \frac{1}{B_m}\frac{\partial u_m}{\partial t} \right)$, $R_{sw} = r_s^2 - r_w^2$, $T_s = \frac{\gamma_m}{k_{ms}}\left(\frac{\partial \varepsilon_v}{\partial t} - \frac{1}{B_m}\frac{\partial u_m}{\partial t} \right)$, $R_s = r_e^2 - r_s^2$, $L_s = r_s^2 \ln\frac{r_s}{r_w}$.

The average pore pressure at certain depth in foundation can be expressed as

$$\bar{u}_m = \frac{1}{\pi(r_e^2 - r_w^2)} \left(\int_{r_w}^{r_s} 2\pi r u_m dr + \int_{r_s}^{r_e} 2\pi r u_m dr \right) \tag{13}$$

Substitute Eqs. (11) and (12) into Eq. (13) we obtained

$$\bar{u}_m = T_s \cdot \left(\frac{L_e r_s^2}{2R_w} - \frac{R_w + 2r_s^2}{8} \right) + T_m \cdot \left(\frac{L_e R_s}{2R_w} - \frac{R_s}{4} \right) + u_p \tag{14}$$

in which, $R_w = r_e^2 - r_w^2$, $L_e = r_e^2 \ln\frac{r_e}{r_w}$.

It can be derived from Eqs. (8) and (10) that

$$-m_v\frac{\partial\bar{u}_m}{\partial t} - \frac{1}{B_m}\frac{\partial u_m}{\partial t} = \beta(\bar{u}_m - u_p)\tag{15}$$

$$\frac{\partial^2 u_p}{\partial z^2} = \frac{K}{B_m}\frac{\partial u_m}{\partial t} + K\cdot m_v\frac{\partial\bar{u}_m}{\partial t} = -K\beta(\bar{u}_m - u_p)\tag{16}$$

in which, m_v is the coefficient of volume compressibility, $K = \frac{k_{ms}R_s + k_m R_{sw}}{k_m k_p r_w}$, $\beta = \frac{8k_m k_{ms}R_w}{4\gamma_m k_m L_e r_s^2 - \gamma_m R_w k_m (R_w + 2r_s^2) + \gamma_m R_s k_{ms}(4L_e - 2R_w)}$.

Assume that $u_p(z,t) = \sum_{k=0}^{\infty} A_1 \sin(\frac{M}{H}z)e^{-\lambda t} - P_0$, and make use of the boundary condition ③ and orthogonality of function system $\sin(\frac{M}{H}z)$ in $[0,H]$. It can be derived from Eqs. (15) and (16) that

$$\bar{u}_m = \sum_{k=0}^{\infty} A_2 e^{-\lambda t}\sin(\frac{M}{H}z) - P_0\tag{17}$$

in which $A_2 = \sum_{k=0}^{\infty}\frac{2P_0}{M}$, $M = \frac{2k+1}{2}\pi, k = 0, 1, 2\cdots$.

Based on the results of Dong (1992), assuming that the sand drain parameters were consistent in saturated and unsaturated zone, the solution of saturated zone ($h \leq z < H$) can be expressed as

$$u_w = \begin{cases} -u_p\sum_{k=0}^{\infty} e^{-B_r t}\frac{2A}{M}\sin\left(\frac{M(z-h)}{H-h}\right) + u_p, & r_w \leq r \leq r_s \\ -u_p\sum_{k=0}^{\infty} e^{-B_r t}\frac{2B}{M}\sin\left(\frac{M(z-h)}{H-h}\right) + u_p, & r_s \leq r \leq r_e \end{cases}\tag{18}$$

in which, $A = -\frac{k_w B_r}{K_{ws}\lambda' F_a}\left(\ln\frac{r}{r_w} - \frac{r^2 - r_w^2}{2r_e^2}\right) + \frac{\lambda' - B_r}{\lambda'}$, $d_e = 2r_e$, $C_h = k_w E_s/\gamma_w$, $B = \frac{B_r}{\lambda' F_a}\left[\ln\frac{r}{r_s} - \frac{r^2 - r_s^2}{2r_e^2} + \frac{K_w}{K_{ws}}\left(\ln s - \frac{s^2 - 1}{2n^2}\right)\right] + \frac{\lambda' - B_r}{\lambda'}$, $B_r = \frac{8C_h}{d_e^2}\left(F_a + \frac{8}{M^2}\frac{n^2 - 1}{n^2}G\right)$, $n = r_e/r_w$, $G = (K_w/K_p)\left(\frac{H-h}{d_w}\right)^2$, $\lambda' = \frac{8C_h}{d_e^2 F_a}$.

Because of $k_{ms}\frac{\partial u_m}{\partial z} = k_{ws}\frac{\partial u_w}{\partial z}$ where $z = h$, it can be derived that $\lambda = \frac{B_m\beta}{\left(\frac{k_{ws}}{k_{ms}} + k_{ms}B_m m_v\right)K\beta\frac{H^2}{M^2} + B_m m_v}$.

Therefore, the final solution of the unsaturated zone ($0 \leq z < h$) can be obtained:

$$u_m = \begin{cases} \sum_{k=0}^{\infty}\frac{MP_0\beta Ce^{-\lambda t}}{M^2 + K\beta H^2}\sin(\frac{M}{H}z) - P_0, & r_w \leq r \leq r_s \\ \sum_{k=0}^{\infty}\frac{MP_0\beta De^{-\lambda t}}{M^2 + K\beta H^2}\sin(\frac{M}{H}z) - P_0, & r_s < r \leq r_e \end{cases}\tag{19}$$

in which, $C = \frac{\gamma_m}{k_{ms}}\left(r_s^2 \ln\frac{r}{r_w} - \frac{r^2-r_w^2}{2}\right) + \frac{R_s\gamma_m \ln r/r_w}{k_m} + \frac{2KH^2}{M^2}$, $D = \frac{\gamma_m}{k_m}\left(r_e^2 \ln\frac{r}{r_w} - \frac{r^2-r_s^2}{2} - L_s\right) + \frac{\gamma_m}{k_{ms}}\left(L_s - \frac{R_{sw}}{2}\right) + \frac{2KH^2}{M^2}$.

5 Analysis of Influence Factors

To study the soil consolidation characteristics under vacuum preloading considering the groundwater table change, the proposed analytical solution of average consolidation degree was used to discuss the some influence factors on the soil consolidation. Three influence factors were discussed herein, namely the drain spacing ratio $n = r_e/r_w$, depth of ground water level h, soil permeability coefficient ratio of undisturbed area and disturbed area k_h/k_s. The time factor can be defined as $T = C_h t/H^2$. The parameters in the baseline case were as follows: $p = 85\,\text{kPa}$, $l = 1.1\,\text{m}$, $r_e = 1.05 \cdot l/2$, $r_w = 0.033\,\text{m}$, $r_s = 0.066\,\text{m}$, $k_p = 540\,\text{m/d}$, $B_m = 3\,\text{MPa}$, $E_s = 3.5\,\text{MPa}$, $k_w = 2 \times 10^{-4}\,\text{m/d}$, $S_r = 0.85$, $h = 2.5\,\text{m}$, $e_0 = 1.1$.

5.1 Effects of Drain Spacing Ratio n

Figure 3 illustrates the influence of drain spacing ratio on the consolidation degree with time factor. It can be seen that with the increase of drain spacing ratio, the soil consolidation rate was reduced. In the early stage of consolidation, the drainage path was short and the effective stress increased rapidly near the drain and ground surface which resulted in the decrease of permeability, thus the far soil drainage channel was hindered and led to the decline of soil consolidation rate (Guo 2015). The discharge of water and gas in unsaturated zone was accelerated effectively due to the smaller drain spacing ratio, and the seepage path and consolidation rate was gradually stable in the late consolidation process.

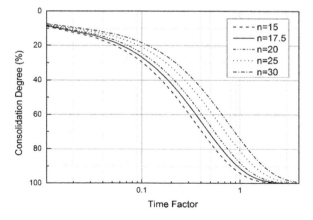

Fig. 3. Effects of n on consolidation behaviors

5.2   Effects of Groundwater Table Change

Figure 4 shows the influence of groundwater table change on the soil consolidation. In Fig. 4, h was the depth of underground water level when pore water pressure was stable, namely the unsaturated zone depth for calculation. It can be seen that the soil consolidation rate decreased with the decrease of groundwater table. Therefore, from the practical point of view, reasonable evaluation of groundwater table change in the process of vacuum preloading is one of the key factors to correctly predict soil consolidation.

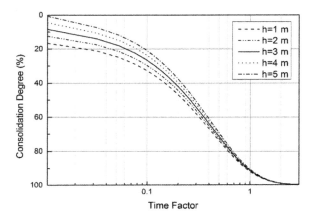

Fig. 4. Effects of h on consolidation behaviors

5.3   Effects of Soil Permeability Coefficient Ratio

Figure 5 illustrates the effect of horizontal permeability coefficient ratio of soil in undisturbed area and smear zone on the consolidation characteristics under vacuum

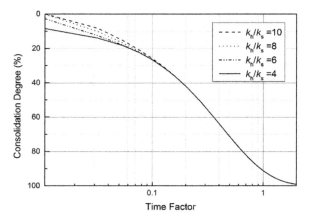

Fig. 5. Effects of k_h/k_s on consolidation behaviors

preloading. It can be seen that the soil consolidation rate decreased with the increase of permeability coefficient ratio k_h/k_s. During the process of vacuum preloading, the insertion of PVD caused additional disturbance to the surrounding soil. Therefore, further study is still needed to figure out the scope and degree of PVD insertion influence on soil in smear zone.

6 Conclusions

(1) Based on the analysis on the field data, it can be concluded that the groundwater table was decreased during vacuum preloading, resulting in an unsaturated zone of soil with high saturation degree would appear above the groundwater table.

(2) Based on the field observation, an analytical solution to calculate the consolidation degree of soil by vacuum preloading considering the groundwater table change was developed using the simplified calculation model of moisture mixed fluid.

(3) The discharge of water and gas in unsaturated zone was accelerated effectively due to the smaller drain spacing, and the soil consolidation rate decreased with the decrease of groundwater table.

Acknowledgements. The authors appreciate the financial support provided by the National Natural Science Foundation of China (NSFC; Nos. 41272294 and 51508408) for this work. And the corresponding author is obliged to the Pujiang Talents Scheme (No. 15PJ1408800) for the continuous support for his research.

References

Dong, Z.: Analytical theory of sand drain consolidation under preloading and vacuum preloading. Port Waterway Eng. **09**, 1–7 (1992)

Xie, K.: Consolidation theory of double-layered ground with vertical ideal drains under equal strain condition. J. Zhejiang Univ. (Nat. Sci.) **05**, 529–540 (1995)

Gong, X., Cen, Y.: Mechanism of vacuum preloading. J. Harbin Univ. C.E. Archit. **35**(2), 7–10 (2002)

Xu, Y., Indraratna, B., Rujikiatkamjorn, C.: Analytical solutions for a single vertical drain with vacuum and time-dependent surcharge preloading in membrane and membraneless systems. Int. J. Geomech. **12**(1), 27–42 (2012)

Dong, Z.: Analysis and calculation of the soil foundation's ground water level and measuring tube's water level under the situations of surcharge preloading and vacuum preloading. Port Waterway Eng. **08**, 15–19 (2001)

Ming, J., Zhao, W.: Study on groundwater level in vacuum preloading. Port Waterway Eng. **01**, 1–6 (2005)

Zhu, J., Li, W., Gong, X.: Vacuum combined pile load preloading in soft ground reinforcement of the underground water level monitoring results analysis. Geotech. Investig. Surv. **05**, 27–30 (2004)

Zhou, Q., Liu, H., Gu, C.: Field tests on groundwater level and yield of water under vacuum preloading. Rock Soil Mech. **11**, 3435–3440 (2009)

Qiu, Q., Mo, H., Dong, Z., Zeng, Q.: Discussion on unsaturated zone in soft ground improved by vacuum preloading. Chin. J. Rock Mech. Eng. **S2**, 3539–3544 (2006)

Cao, X., Yin, Z.: Simplified computation of two-dimensional consolidation of unsaturated soils. Rock Soil Mech. **09**, 2575–2580 (2009)

Su, W., Xie, K.: Analytical solution of 1D consolidation of unsaturated soil by mixed fluid method. Rock Soil Mech. **08**, 2661–2665 (2010)

Bishop, A.W., Blight, G.E.: Some aspects of effective stress in saturated and unsaturated soils. Geotechnique **13**(3), 177–196 (1963)

Childs, E.C., Collis-George, N.: The permeability of porous materials. Proc. Roy. Soc. A **201** (1066), 392–405 (1950)

Fredlund, D.G., Rahardjo, H.: Soil Mechanics for Unsaturated Soils. Wiley, New York (1993)

Guo, B.: Theory and New Technology Study on Using Vacuum Preloading to Improve Recently Filled Dredge. Tianjin University, Tianjin (2015)

DEM Analysis of the Effect of Grain Size Distribution on Vibroflotation Without Backfill

Mingjing Jiang[1,2,3(✉)], Huali Jiang[1,2,3], and Banglu Xi[1,2,3]

[1] State Key Laboratory of Disaster Reduction in Civil Engineering,
Tongji University, Shanghai, China
mingjing.jiang@tongji.edu.cn
[2] Department of Geotechnical Engineering, Tongji University, Shanghai, China
[3] Key Laboratory of Geotechnical and Underground Engineering
of Ministry of Education, Tongji University, Shanghai, China

Abstract. Vibroflotation without backfill is a widely-employed ground improvement technique in geotechnical engineering, which has been proven to be a quite effective method for granular sands. However, some mechanics still remain unclear, such as the effect of the grain size distribution (GSD) of the ground. Hence, firstly, two numerical grounds were generated using the Distinct Element Method (DEM), which only differed in the GSD while other parameters were the same. Then, the vibroflotation without backfill was simulated in the two numerical grounds. Finally, the effect of the GSD on vibroflotation without backfill was studied by comparing the relative change of void ratio during vibroflotation and presenting the micro information, i.e., particle velocity field. The results show that a more efficient densification, a better treatment and a larger affected area can be achieved by the vibroflotation without backfill in the test ground with bigger uniformity coefficient.

Keywords: Vibroflotation without backfill · Distinct Element Method
Grain size distribution

1 Introduction

Vibroflotation without backfill is a widely-employed ground improvement technique, which can significantly enhance the strength and bearing capacity of unsaturated granular soils through vibration and water jetting. Because of its simple process, low cost and good treatment effect, vibroflotation has been widely applied in engineering practices successfully since it was firstly proposed by Steuerman in 1936 [1, 2].

In last decades, lots of researches about vibroflotation were carried out with focus mainly on the densification mechanism, construction technology and applicability. Webb and Hall [3] performed vibroflotation tests on clayey sands and the results showed that the vibroflotation could obtain good treatment effect when the fine content of the ground reached 30%. Slocombe et al. [4] investigated the densification of vibro method on granular soils and revealed that vibroflotation could enhance the ground with fine content more than 45% by improving the power and construction technology. However, Saito [5] found that vibroflotation could not improve the ground with fine content more

L. Li et al. (Eds.): GSIC 2018, *Proceedings of GeoShanghai 2018 International Conference: Ground Improvement and Geosynthetics*, pp. 41–47, 2018.
https://doi.org/10.1007/978-981-13-0122-3_5

than 20%. Such different conclusions imply that there still remain some arguments on the applicability of vibroflotation, which should be caused by the different grain size distribution (GSD) of the ground, which constitutes the strong motivation of this study.

Physical model tests and numerical simulations are usually employed to investigate vibroflotation. Physical model tests can provide reliable results, but it is expensive, time-consuming and the micro information can hardly be observed for further analysis. Hence, numerical simulation techniques, which can provide both reasonable results and micro information, is a promising method. Since the Finite Element Method, which has been widely used in geotechnical engineering, works poorly in the problems related to large deformation and soil failures, the Distinct Element Method (DEM) [6], which treats the soil as non-continuous aggregates, is an alternative tool to simulate the vibroflotation process. Jia et al. [7] performed DEM simulations of vibroflotation without backfill and found that the simulation results agreed well with the experimental data. Jiang et al. [8] employed DEM to study the macro-micro mechanics of vibroflotation in dry and wet vibration conditions. These researches prove that DEM is an effective method to simulate the vibroflotation process. In addition, different GSDs can be easily taken into account in DEM simulations. Hence, the DEM is employed to simulate the vibroflotation process in this study.

In this paper, the effect of GSD on the treatment effect of vibroflotation without backfills is studied with DEM. Firstly, two DEM test grounds which differed only in the GDS were generated. Then, the process of vibroflotation without backfills was simulated in the two grounds. Finally, the effect of GDS on vibroflotation was analyzed by examining the relative change of the void ratio and presenting the particle velocity field.

2 DEM Simulations

2.1 Contact Model

Figure 1 schematically illustrates the mechanical response of the contact model [9] used in this study. The contact model consists of the normal, tangential and rolling contact components, which resist the normal force F_n, shear force F_s and rolling moment M, respectively. The forces or moment can be calculated as follows:

$$F_n = k_n \times u_n \tag{1}$$

$$F_s = k_s \times u_s \tag{2}$$

$$M = \begin{cases} K_m \theta = \frac{\theta \times k_n \times r^2 \times \beta^2}{12}, & \theta \leq \theta_r, \\ M_0 = \frac{1}{6}(F_n \times r \times \beta), & \theta > \theta_r. \end{cases} \tag{3}$$

where k_n, k_s and k_m are the normal, tangential and rolling contact stiffness, respectively; u_n is the relative normal displacement, u_s is the relative tangential displacement and θ is the relative rotation angle; β is the rolling resistance coefficient, θ_r is critical relative

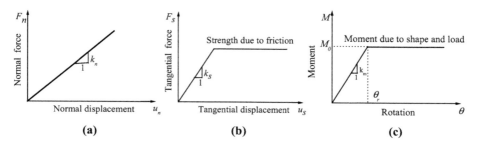

Fig. 1. The mechanical response of the contact model: (a) normal contact model; (b) tangential contact model; (c) rolling contact model [9].

rotation angle that separates the elastic from the plastic case, M_0 is the critical moment corresponding to θ_r and r is the average radius of the two contact particles, namely:

$$r = \frac{2r_1 r_2}{r_1 + r_2} \tag{4}$$

where r_1 and r_2 are the radii of two contact particles, respectively.

2.2   Simulation Conditions

Previous researches reveal that vibroflotation shows good treatment effects on the sands within a certain range of GSD, as shown in Fig. 2(a) [1]. Figure 2(a) shows that the uniformity coefficient C_u and curvature coefficient C_c of the GSD curves fall into the ranges of 6.41–6.49 and 1.10–2.05, respectively. In DEM simulation, a large range of grain size will result in computational inefficiency. Thus, two GSDs with small values of C_u, which reflects the range of grain size, were selected in this study, as shown in Fig. 2(b). The GSD-A ranges from 6–9 mm with C_u being 1.3 and C_c being 1.1, while

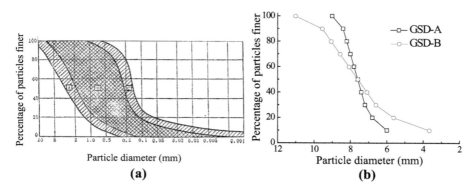

Fig. 2. The grain size distributions treated by vibroflotation without backfill: (a) commonly used; (b) in this study.

the GSD-B ranges from 3.7–11 mm with C_u being 2.2 and C_c being 1.5. Note that the values of mean diameter d_{50} of the two GSDs are both 7.6 mm.

Then, biaxial compression test simulations were performed to obtain the mechanical properties of the two sands with the selected GSDs. 10000 particles were used to generate a homogeneous sample for each sand using the Multi-layer under-compaction method (UCM) [10]. Table 1 presents the parameters used in the DEM simulations. The biaxial compression test results show that the measured internal friction angle of the sand with GSD-A (i.e., $\varphi = 24°$) is slightly higher than that of the sand with GSD-B (i.e., $\varphi = 22°$).

Table 1. Model parameters used in the simulations.

Parameters	Value
Initial void ratio	0.24
Particle density (kg/m^3)	2600
Inter-particle normal stiffness (N/m)	1.5×10^8
Inter-particle tangential stiffness (N/m)	1.0×10^8
Rolling resistance coefficient	0.4
Friction coefficient between particles	0.5
Normal stiffness of boundaries (N/m)	1.5×10^8
Tangential stiffness of boundaries (N/m)	1.0×10^8
Friction coefficient of boundaries	0

2.3   The Process of Vibroflotation

Figure 3 provides the schematic diagram of the test ground, the length and width of which are 5 m and 2.5 m respectively. Two test grounds, whose contact parameters and compacting method were the same as the samples used in the biaxial compression tests, were generated with the selected GSDs. The test grounds were then consolidated under 2 g gravity filed, after which a square vibrator with length $d = 0.2$ m was generated at the center of each test ground and vibrated in the horizontal direction periodically. Figure 4 shows the periodical variation of the loading velocity of the vibrator, where

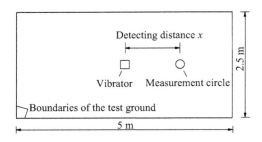

Fig. 3. The schematic diagram of the ground.

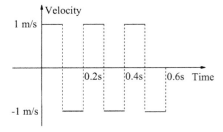

Fig. 4. The loading velocity of the vibrator.

the period T is 0.2 s and the maximum velocity v_{max} is 1 m/s. When the velocity is positive, the vibrator moves right, otherwise, it moves left. For simplification, the penetration process of the vibrator and the influence of excess pore water pressure are ignored in this study. In addition, to detect the change of the void ratio, a series of measurement circles with different detecting distances x (i.e., the distance between the measurement circle and the vibration point.) were arranged in the test grounds.

3   Simulation Results

3.1   The Distribution of Void Ratio

Figure 5 presents the relative change of the void ratio e_{re} in the measurement circle at the detecting distance $x = 1.1$ m during the vibroflotation process. The relative change of the void ratio is calculated by the following equation:

$$e_{re} = \frac{e_{ini} - e_{mear}}{e_{ini}} \tag{5}$$

where e_{mear} is the measured void ratio during vibroflotation and e_{ini} is the initial void ratio. A positive value of e_{re} means that the soils are compacted while a native value means that the soils are loosened. Figure 5 shows that the relative changes of void ratio in two test grounds evolve in a similar tendency. The soils are compacted when the vibrator goes right and loosened when the vibrator goes left. As the vibrator vibrates periodically, the void ratio also fluctuates periodically. Nevertheless, after several vibroflotation loops, the relative change of the void ratio tends to be a constant. However, in our simulations, e_{re} in the test ground with GSD-B (Ground B) takes about 1.2 s ($6T$) to be stable, while the e_{re} in test ground with GSD-A (Ground A) still fluctuates at that time, which implies that a more efficient densification can be achieved in Ground B.

Figure 6 provides the e_{re} with different detecting distance at the time of $0.5T$ when the vibrator firstly reached the rightmost place. Figure 6 shows that e_{re} in two test

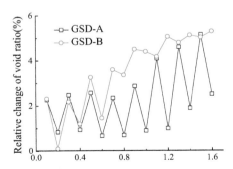

Fig. 5. The relative change of void ratio with time.

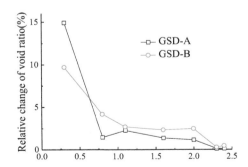

Fig. 6. The relative change of void ratio with distance.

grounds evolves in a similar trend: e_{re} decreases with the increasing detecting distance x, which implies that the densification effect of the vibroflotation weakens with the increasing distance from the vibration point. When the detecting distance is long enough, the soils seem unaffected by the vibroflotation. Moreover, e_{re} in Ground B is slightly larger than that in Ground A, which means that a better treatment effect can be obtained in ground with larger uniformity coefficient.

3.2 The Particle Velocity Field

Figure 7 illustrates the particle velocity field of the test grounds at the time of $0.5T$. Figure 7 shows that there exits an obvious affected area in the test ground and the particle velocity decreases rapidly near the boundary of the affected area. Comparison of the affected areas in the two grounds shows that a larger affected area can be observed in Ground B, which demonstrates that the vibroflotation has a larger affected area in ground with larger uniformity coefficient. The comparison, from the micro aspect, illustrates the reason why a more efficient densification and better treatment effect can be obtained in the ground with larger uniformity coefficient.

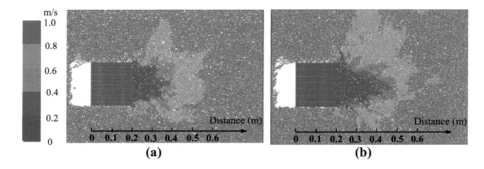

Fig. 7. The particle velocity field: (a) Ground A; (b) Ground B.

4   Conclusions

This paper focuses on investigating the effect of grain size distribution (GSD) on the treatment effect of vibroflotation without backfill using DEM. The process of vibroflotation was simulated in two test grounds with different GSDs, where the relative change of void ratio and particle velocity field were analyzed. The main conclusions can be drawn as follows:

(a) A more efficient densification and better treatment effect can be obtained in the ground with larger uniformity coefficient;
(b) A larger affected area can be observed in the ground with bigger uniformity coefficient from the particle velocity field, which confirms the better treatment effect in the ground.

Acknowledgement. The research was financially supported by the National Natural Science Foundation of China with Grant Nos. 51639008 and 51579178, which are sincerely appreciated.

References

1. Mitchell, J.K.: In-place treatment of foundation soils. J. Soil Mech. Found. Div. **96**(SM1), 73–110 (1970)
2. Brown, R.E.: Vibroflotation compaction of cohesionless soils. J. Geotech. Geoenviron. Eng. **103**(GT12), 1437–1451 (1977)
3. Webb, D.L., Hall, R.I.: Effects of vibroflotation on clayey sands. J. Soil Mech. Found. Div. **97**(SM6), 1365–1378 (1969)
4. Slocombe, B.C., Bell, A.L., Baez, H.U.: The densification of granular soils using vibro methods. Géotechnique **50**(6), 715–725 (2000)
5. Saito, A.: Characteristics of penetration resistance of a reclaimed sandy deposit and their change through vibratory compaction. Soils Found. **17**(4), 31–43 (1977)
6. Cundall, P.A., Strack, O.D.: A discrete numerical model for granular assemblies. Geotechnique **29**(1), 47–65 (1979)
7. Jia, M.C., Wang, L., Zhou, J.: Meso-mechanics simulation on vibrocompaction of sand with two-dimension Particle Flow Code. J. Hydraul. Eng. **4**, 421–429 (1979). (in Chinese)
8. Jiang, M.J., Liu, W.W., He, J., et al.: A simplified DEM numerical simulation of vibroflotation without backfill, p. 012044. IOP Publishing, Warwick (2015)
9. Jiang, M.J., Yu, H.S., Harris, D.: A novel discrete model for granular material incorporating rolling resistance. Comput. Geotech. **32**(5), 340–357 (2005)
10. Jiang, M.J., Konrad, J., Leroueil, S.: An efficient technique for generating homogeneous specimens for DEM studies. Comput. Geotech. **30**(7), 579–597 (2003)

Steel Drilled Displacement Piles (M-Piles) – Overview and Case History

Antonio Marinucci[1(✉)] and Stephen E. Wilson[2]

[1] V2C Strategists, LLC., Brooklyn, NY 11211, USA
antmarinucci@gmail.com
[2] M-Pile Sales, LLC., Salt Lake City, UT 84104, USA
swilson@m-pile.com

Abstract. Drilled displacement (DD) piles are cast-in-place piles that are formed with little or no soil removal, where the drilling tool displaces the soil radially outward into the formation, and can be used for ground improvement and/or for structural foundation systems. M-Piles are a type of steel DD pile that are constructed using a conventional rotary drill rig to supply downward thrust and rotation to install into the ground a permanent steel pipe connected to a sacrificial drill bit. Steel reinforcement and concrete can be placed within the open inner space of the steel shaft to provide additional rigidity and structural strength. There are many benefits to using DD piles, including minimal amount of soil removal, low ground vibrations during installation, larger unit values of side shear, and a stiffer pile response to loading. In general, the use of DD piles where the in-situ soils can be displaced and compacted. The use of M-Piles is applicable in very loose (i.e., running) -to-medium dense granular soils and in very soft-to-firm cohesive soils. The steel shaft provides support to the unstable soil so that the integrity of the supporting ground and the performance of the completed pile will not be compromised during installation. This paper will provide a general overview of steel DD piles (M-Piles), how a pile is constructed, applicability for use, and benefits afforded from using these piles. This paper will also present and discuss project conditions and test results via a mini case history where the M-Pile technique was implemented for the Bachelors Enlisted Quarters on Coronado Island in southern California, USA.

Keywords: Steel drilled displacement piles · M-Piles · Applicability
Benefits · Case history

1 Introduction

Drilled displacement (DD) piles refer to a specialized technology in which a bored pile is constructed using a process where rotation and downward thrust from a drill rig are applied to advance a specially designed tool into the ground. During installation, the in-situ soil is displaced radially outward into the surrounding formation, thereby resulting in a limited amount of drill spoils returning to the ground surface. DD piles are well suited for a wide spectrum of soil conditions, ranging from sandy gravel to clay, with the caveat being that the soil must be able to be both displaced and compacted. Brown [7] explained, "The energy required to install the pile is related to the

© Springer Nature Singapore Pte Ltd. 2018
L. Li et al. (Eds.): GSIC 2018, *Proceedings of GeoShanghai 2018 International Conference: Ground Improvement and Geosynthetics*, pp. 48–58, 2018.
https://doi.org/10.1007/978-981-13-0122-3_6

resistance of the soil to the displacement, and so the piles are often installed to a depth that is controlled by the capabilities of the drilling rig." Modern hydraulic drilling/piling rigs must be capable of delivering high torque ($\geq 370,000$ ft-lb (≥ 500 kN-m)) and large crowd forces (100,000 lb (450 kN)) to install DD piles properly to the desired depths required by the design.

Paniagua [11] provided a detailed history of the evolution of DD piles and the principal advancements realized during the development of the three generations of tooling. Basu et al. [5] presented a comprehensive overview for many of the different tools and techniques used to construct concrete DD piles in Europe and North America. DD piles can be classified as either partial or full displacement type according to the installation method and/or to the type and shape of tooling used to create the cylindrical or screw shaped steel or concrete pile (Fig. 1).

(a)   (b)   (c)

Fig. 1. Photographs of (a) a fabricated steel DD M-Pile with tip welded to steel casing and (b, c) concrete DD piles: cylindrical-shaped and screw-shaped [8]

For typical concrete DD piles, concrete is injected and steel reinforcement (if required) is inserted to fill the created hole and to provide structural stiffness. Methods comprising modern types of concrete DD piles include Omega pile, Berkel APGD pile, Menard CMC, Trevi Discrepiles, and Bauer FDP System. Depending upon the method and equipment, concrete DD piles range in diameter from about 300 mm to 800 mm (12- to 32-in.) and to a maximum depth of about 35 m (115 ft), as reported in the published literature and commercial brochures.

A steel DD pile (i.e., M-Pile) is constructed using a conventional rotary drill rig to supply downward thrust and rotation to install a permanent steel pipe that is connected to a sacrificial, proprietary displacement-drilling tip (to loosen the soil during advancement of the pile). If required per design, steel reinforcement bars and concrete may be placed within the open inner space of the steel pipe to provide additional rigidity and increased structural strength. M-Piles are about 325 mm (12.75 in.) in outside diameter with a wall thickness of about 9.5 mm (3/8 in.), and have been installed to a maximum depth of about 27 m (90 ft). On both commercial and private projects, M-Piles have been used as structural foundation elements, for ground improvement, and have been incorporated in the support of excavation design (i.e., as the soldier beam in a soldier beam-and-lagging system).

This paper will focus on steel DD piles, and will provide an overview of this type of pile, applicability of its use, components, and installation. This paper will also present a mini case history (i.e., project conditions and static and dynamic load test results) for a project on Coronado Island, California, USA.

2   Generalized Construction Procedure

During installation, an M-Pile is connected to the drill rig, positioned and plumbed at the desired location, and is rotated clockwise and penetrates the ground using the single rotary drive and crowd force provided by the drill rig. The drilling tip (Fig. 1a) is used to loosen and displace the soil during the advancement of the pile. The drilling phase continues until the desired depth is achieved, whereby the maximum achievable depth is limited by the capabilities of the drill rig. When required, lengths of steel pipe can be added by splicing and welding to achieve greater depths. Once the desired depth has been achieved, the pile is disconnected from the drilling rig, where the drill can move onto the next location to continue pile installation.

The sequencing of the placement of the infill concrete and steel reinforcement, when required, can occur concurrently or subsequently to the installation of the steel pipe. The concrete mix is typically a structural concrete mix composed of Portland cement, aggregate, water, and additives and admixtures (e.g., fly ash, water reducer, and plasticizer). Admixtures affect and control the rate of hydration (for workability and set time) and water reducers (e.g., plasticizers) affect the amount of water needed for fluidity and flowability to ensure the fresh concrete can get to its intended location without clogging the lines. The concrete infill can be placed using free-fall or tremie concrete placement methods.

3   Applicability and Installation-Induced Changes

Steel DD piles are well suited for use in a wide range of soil conditions ranging from sandy gravel to clay, with the one caveat being that the soil has to able to be displaced and compacted. The soil surrounding a DD pile will undergo changes to its stress state (e.g., change in void ratio) as a function of the soil type, original stress state and consistency, shape of the tool, and installation method.

The use of steel DD piles is applicable in very loose (running) to medium-dense cohesionless soil conditions, as long as the relative density (D_r) is less than about 65%, CPT tip resistance (q_c) is less than about 14 MPa (2,000 psi), and/or SPT N-values are less than about 30–35 blows/0.3 m (per ft). NeSmith [9] reported that installation in dense cohesionless soils could be difficult and uneconomical. During installation, the void space is decreased and the soil structure is reorganized. In partially saturated or fully saturated cohesionless soils, the installation of the pile may generate excess pore water pressures in the soil surrounding the pile. For cohesionless soils with minor fines content ($\leq 15\%$), the dissipation of induced excess pore water pressures will be relatively rapid and the beneficial effects of the installation are realized relatively soon after construction is complete. For cohesionless soils with appreciable fines (>15%), the dissipation will require time, which will depend on the length of the drainage path.

The use of steel DD piles is suitable in very soft to stiff (yet displaceable) fine-grained and cohesive soils, where the undrained shear strength (S_u) does not exceed about 100 kPa (2,000 psf). During installation, cohesive soils will undergo plastic deformation as they are compacted; however, stiff-to-very stiff cohesive soils are difficult to compact, rendering this method uneconomical in these soil conditions. In partially saturated or fully saturated conditions, the induced excess pore water pressures generated during installation will require time to dissipate, which may require some time to realize the increase in shear strength as the soil undergoes consolidation within the affected zone. In sensitive soils, the installation-induced disturbance may result in remolding of the soil and formation of residual shear planes, which could be detrimental to the soil structure, shear strength, and performance of the steel DD pile.

4   Benefits and Advantages

As described in the technical literature (e.g., [5, 6, 10]), DD piles, in general, provide numerous benefits, including:

- Larger values of side resistance are realized due to the ground improvement induced by the installation process (i.e., compaction/densification of the soil), which results in a comparatively stiffer load-displacement response for a DD pile than that of a similarly sized non-displacement pile. Therefore, the DD pile is able to achieve a given load resistance at a shorter length resulting in a lower cost (per ton of load).
- An environmentally friendly construction approach, whereby only minimal amount of drill spoils return to ground surface, which lowers the risk associated with transport and disposal (especially contaminated material).
- Minimal ground vibrations are generated during installation.
- Cleanliness of the working platform resulting from minimal drill spoils returning to the surface.

Steel DD piles can be installed in very loose or very soft soils because an open borehole is not required because the steel pipe provides the support to maintain the diameter. Moreover, there is increased flexibility with steel DD piles as additional lengths of steel pipe can be spliced to accommodate achieving greater depths due to changing conditions realized during construction. Furthermore, high production rates can be realized, as much as 300 to 400 linear meters of pile installed per shift (1,000 to 1,300 linear ft), since the steel pipe is left in the ground during one operation; the steel reinforcement/concrete infill can be placed during a separate operation, which allows the drill to move to the next location to continue installing the M-piles.

5   Case History – P-730 Bachelors Enlisted Quarters Project

A new 7-story mid-rise structure (P-730 Bachelors Enlisted Quarters) was constructed to house enlisted personnel on Coronado Island across the bay from San Diego in southern California, USA (delineated by red circle on Fig. 2a). For this portion of the project, the foundation system was installed by the geotechnical specialty contractor Magco Drilling. Given the location and sensitivity of the site and its surroundings, the design criteria mandated that the proposed foundation system minimally affect base operations, mitigate the detrimental effects of the potentially liquefiable sands and silty sands, produce no vibration at the site because of its sensitivity as an active base, and minimize the costly disposal of contaminated soils.

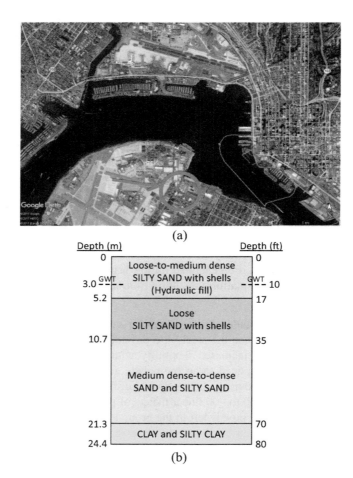

Fig. 2. (a) Map view of approximate project area on Coronado Island (image from Google Earth), and (b) generalized subsurface profile for proximity around the proposed building

The site investigation and laboratory testing program included 6 cone penetration tests (CPTs), 4 soil borings with sampling and standard penetration tests (SPTs), and laboratory tests on the soil samples for classification, compaction parameters, and to determine resistance/corrosion characteristics of the in-situ soils. The generalized subsurface profile (Fig. 2b) consists mainly of silty sands, with a clay and silty clay layer beginning at a depth of about 21.3 m (70 ft). The upper 10.7 m (35 ft) of the deposit is composed of loose to medium dense silty sand, which is then underlain by a deposit of about 10.7 m (35 ft) of medium dense to dense sand and silty sand. The lowermost layer sampled consists of clay and silty clay. At this site, the groundwater table was located at a depth of approximately 3 m (10 ft) from the ground surface.

The purpose of the foundation system was three-fold: to support the expected vertical loading from the proposed structure, to resist ground motion due a design seismic event, and to provide resistance and support against potential liquefaction-induced deformations. As part of the original foundation design package, the designer offered two foundation options: a deep foundation system consisting of augered cast-in-place (ACIP) piles or a ground modification system consisting of vibratory stone columns. For the ACIP pile option, the piles would be 405 to 510 mm (16 to 20 in.) in diameter and would be installed on a center-to-center spacing of about 0.9 m (3 ft) or 2.5 pile diameters to a minimum depth of about 10.7 m (35 ft). In addition, full-scale load testing would be required for any ACIP pile element with a vertical design load of at least 356 kN (80 kips), and would include a lateral test load of 22 to 45 kN (5 to 10 kips) each. For the vibratory stone column option, the columns would be about 915 mm (36 in.) in diameter and would be installed on a center-to-center triangular spacing of about 2.4 m (8 ft) or 2.67 pile diameters to a minimum depth of about 15.2 m (50 ft).

The specialty contractor offered an alternative foundation system option for consideration. For this option, steel DD piles (M-Piles) with a diameter of 324 mm (12.75 in.) and internal steel reinforcement and concrete infill would be installed a depth of 12 to 18 m (40 to 60 ft), as determined based on the loading requirements. For a factor of safety of 2.0, the M-Piles were designed for an axial compression and tension resistance of 1,068 kN (240 kips) and 512 kN (115 kips), respectively, for a pile embedment of about 18 m (60 ft). The upper 10.4 m (34 ft) of side resistance was neglected in the determination of the allowable axial resistance to address the potential zone of liquefiable soil and the potential loss of axial resistance due to liquefaction during the design seismic event.

Each of the piles was designed for an axial compression load (i.e., design load, DL) of 1,112 kN (250 kips). The schematic shown in Fig. 3a provides a plan view of the structure along with the approximate locations of the steel DD piles. In total, 273 displacement piles were installed at this site. As shown in Fig. 3b, the schematic provides a cross-sectional view of the pile cap/grade beam attached to and supported by an M-Pile. Internal steel reinforcement was used in the uppermost 1.8 m (6.0 ft) of the pile with an approximate 0.6 m (2.0 ft) embedment into the pile cap/grade beam to develop and provide adequate shear and compression load transfer between the structural elements. The internal steel reinforcement (Fig. 3c) consisted of a rebar cage comprising 5 ea, No. 29 (#9) longitudinal rebar that were bent on one side and transverse shear reinforcement consisting of a No. 10 (#3) rebar spiral at a pitch of about 76 mm (3 in.).

The design assumptions of the axial resistance in compression were confirmed via full-scale load testing comprising three static compression tests, two static tension tests, and four dynamic tests. The project specifications required that the piles be loaded to a test load (TL) of 2,670 kN (600 kips) in compression, which was approximately 2.4*DL, and to 1,112 kN (250 kips) in tension. Linear variable displacement transducers (LVDTs) were used to measure the displacements at the top of the test piles, and the loads were measured and recorded using both a calibrated load cell and manually, to provide backup and confirmation. The piles had two levels of strain gages, one at 0.46 m (1.5 ft) below grade and one at 10.7 m (35 ft) below grade, to evaluate the behavior within the potentially liquefiable zone.

The quick load testing procedures were used for the compression and tension load tests in accordance with ASTM D1143 [1] and ASTM D3689 [2], respectively. The load-displacement responses for the test piles determined using the results of the static axial compression and tension tests are shown in Fig. 4. The project specifications stated that, when a pile does not exhibit a plunging failure, the failure load would be determined using the Davisson failure criterion (shown in Fig. 4a and b via the dashed linear lines). The Davisson failure criterion is the point when the applied load intersects a curve that is defined as the elastic elongation of pile plus 0.15-in. (3.8 mm) plus 1% of the pile diameter. Test piles P-1, P-5, and P-6 did not exhibit a plunging failure. For P-1 and P-6 (Fig. 4a), the ultimate compressive load resistance was determined to be (a minimum of) 2,670 kN (600 kips). However, P-5 intersected its Davisson failure curve (Fig. 4b) at a compression load (i.e., ultimate compressive load resistance) of approximately 2,580 kN (580 kips). As shown in Fig. 4c, the test piles (P-1 and P-5) were able of sustain an axial tension test load of 1,557 kN (350 kips) without obtaining a clear geotechnical failure. Therefore, based on the results of the static axial compression and tension testing, the project criteria were satisfied.

The dynamic tests were performed by GRL Engineers, Inc. in accordance with ASTM D4945 [3] and ASTM D7383 [4] using a pile driving analyzer (PDA), the Case pile wave analysis program (CAPWAP), and the APPLE V drop hammer system. Though the piles would eventually be filled with steel reinforcement and concrete, only the outer steel pipe was in place at the time of the dynamic testing. In short, the CAPWAP process iteratively determines the soil model unknowns using signal matching and a numerical analysis procedure by analyzing the measured force and velocity data to solve for soil resistance parameters and their distribution along the length of a pile, which combines the pile and soil models of the wave equation with field measurements of the Case Method. The impact hammer used for the high-strain dynamic testing was the GRL APPLE V drop hammer system, with a ram weight of approximately 7,260 kg (8 tons) and a ram stroke (i.e., drop height) of about 762 mm (30 in.).

The dynamic measurements were obtained using two accelerometers and two strain transducers attached on opposite sides of the piles. The recorded analog signals from the gages were conditioned, digitized, and processed using the PDA, where the measurements were monitored during the high-strain dynamic impacts during the testing for indications of notable pile impedance changes and possible damage to the piles.

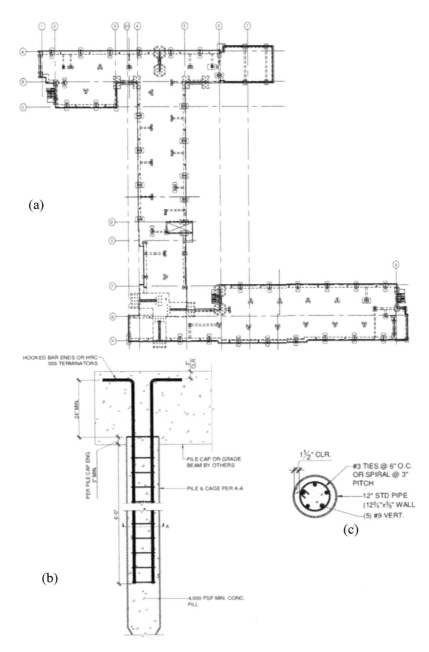

Fig. 3. (a) plan view of structure and M-Pile locations, (b) connection detail for M-Pile and pile cap/grade beam, and (c) cross-section A-A detailing internal steel reinforcement for pile

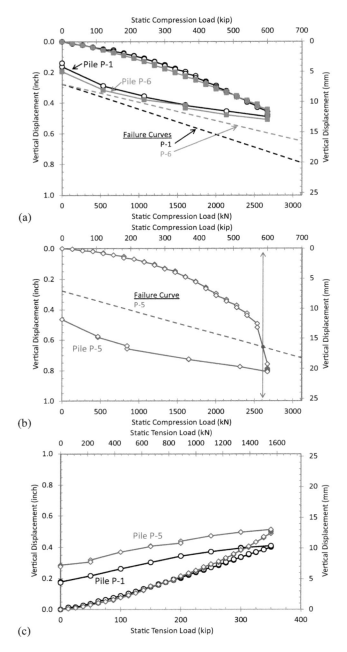

Fig. 4. Static axial load-displacement responses with Davisson Failure Criterion: (a) piles P-1 & P-6 (compression), (b) P-5 (compression), and (c) P-1 & P-5 (tension)

Based on the measurements, no indications of pile damage were observed. Ultimate axial compressive resistance is defined when a permanent set approaches 5 mm (0.2 in.) during the dynamic testing. For these four dynamic tests, the smaller recorded sets (ranged from 0.05 to 0.13 in., as shown in Table 1) imply that more soil resistance may have been available but were not observed or measured; therefore, the resistance was a mobilized resistance and not an ultimate resistance for the four tested piles. Based on the CAPWAP analyses for the four test piles, the axial compressive resistance due to the high-strain dynamic loading ranged from 2,935 to 3,605 kN (660 to 810 kips), as listed in Table 1. Moreover, it was recommended that the axial resistance in tension (or uplift) be equal to 70% to 80% of the axial compressive resistance. Based on the results of the high-strain dynamic compression testing, the project criterion was satisfied.

Table 1. Testing details, measurements, and computed axial compressive resistance based on the results of the high-strain dynamic testing

Pile No.	Pile Embed. m [ft]	Testing details and measurements				Resistance or capacity		
		Drop Height mm [in]	Pile Set mm [in]	Avg Max Stress		End MN [kip]	Side MN [kip]	Total MN [kip]
				Compr. MPa [ksi]	Tension MPa [ksi]			
158	13.4 [44]	762 [30]	3.0 [0.12]	317 [46]	4.8 [0.7]	0.8 [180]	2.5 [550]	3.3 [730]
173	15.5 [51]	762 [30]	1.0 [0.05]	331 [48]	4.1 [0.6]	0.8 [180]	2.8 [630]	3.6 [810]
188	14.6 [48]	762 [30]	2.0 [0.07]	269 [39]	4.1 [0.6]	0.6 [140]	2.4 [540]	3.0 [680]
209	14.3 [47]	762 [30]	3.0 [0.13]	283 [41]	6.9 [1.0]	0.5 [120]	2.4 [540]	2.9 [660]

6 Conclusions

The use of DD piles has increased significantly during the past two decades as a result of various factors, including advancements in tooling (e.g., increased diameters, increase production rates) and equipment capabilities (e.g., greater torque and pulldown force). As a result, DD piles have been used as structural foundations, for ground improvement, and in support of excavation systems. As long as the soil can be displaced and compacted, the technique is best suited for loose-to-medium dense cohesionless soils but can be used in a wide range of ground conditions ranging from soft-to-firm ground conditions and from sandy gravel to clay. During construction, the void space (porosity) decreases, and the soil structure is reorganized. The various benefits of DD piles were presented, and include enhanced unit side friction or resistance, minimal drill spoils, environmentally friendly, minimal ground vibrations, cleaner work area, and, ultimately, lower cost per kiloNewton (kN) or ton of load.

As a design alternative, 273 steel DD piles (M-Piles) with a diameter of 324 mm (12.75 in.) were installed to a depth ranging from 12 to 18 m (40 to 60 ft) to support a new 7-story mid-rise structure on Coronado Island. The proposed foundation system was designed to provide the required support for the structure as well as providing adequate resistance in the event of potential liquefaction of the loose-to-medium dense silty sands in the upper 10.7 m (35 ft) of the deposit. Full-scale static and dynamic load testing was performed on sacrificial piles (only the outer steel pipe was in place at the time of the dynamic testing) embedded to a depth of about 18 m (60 ft) to a test load of 2,670 kN (600 kips) in compression, which was approximately 2.4 times the design load, and to 1,112 kN (250 kips) in tension. Based on the compressive load testing, the axial resistance in compression ranged from 2,580 to 2,670 kN (580 to 600 kips) due to static loading and from 2,935 to 3,605 kN (660 to 810 kips) due to the high-strain dynamic loading. Therefore, using the results of the load testing program, the pile lengths were shortened based on the anticipated specific pile loads and the depth of the liquefiable layer. Ultimately, the foundation alternative design and the full-scale testing provided a savings in both material and time to the project, even when including the costs for the testing.

References

1. ASTM: Standard Test Methods for Deep Foundations Under Static Axial Compressive Load. ASTM D1143/D1143M-07, ASTM International, West Conshohocken, PA (2013). www.astm.org
2. ASTM: Standard Test Methods for Deep Foundations Under Static Axial Tensile Load, ASTM D3689/D3689M-07, ASTM International, West Conshohocken, PA (2013). www.astm.org
3. ASTM: Standard Test Method for High-Strain Dynamic Testing of Deep Foundations, ASTM D4945, ASTM Intern'l, West Conshohocken, PA (2012). www.astm.org
4. ASTM: Standard Test Methods for Axial Compressive Force Pulse (Rapid) Testing of Deep Foundations, ASTM D7383, ASTM International, West Conshohocken, PA (2010). www.astm.org
5. Basu, P., Prezzi, M., Basu, D.: Drilled displacement piles – current practice and design. DFI J. 4(1), 3–20 (2010)
6. Bottiau, M.: Recent evolutions in deep foundation technologies. In: Proceedings of the DFI/EFFC 10th International Conference on Piling and Deep Foundations, Amsterdam, Netherlands (2006)
7. Brown, D.A.: Recent advances in the selection and use of drilled foundations. In: Hryciw, R. D., Athanasopoulos-Zekkos, A., Yesiller, N. (eds.) Proceedings of the GeoCongress 2012: State of the Art and Practice in Geotechnical Engineering. Geotechnical Special Publication No. 225. Sponsored by Geo-Institute of the ASCE (2012)
8. Marinucci, A., Chiarabelli, M.: The use of displacement piling technology in soft soil conditions. In: SMIG-DFI Conference, Mexico City, Mexico (2015)
9. NeSmith, W.M.: Design and installation of pressure-grouted displacement piles. In: Proceedings of the Ninth International Conference on Piling and Deep Foundations, Nice, France, pp. 561–567 (2002)
10. NeSmith, W.M.: Application of augered, Cast-in-Place Displacement (ACIPD) piles in New York City. In: Proceedings of the Deep Foundation Institute (DFI) Augered Cast-in-Place Piles Committee Specialty Seminar, McGraw-Hill Building, New York, NY, pp. 77–83 (2004)
11. Paniagua, W.I.: Construction of drilled displacement and auger cast in place piles. In: Proceedings of the International Symposium: Rigid Inclusions in Difficult Soft Soil Conditions, TC36, México DF (2006)

Field Investigation of Highway Subgrade Silty Soil Treated with Lignin

Tao Zhang[1,2(✉)], Songyu Liu[2], Guojun Cai[2], and Longcheng Duan[1]

[1] Faculty of Engineering, China University of Geosciences,
Wuhan 430074, Hubei, China
[2] Institute of Geotechnical Engineering, Southeast University,
Nanjing 210096, Jiangsu, China
zhangtao_seu@163.com

Abstract. Lignin is a by-product of paper or timber industry, and it has not been fully utilized all over the world. Improper disposal of lignin would pose significant risk to public health and surrounding environment. A field test was conducted to investigate the road performance of problem silty soil treated with lignin in highway subgrade applications. Quicklime, a traditional soil stabilizer, was selected as a control binder for comparison purpose. A series of field tests, including California Bearing Ratio (CBR) test, resilient modulus (E_p) test, and Benkelman beam deflection test were performed to explore the evolution of mechanical properties of lignin treated silty soil during the curing period. The effects of additive content and curing time on the bearing capacity of the treated soil were also investigated. The field test results reveal that lignin possesses a good ability to improve the bearing capacity of the silty soil. 12% lignin treated soil exhibits higher values of CBR, E_p, and lower value of resilient deflection as compared with those of 8% quicklime treated soil. The use of lignin as a stabilization chemical mixture for silty soil may be one of the viable answers to the reuse of organic by-product in geotechnical engineering.

Keywords: Silty soil · Recycled materials · Stabilization · Strength

1 Introduction

Silty soil possesses poor engineering properties, such as low strength and stiffness, and difficult to compaction, making them not permitted to be used directly in road subgrade. Silty soil subgrade can easily cause excessive settlements under traffic loading if effective improvement is not implemented [1]. According to the requirements specified in the Specifications for Design for Highway Subgrades (in Chinese), silty soil should be removed or employed in specified sites after testing. It is a fact that traditional soil stabilizers, i.e., Portland cement, lime, fly ash, and gypsum, would have a negative effect on the surrounding environment including threat on the safety of ground water and reduction in water/nutrients holding capacity of soils [2]. Furthermore, the brittle performance of the stabilized soils usually affect the stability of structures. Consequently, it is necessary to explore an environmentally friendly and cost-effective stabilizer for silty soils.

© Springer Nature Singapore Pte Ltd. 2018
L. Li et al. (Eds.): GSIC 2018, *Proceedings of GeoShanghai 2018 International Conference: Ground Improvement and Geosynthetics*, pp. 59–67, 2018.
https://doi.org/10.1007/978-981-13-0122-3_7

Under the situation of sustainable development, the utilization of by-product for stabilizing soils is strongly encouraged all over the world. Lignin is a by-product from paper and timber industry, which shows a great potential in stabilizing both cohesive and non-cohesive soils [3, 4]. The engineering properties of lignin treated various soils have been systematically investigated by earlier researchers. The results indicate that lignin treated clay or silty sand presents a higher compressive strength and better durability as compared to the untreated ones [5, 6]. In addition, the swelling potential are decreased after lignin treatment [7]. The authors have conducted a series of laboratory tests on lignin treated silty soil to study the evolution of mechanical behaviors during the curing period and to explore the mechanism of lignin stabilization. For silty soils deposited in eastern Jiangsu province, the optimum lignin content is approximately 12%, while the additive content higher than this optimum content, the soil compressive strength presents a decreasing trend [8]. The soil particles were coated and connected with a kind of cementing agent, which is produced from lignin and clay minerals, and then a stronger soil structure was generated [9, 10]. Although there have been several studies on the engineering behaviors of lignin treated soils, these were predominantly conducted in laboratory scale only. Studies on the field trials scale to investigate the mechanical properties of lignin treated soils have been noticed to be quite limited.

In view of the above, a field trial was conducted on highway subgrade to investigate the road performance of lignin treated silty soil in this study. The quicklime, as a traditional soil stabilizer, was selected for comparison purpose. A series of field tests, including California Bearing Ratio (CBR), resilient modulus (E_p), and Benkelman beam deflection were performed for evaluating the mechanical properties of the treated silty soil. Based on the testing results, the effects of curing time, lignin content, and stabilizer type on CBR, E_p, and resilient deflection (H_r) were studied. The heavy metal concentration within the stabilized soils was also measured and discussed. A great effort has been made in this study to verify the applicability of using lignin treated silty soil as a highway subgrade fill material.

2 Field Tests

2.1 Site Description

The field tests were conducted on Ramp C of Fu-Jiang highway, which was located in Yancheng city, Jiangsu Province. The static cone penetration test was firstly carried out at location CK 1+088 as per ASTM (2005) D3441 for obtaining the detailed soil profiles, and the test results were shown in Fig. 1. The in-situ soil in Ramp C was predominantly composed of silts and silt clays.

Figure 2 shows the plan view of Ramp C, which was divided into three Sections, i.e., Section A (CK0+800 to CK0+850, 50 m in length), Section B (CK0+850 to CK0+899, 49 m in length), and Section C (CK0+899 to CK0+950, 51 m in length). The width of upper pavement and subgrade base are 15 m and 24.5 m, respectively. According to the specifications for design of highway subgrades, this subgrade was divided into two zones, namely, 96% compaction degree zone and 94% compaction degree zone.

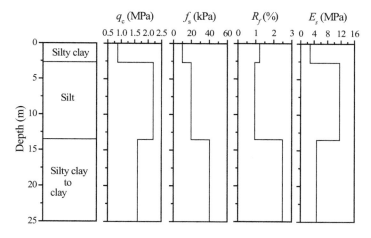

Fig. 1. Soil profiles of the testing site at CK 1+088 from static cone penetration test.

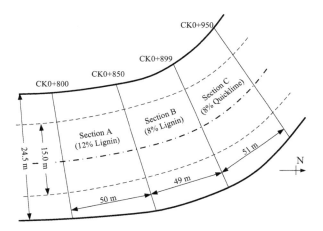

Fig. 2. Plan view of the test Sections for different stabilizers treated silty soil in Ramp C.

2.2 Materials and Methods

In all three Sections, the subgrade was filled with soils excavated from the nearby region, which was stockpiled on the ground under natural conditions for about one week for reducing moisture content prior to compaction procedure. Eight samples were collected from site and taken to laboratory for engineering property testing. Table 1 lists the basic engineering and physical properties of the filled soil. According to the Unified Soil Classification System (ASTM (2011a) D2487), this soil is classified as low plasticity silt (ML). The 12% lignin, 8% lignin, and 8% lignin and quicklime were employed to treat the filled soils in Section A, Section B, and Section C, respectively. The designed contents were calculated by dry weight of filled soil. Lignin, exhibited a yellow-brown powder with a smell of fragrance, is a by-product from a paper factory. This material was the same as that used by the authors in laboratory tests. Quicklime was classified as a high-calcium lime as per ASTM (2011b) C51.

Table 1. Basic engineering properties of the tested soil.

Property	Characteristic
Natural moisture content, w_n (%)	28.5
Specific gravity, G_s	2.71
Grain size distribution (%)[a]	
Clay (<0.005 mm)	10.8
Silt (0.005–0.075 mm)	80.1
Sand (0.075–2 mm)	9.1
Liquid limit, w_L (%)	32.4
Plasticity limit, w_P (%)	23.6
Plasticity index	8.8
Optimum moisture content (%)	16.1
Maximum dry unit weight (γ_{dmax}), (kg/m^3)	1720
pH[b]	8.74

[a]Measured using a laser particle size analyzer Mastersize 2000.
[b]Measured as per ASTM D4972 (2013).

The CBR test was conducted by setting the truck load as 60 kN and fixing the load penetration rate as 1 mm/min, as detailed by China MOT (2008) JTG E60-2008. The E_p of filled soil was determined by portable falling weight deflectometer (PFWD), which is non-destructive testing device widely used in subgrade detection. The testing method of PFWD can refer to the reports in the literature. The Benkelman beam test was conducted for measuring the surface deflections under a given static loading. The testing truck with a load of 100 kN and a tire pressure of 0.7 MPa was employed in this field test. The concentrations of heavy metals within soil samples which were collected from the field site were measured, as per US EPA (1996) methods, by the Center of Environmental Monitoring in Nanjing city.

3   Results and Discussion

3.1   CBR Test

Figure 3 presents the variation of CBR tested after 0 day, 8 days, and 15 days curing for all testing points in the 96% compaction degree zone, and the average values are marked on it. Generally, the CBR values increase with an increase in curing time in three Sections. At 0 day of curing, the average CBR value in Section A is higher as compare to that of Section B and Section C. In addition, the average values in both Section B and Section C are approximately identical (40.3% and 43.7% for Section B and Section C, respectively) immediately after the construction completion (i.e., 0 day curing). The reason for higher CBR value of the lignin treated silt in Section A at 0 day of curing is that the Section A located on the straight road, which is convenient for the compaction of subgrade soils. Moreover, the hydration reaction rate of quicklime is very rapid, which leads to higher CBR value in the silt treated by quicklime (Section C) relative to that treated by lignin (Section B) with the same additive dosage (8%) at 0 day curing. As curing time increases from 0 day to 8 days, the average CBR values increase dramatically from 50.6% to 87.7% for 12% lignin stabilized silt (Section A), from 40.3% to 70.6% for 8% lignin treated silt (Section B), and from 43.7% to 76.3% for 8% quicklime treated silt (Section C), respectively. When the curing time increases to 15 days, the variation of CBR in all Sections shows a similar trend with that of curing with 8 days.

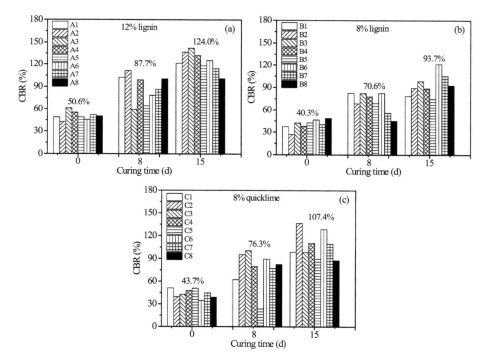

Fig. 3. Variation in CBR values with curing time and additive content: (a) Section A, 12% lignin stabilized soil; (b) Section B, 8% lignin stabilized soil; and (c) Section C, 8% quicklime stabilized soil.

3.2   Resilient Modulus Test

Figure 4 depicts the variations of E_p with the curing time in each test Section. The average magnitude of E_p in Section C (30.4 MPa) is slightly higher than those of Section A (29.8 MPa) and Section B (23.9 MPa) momentarily right after the compaction. After

8 days curing, the E_p of stabilized silt in all three Sections shows an increasing trend. When the curing time up to 15 days, the average values of E_p increase dramatically from 29.8 MPa to 66.4 MPa for 12% lignin treated silt, from 23.9 MPa to 43.2 MPa for 8% lignin treated silt, and from 30.4 MPa to 55.8 MPa for 8% quicklime treated silt, respectively. The variation in E_p with curing time is consistent with CBR, which has been described earlier. The reason for the E_p difference between Section B and Section C is the diverse stabilization mechanisms of these two soil stabilizers, which would result in a dramatically different in the type of cementation materials and the intensity of reactions. Overall, the resilient modulus properties of lignin treated silt can not compare well with quicklime treated silt under the same additive content and curing time. Nevertheless, when lignin content increases to 12%, a more superior stiffness performance is obtained relative to 8% quicklime treated silt after 15 days of curing.

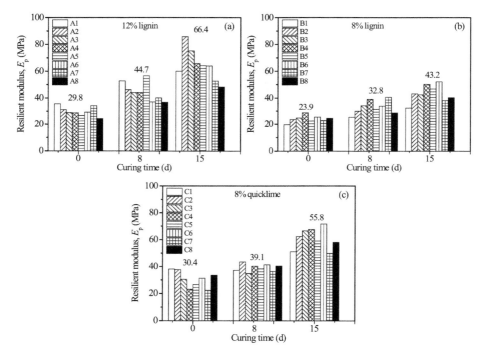

Fig. 4. Variation in E_p values with curing time and additive content: (a) Section A, 12% lignin stabilized soil; (b) Section B, 8% lignin stabilized soil; and (c) Section C, 8% quicklime stabilized soil.

3.3   Benkelman Beam Test

Figure 5 shows the effect of curing time on H_r in all Sections. It is noteworthy that this Figure presents the average values and standard deviations for all 60 testing points in three Sections to investigate the effects of soil stabilizers (viz., lignin and quicklime) and curing time on the H_r of treated silt. It was calculated that the coefficient of variation (C.V) of measured H_r values for all Sections were less than 6%, demonstrating that the average values can represent the individual measuring data. Consequently, average values with standard deviations were selected in the Figure to quantitatively analyse of the mechanical properties of the stabilized subgrade silt.

At curing time of 0 day (momentarily after the compaction), the initial H_r in Section A is 3.29 mm, which is slightly higher than that in Section B (3.00 mm) or in Section C (2.75 mm). As the curing time increases from 0 day to 8 days, the H_r in each Section shows a dramatical reduction. The average H_r values in Section A, Section B, and Section C are 1.91 mm, 2.13 mm, and 1.11 mm, respectively. At the subsequent curing of 7 days, the H_r values of each Section are further reduced. The H_r values of both lignin and quicklime stabilized silts are lower than 1 mm after 15 days curing, and all the Sections exhibit approximately similar deflections (e.g., 0.33 mm in Section A, 0.56 mm in Section B, and 0.47 mm in Section C). This indicates that after treatment, the strength and stiffness of the compacted subgrade silt are improved. Overall, the 12% lignin treated silt displays greater deformation resistance behaviors compared to 8% lignin and 8% quicklime treated silt. These observations are also consistent with the results of CBR and resilient modulus tests mentioned above.

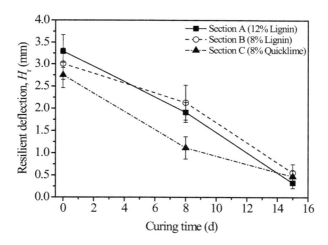

Fig. 5. Effect of curing time on H_r of both lignin and quicklime treated silt.

3.4   Heavy Metal Concentration

The total concentrations of major heavy metals within the stabilized silt were measured to evaluate the effects of lignin stabilized subgrade soils on the surrounding environment, and the analyzing results were shown in Table 2. The background limit values listed in the table were specified by China Soil Environmental Quality Standard (GB 15618–1995) (China MEP, 1995). It can be observed that the concentrations of major heavy metals within natural silt and 8% quicklime stabilized silt (15 days of curing) are lower than the background limit value. This indicates that the quality of those stabilized soils satisfy the requirements of first grade soil quality standards. However, the concentrations of Cu and Zn within the 12% lignin stabilized silt achieve 38.1 mg/kg and 118.5 mg/kg, respectively, which slightly exceed the background values (i.e., Cu is 35 mg/kg and Zn is 100 mg/kg). This phenomenon may attributed to the fact that the used lignin here is a kind of industrial by-product, which would possess some contaminants during the production process. Moreover, the heavy metal concentrations of 12% lignin stabilized silt satisfy the corresponding requirements of second grade soil quality standards in China (limit values of Cu and Zn are 100 mg/kg and 300 mg/kg (pH > 7.5), respectively.). Therefore, for the highway subgrade construction, the lignin used in this investigation is an environmentally friendly by-product that poses a negligible risk to human health.

Table 2. Heavy metal concentrations within the natural silt, 12% lignin stabilized silt, and 8% quicklime stabilized silt (mg/kg).

Heavy metal	Background value	Natural silt	12% Lignin	8% Quicklime
Cu	35	29.6	38.1	28.2
Zn	100	89.7	118.5	91.8
Ni	40	27.5	26.8	27.8
Cr	90	40.9	39.9	41.1
Pb	35	11.2	11.7	15.6
Cd	0.20	0.106	0.118	0.175
Hg	0.15	0.067	0.079	0.081
As	15	12.7	12.8	19.3

4   Conclusions

This study provides a field testing program conducted to investigate the road performances of lignin treated silty soil employed as highway subgrade filling material. Based on the results reported, the following conclusions can be drawn:

By-product lignin had the capacity for improving the engineering properties of silty soil and its treated silt could be used properly in highway subgrade construction. The 12% lignin treated silty soil possessed a superior mechanical behaviors compared to the 8% quicklime treated soil. The strength and modulus of lignin treated silt were slightly lower than that of quicklime treated one with the same additive content.

After 15 days of curing, 12% lignin treated silt presented the best mechanical performances relative to other treated soils in both Section B and Section C. Even if the stabilization mechanisms of lignin and quicklime treated soils were different, the 12% lignin treated silt still presented higher E_p and lower H_r than 8% quicklime stabilized silt. The lignin stabilized soils pose a negligible threat on surrounding environment.

References

1. Zhu, Z.D., Liu, S.Y.: Utilization of a new soil stabilizer for silt subgrade. Eng. Geol. **97**(3), 192–198 (2008)
2. Chen, Q., Indraratna, B., Carter, J., Rujikiatkamjorn, C.: A theoretical and experimental study on the behaviour of lignosulfonate-treated sandy silt. Comput. Geotech. **61**, 316–327 (2014)
3. Ceylan, H., Gopalakrishnan, K., Kim, S.: Soil stabilization with bioenergy coproduct. Transp. Res. Rec. **2186**, 130–137 (2010)
4. Santoni, R.L., Tingle, J.S., Webster, S.L.: Stabilization of silty sand with nontraditional additives. Transp. Res. Rec. **1787**, 61–70 (2002)
5. Indraratna, B., Muttuvel, T., Khabbaz, H., Armstrong, R.: Predicting the erosion rate of chemically treated soil using a process simulation apparatus for internal crack erosion. J. Geotech. Geoenvironmental Eng. **134**(6), 837–844 (2008)
6. Kim, S., Gopalakrishnan, K., Ceylan, H.: Moisture susceptibility of subgrade soils stabilized by lignin-based renewable energy coproduct. J. Transp. Eng. **138**(11), 1283–1290 (2011)
7. Puppala, A.J., Hanchanloet, S.: Evaluation of a new chemical (SA-44/LS-40) treatment method on strength and resilient properties of a cohesive soil. In: 78th Annual Meeting of the Transportation Research Board, Washington, DC, Paper No. 990389 (2009)
8. Zhang, T., Liu, S., Cai, G., Puppala, A.J.: Experimental investigation of thermal and mechanical properties of lignin treated silt. Eng. Geol. **196**, 1–11 (2015)
9. Zhang, T., Cai, G., Liu, S., Puppala, A.J.: Engineering properties and microstructural characteristics of foundation silt stabilized by lignin-based industrial by-product. KSCE J. Civil Eng., 1–12 (2016)
10. Cai, G., Zhang, T., Liu, S., Li, J., Jie, D.: Stabilization mechanism and effect evaluation of stabilized silt with lignin based on laboratory data. Mar. Georesour. Geotechnol. **34**(4), 331–340 (2016)

Investigation of Ground Displacement Induced by Hydraulic Jetting Using Smoothed Particle Hydrodynamics

Pierre Guy Atangana Njock[1,2] and Shuilong Shen[1,2,3(✉)]

[1] State Key Laboratory of Ocean Engineering, School of Naval Architecture, Ocean, and Civil Engineering, Shanghai Jiao Tong University, Minhang District, Shanghai 200240, China
slshen@sjtu.edu.cn
[2] Collaborative Innovation Center for Advanced Ship and Deep-Sea Exploration (CISSE), Shanghai Jiao Tong University, Shanghai 200240, China
[3] Department of Civil and Construction Engineering, Centre for Sustainable Infrastructure, Swinburne University of Technology, Hawthorn, VIC 3122, Australia

Abstract. The few methodologies available for estimating the displacement induced by the action of a fluid jet on the soil are almost all based on the cavity expansion theory, which the adaptiveness to model this phenomenon can be questioned due to some significant drawbacks. This paper investigates the suitability of the SPH approach to investigate the ground displacement induced by jet grouting. A simulation of a water jet impinging on a soil mass is performed using AUTODYN-2D. In this simulation, the soil is modelled as a granular elastoplastic material, whereas the water is assumed to perform as a Newtonian fluid. The interaction between these two bodies is therefore investigated in order to demonstrate the advantage of the SPH method over others. In particular, the strain and pressure variations in the soil are exploited to clearly expose the limitations of the cavity expansion based approaches. It is apparent that the SPH approach simulates the fluid-soil interaction more realistically and thus can be used to investigate the aforementioned mechanism. The results also highlight some computation instabilities that occurred during the simulation. It is therefore recommended to adopt coding as solving technique, notably to have a greater flexibility in dealing with issues of this type.

Keywords: Jet grouting · Smoothed particle hydrodynamics
Ground displacement · Water-soil interaction

1 Introduction

The ground displacement induced by high pressure fluid jet has become a central issue in jet grouting practice and has generated considerable recent research interest. Initially registered in the UK in 1950s [1], the jet grouting method consists of creating in-situ rigid elements in the soil, along with the successive processes of drilling, jetting and grouting. However, during the construction of jet grouting column, a large amount of

© Springer Nature Singapore Pte Ltd. 2018
L. Li et al. (Eds.): GSIC 2018, *Proceedings of GeoShanghai 2018 International Conference: Ground Improvement and Geosynthetics*, pp. 68–75, 2018.
https://doi.org/10.1007/978-981-13-0122-3_8

grout is injected into the ground via a small diameter nozzle. For the single fluid system for example, the injection pressure even varies from 40 to 70 MPa [2, 3]. Moreover, it was demonstrated that the injection of cement slurry at high pressure tends to cause detrimental effects to the surrounding subsoil. In densely urbanized areas, these effects adversely affect the constructed structures in the vicinity of the implementation zone [4]. Few approaches have been dedicated to the assessment of fluid jet induced displacement in soil, and were regularly found acceptable [4–9]. However, these methods are largely established on the basis of the cavity expansion theory, which the appositeness to describe the mechanism of relevance in this paper can be questioned owing to some intrinsic limitations. The exploration of a new approach such as Smoothed Particle Hydrodynamics (SPH), which incorporates numerous desirable advantages, thus seems legitimate.

This paper analyzes the effect of a water jet impinging on a soil mass modeled as granular material (single phase), using the SPH method. The main purpose underlying this approach is to investigate the suitability of the SPH approach to model and evaluate the ground displacement observed during the jet grouting procedure. The fundamentals of the Smoothed Particle Hydrodynamics formalism are first presented, followed by a detailed methodology of the analysis of water-soil interaction. The results of the simulation (obtained using the commercial software ANSYS-Autodyn) tend to confirm the aptitude of the SPH method to simulate jet grouting mechanism.

2 Fundamentals of Smoothed Particle Hydrodynamics (SPH)

2.1 Definition

The smoothed particle hydrodynamics (SPH) is an adaptive meshfree particle method that harmonically combines particle approximation and Lagrangian formulation. Its adaptive nature is a remarkable advantage that provides a certain ease in handling large deformations problems [10]. Originally, this meshless method was developed independently by [11] and [12] for astrophysics purposes; and thenceforth, has been extended to a wide range of applications owing to a number of advantages detailed in [13]. In this approach, physical phenomena are translated into mathematical models (conventionally in the form of governing equations) and then expressed in SPH form.

2.2 SPH Formulation

The prominent idea behind the SPH formulation is that numerical results rely on an interpolation process to describe both the space and time evolution of the problem investigated. Basically, the domain of interest is discretized into a finite number of particles, each of which independently carries the material properties. The material properties at a given point x (particle i) is subsequently determined through an interpolation process with respect to its neighboring points x' (particles j), and within a support domain Ω (see Fig. 1). Noticeably, the particles have a spatial distance h (usually termed smoothing length), over which their properties are approximated by using a "Kernel" or smoothing function W. This smoothing function thus plays a crucial role in the computation process, as it not only determines the interpolation pattern of field variables, but also the stability, the accuracy and the efficiency of this interpolation. More details related to the definition or the choice of the smoothing function can be found in [13]. Usually, the aforementioned interpolation is based on the integral representation of a field function that can be a density, a velocity, an energy functions, etc., depending on the problem to be solved. Yet, one must understand that the solution scheme to the discretized governing equations (that define the problem) should be formulated following three principal steps:

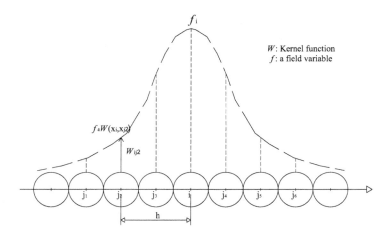

Fig. 1. Example of a field variable approximated at a given particle i.

(i) The first step also called the Kernel approximation consists of representing a given field function and its derivatives in a continuous integral representation as shown by the following Eqs. (1) and (2).

$$f(x) = \int_{\Omega} f(x') \, W(x - x', h) \, dx' \tag{1}$$

$$\frac{\partial f(x)}{\partial x} = -\int_{\Omega} f(x') \, \frac{\partial W(x - x', h)}{\partial x'} \, dx' \tag{2}$$

(ii) In the second step (particle approximation), the field variables at a given particles are approximated using the information of its neighboring particles, Eqs. 3 and 4. This step is clearly illustrated by Fig. 1.

$$f(x_i) = \sum_{j=1}^{N} \frac{m_j}{\rho_j} f(x_j) \, W_{ij} \tag{3}$$

$$\frac{\partial f(x_i)}{\partial x} = \sum_{j=1}^{N} \frac{m_j}{\rho_j} f(x_j) \cdot \frac{\partial W_{ij}}{\partial x_i} \tag{4}$$

(iii) Finally, a numerical computation technique is adopted to solve the resulting equations through an "updated Lagrangian scheme". However, the method used must be sufficiently robust to handle challenging issues such as the boundary conditions, numerical instabilities and time integration. The commonly used approaches are coding (FORTRAN, MATLAB, DualSPHysics) and simulations using commercial software (AUTODYN, LS-DYNA, etc.)

3   Investigation Methodology

The concept of investigating the ground displacement induced by jet grouting using the SPH approach and based on the simulation of water-soil interaction stems from [14]. As illustrated by Fig. 2, the present simulation intended to describe a soil mass response due to the action of a water jet impinging on it. AUTODYN-2D SPH-code was used in this case to model the relevant mechanisms. AUTODYN-2D is an ANSYS software product that is accessible from ANSYS Workbench (under software license). This software possesses a rich materials library that also includes their constitutive models and equations of states (EOS). Taking advantage of this feature, the soil was modelled as a granular material with the following properties: reference density $\rho = 1674$ kg/m^3; bulk modulus $K = 125$ MPa and poisson's ratio $\upsilon = 0.3$. Its behavior was assumed elastoplastic under the action of water jet. Additionally, the water was modelled as a Newtonian fluid with a virtual sound of speed $C = 483$ m/s, and a density $\rho_{fo} = 1000$ kg/m^3. The initial velocity of the water jet was however taken as 400 m/s.

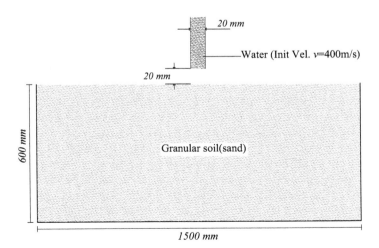

Fig. 2. Simulation arrangement

The water jet was discretized by 750 particles, while 56250 particles were used for the soil mass. The shape and packing of the particles in these two bodies are depicted in the Fig. 2. Moreover during the calculation, the relevant planes were modelled by "rigid wall" boundary conditions that are available in AUTODYN as velocity boundaries conditions.

4   Simulation Results and Discussion

Three typical variables were investigated for the purposes of this study namely, velocity vectors, pressure and strains variations. These results are respectively presented in Figs. 3, 4 and 5. However, it is crucial to understand the limitations of the cavity expansion models while discussing these results.

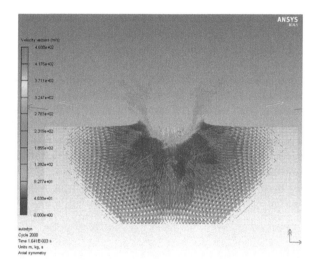

Fig. 3. Velocity vectors of SPH particles after the water jet impact

There are several reasons why the pertinence of the cavity expansion based models to realistically approximate the ground displacement induced by jet grouting can be questioned. One reason is that the cylindrical cavity expansion theory cannot simulate the incremental construction process of jet grouting. As reported by [6], during the jetting action, the radial pressure is non-uniformly distributed to the surrounding soil; however, the cylindrical cavity expansion theory cannot fulfill the non-uniform strain state of the soil undergoing the action of the fluid jet. In other words, the deformation of the soil is solely assumed in one direction and the other relevant effects (e.g. the effect of upward movement on lateral displacement) are ignored. Another important reason is that the cavity expansion models treat the soil as a continuum; and consequently, they are unable to simulate some fundamental behaviors of the soil that arise from its

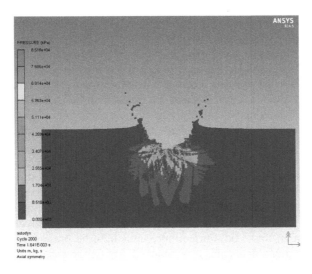

Fig. 4. Pressure distribution in the soil

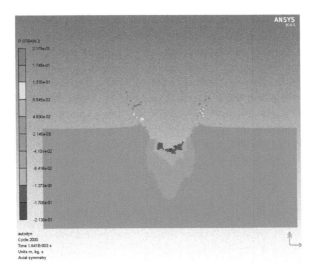

Fig. 5. Stain distribution in the soil during the jetting

particulate nature. Indeed, these models neglect the influence of the hydraulic fracturing of soil particles on the final ground displacement. Importantly, it has been demonstrated that the mechanical agitation in the ground generated by the fluid pressure is the principal mechanism inducing ground displacement [15].

Conversely, the SPH simulation allows picturing some crucial soil responses to fluid jet. As illustrated by the Fig. 3, which presents the particles velocity vectors after the water jet impact, it can be seen that the water jet affects the surrounding mass in a non-uniform fashion. This effect regularly fades as the distance of the particles from the immediate impact area becomes larger. It can therefore be assumed that the stress in the soil decrease as the particles get rearranged in a compact way. This assumption is confirmed by Fig. 4 that presents pressure distributions in the soil. It is obvious that the farther from the impact area, the smaller the pressure in the soil. This pressure is also develops non-uniformly in the soil. More to the point, it is important to pay attention to the erosion of the soil by fluid jet. This erosion is the result of hydraulic fractures of soil particles, and represents a crucial phenomenon in jet grouting as it determines the strains distribution in the soil (see Fig. 5). It can be seen that the most affected zones are the ones located in the vicinity of the water jet. Interestingly, the development of strain in the soil is non-uniform and strains evolution can be observed in all directions. Though, one should notice that unexpectedly in the Fig. 4, the contours present some irregularities that can be ascribed to computation instabilities, given that the soil is assumed isotropic in this case. Furthermore, although the present study only investigates the case of granular media, it is believed that in the case of saturated soil, the contribution of pore water pressure could be a decisive parameter.

5 Conclusion

The present study investigated the suitability of the smoothed particle hydrodynamics (SPH) technique to simulate the mechanism of ground displacement during the jet grouting procedure. Therefore the approach adopted in this paper was the simulation of water-soil interaction using the software AUTODYN-2D. The results obtained allow affirming this approach is more realistic, and might overcome the limitations of the cavity expansion based methods. In addition, this exploration has also permit to underline some particles instabilities issues during the computation. We therefore recommend implementing this approach using a coding solving technique, which offers more confidence in controlling the relevant parameters for both single and double phase materials.

References

1. Essler, R., Yoshida, H.: Jet grouting — from Ground Improvement, 2nd edn., pp. 160–196. Taylor and Francis (2004)
2. Lunardi, P.: Ground improvement by means of jet grouting. Ground Improv. 1, 65–85 (1997)
3. Burke, G.K.: Jet grouting systems: advantages and disadvantages. In: GeoSupport 2004: Drilled Shafts, Micropiling, Deep Mixing, Remedial Methods, and Specialty Foundation Systems, Orlando, pp. 875–886 (2004)
4. Wang, Z.F., Shen, S.L., Ho, E.C., and Kim, Y.H.: Investigation of field installation effects of horizontal Twin-Jet grouting in Shanghai soft soil deposits. Can. Geotech. J. 50(3), 288–297 (2013). https://doi.org/10.1139/cgj-2012-0199

5. Shen, S.L., Wang, Z.F., Cheng, W.C.: Estimation of lateral displacement induced by jet grouting in clayey soils. ICE Geotech. **67**(7), 621–630 (2017). https://doi.org/10.1680/geot./16-P-159
6. Chai, J.C., Miura, N., Koga, H.: Lateral displacement of ground caused by soil–cement column installation. J. Geotech. Geoenviron. Eng. **131**(5), 623–632 (2005)
7. Chai, J.C., Carter, J.P., Miura, N., Zhu, H.H.: Improved prediction of lateral deformations due to installation of soil–cement columns. J. Geotech. Geoenviron. Eng. **135**(12), 1836–1845 (2009)
8. Wang, Z.F., Bian, X., Wang, Y.Q.: Numerical approach to predict ground displacement caused by installing a horizontal jet grout column. Mar. Georesour. Geotechnol. (2016). https://doi.org/10.1080/1064119X.2016.1273288. Published Online
9. Wu, Y.D., Diao, H.G., Ng, C.W.W., Liu, J., Zeng, C.C.: Investigation of ground heave due to jet grouting in soft clay. J. Performance Constructed Facil. **30**(6) (2016)
10. Liu, G.R., Liu, M.B.: Smoothed Particle Hydrodynamics: A Meshfree Particle Method. World Scientific, Singapore (2003)
11. Lucy, L.: A numerical approach to the fission hypothesis. Astron. J., 82–1013 (1977)
12. Gingold, R.A., Monaghan, J.J.: Smoothed particle hydrodynamics - theory and application to non-spherical stars. Mon. Not. R. Astron. Soc. **181**, 375–389 (1977)
13. Liu, M.B., Liu, G.R.: Smoothed particle hydrodynamics (SPH): an overview and recent developments. Arch. Comput. Methods Eng. **17**, 25–76 (2010). https://doi.org/10.1007/s11831-010-9040-7
14. Atangana Njock, P.G., Shen, J.S., Modoni, G., Arulrajah, A.: Recent Advances in horizontal jet grouting (HJG): an overview. Arab. J. Sci. Eng., 1–18 (2017). https://doi.org/10.1007/s13369-017-2752-3
15. Miura, N., Shen, S.-L., Koga, K., Nakamura, R.: Strength change of clay in the vicinity of soil-cement column. J. Geotech. Geoenviron. Eng. **596**(43), 209–221 (1998)

Monte-Carlo Simulation of Post-construction Settlement After Vacuum Consolidation and Design Criterion Calibration

Wei He[1(✉)], Mathew Sams[1], Barry Kok[1], and Pak Rega[2]

[1] Geoinventions Consultancy Services, Brisbane, QLD 4119, Australia
[2] Indonesia Port Corporation, Jakarta Utara, Jakarta 14310, Indonesia

Abstract. Reliability-based design is required to minimise risk induced by soil properties variation and laboratory tests discrepancy in geotechnical engineering. A procedure was proposed to analyse probability of post-construction settlement (PCS) after vacuum consolidation, and to calibrate the design criteria to achieve a target reliability index. A Monte-Carlo simulation based on analytical solution of vacuum consolidation was developed to incorporate both primary and secondary consolidation settlement. The reduction of secondary consolidation coefficient during construction was considered in the method. This design and analysis approach were applied in the design review of Kalibaru port, Indonesia. Statistical analysis on soil properties was performed based on comprehensive investigations. The original design was reviewed by using both deterministic analysis with FEM and reliability-based analysis with the proposed method. Lastly, the coefficient of variation (COV) of 1.164 was found for PCS, and design criteria were calibrated to target different levels of P_e, from 6.7% to 25%.

Keywords: Reliability-based geotechnical design · Reclamation
Post-construction settlement (PCS) · Monte-Carlo simulation

1 Introduction

In a geotechnical design, significant uncertainties exist in the process of defining geomaterial properties, which need to be evaluated via reliability analysis. In Kalibaru port, Indonesia, prefabricated vertical drains (PVDs) with preloading and vacuum is proposed to improve the soft ground at a 900×2600 m site. Comprehensive investigations were performed to mitigate potential risk. However, there was no similar case using reliability-based design (RBD) could be found in literature. Therefore, a RBD procedure and analysis method need to be developed.

Several reliability-based analysis approaches have been developed for geotechnical structures [1], such as the first-order reliability method (FOSM), the second-order reliability method (SORM), and numerical simulations. Monte-Carlo simulation is a numerical process of repeatedly calculating a performance function, in which the variables within the function are random or contain uncertainty with prescribed probability distributions. A large number of outputs can be obtained and used in statistical analysis for directly estimating the probability of failure (P_f), or the

© Springer Nature Singapore Pte Ltd. 2018
L. Li et al. (Eds.): GSIC 2018, *Proceedings of GeoShanghai 2018 International Conference: Ground Improvement and Geosynthetics*, pp. 76–88, 2018.
https://doi.org/10.1007/978-981-13-0122-3_9

probability of exceedance (P_e). In this way, conventional deterministic modelling can be extended to reliability analysis without complex concept and algorithms.

In this paper, RBD procedure based on Monte-Carlo simulation was developed and applied in the Kalibaru port. The mean value of PCS and COV were obtained, then the design criterion was calibrated to achieve reliability index of 1.5. This also provides a detailed case study for future engineering practice.

2 Analytical Solution of Vacuum Consolidation Combined with Preloading

PVDs with vacuum and preloading have been widely applied to accelerate the consolidation of soft ground all over the world. The successful applications include Port of Brisbane, Ballina Bypass, and Sunshine Coast Motorway, in Australia [2, 3]; Tianjin Port and Wenzhou Reclamation, in China [4, 5]; Philadelphia International Airport, in USA [6]; North South Expressway, in Malaysia [7]; Second Bangkok International Airport, in Thailand [8]; Shin-Moji Oki Disposal Pond, in Japan [9]. In these projects, fill preloading was combined with vacuum to avoid excessively high embankment and a lengthy preloading period to achieve the same amount of consolidation degree.

A typical cylindrical element of a PVD with preloading and vacuum is shown in Fig. 1. The PVD has the equivalent radius of r_w, and the influence radius of r_e. A smear zone with the radius of r_s is formed during vertical drains installation with a steel mandrel which significantly remoulds its immediate vicinity. Studies showed that, the radius of the smear zone is about 2.5 times the equivalent radius of the mandrel, and the lateral permeability within the smear zone is 61%–92% of the outer undisturbed zone [10]. Preloading, p_0, is applied on the ground surface, and vacuum pressure, p_s, is applied via pump connected to PVDs. Experience has shown that the vacuum pressure applied in field through PVDs may decrease with depth. Let the decreasing rate is λ, then the suction propagated to the toe is $p_s - \lambda l$ when the length of PVDs, l, is not sufficient to reduce the suction down to zero. This pressure loss rate, λ, was found up to 3 kPa/m in experiments [11].

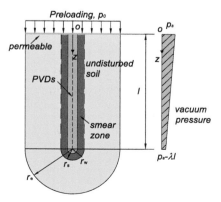

Fig. 1. Illustration of PVDs with preloading and vacuum pressure [2]

Settlement development by using PVDs with preloading and vacuum incorporates both primary and secondary consolidation.

Primary consolidation. The primary consolidation of soil with PVDs is dominated by radial drainage. The theory of radial drainage and consolidation has been developed by many researchers [12, 13]. Consider a thin layer with ΔH in thickness, the excess pore pressure during vacuum consolidation can be solved by:

$$\frac{\bar{u}_{h,t}}{u_0} = \left(1 + \frac{p_s}{u_0}\right) e^{\left(\frac{-8c_h t}{\mu d_e^2}\right)} - \frac{p_s}{u_0} \tag{1}$$

where, $\bar{u}_{h,t}$ is excessive pore pressure at depth of h and time of t; u_0 is the initial excessive pore pressure induced by preloading, p_0; c_h is the coefficient of consolidation; d_e is the diameter of the influencing zone, and $d_e = 2r_e$; μ is a factor as follow:

$$\mu = \frac{n^2}{n^2-1}\left[\ln\left(\frac{n}{s}\right) + \frac{k_h}{k_s}\ln(s) - \frac{3}{4}\right]$$
$$+ \frac{s^2}{n^2-1}\left(1 - \frac{s^2}{4n^2}\right) + \frac{k_h}{k_s}\frac{1}{n^2-1}\left(\frac{s^4-1}{4n^2} - s^2 + 1\right) \tag{2}$$

where, $n = r_e/r_w$, $s = r_s/r_w$.

The degree of consolidation is:

$$U_{h,t} = \frac{1 - \bar{u}_t/u_0}{1 - u_\infty/u_0} = \frac{u_0 - \bar{u}_t}{u_0 - p_s} \tag{3}$$

Thus, primary consolidation can be obtained by:

$$s_t = s_p U_{h,t} \tag{4}$$

where, s_p is the ultimate primary consolidation computed by,

$$s_p = \frac{C_c}{1+e_0}\Delta H \log\left(\frac{\sigma'_{z,t}}{\sigma'_{z,0}}\right) \tag{5}$$

where, C_c is compression index; e_0 is initial void ratio; $\sigma'_{z,t}$ is vertical effective stress at time t; $\sigma'_{z,0}$ is initial vertical effective stress at time 0.

Except for loading process, the recompression index, C_r, is smaller than C_c.

Secondary consolidation. Secondary consolidation plays an important role in long-term settlement. If the time to reach the end of primary consolidation is relatively short which benefits from vacuum consolidation and reloading, the time-dependent settlement is basically controlled by the secondary consolidation [14].

Secondary consolidation, s_s, is given by the formula:

$$s_s = \frac{\Delta H}{1 + e_0} C_\alpha \log\left(\frac{t}{t_{95}}\right) \tag{6}$$

where, C_α is the coefficient of secondary consolidation; t_{95} is the time when primary consolidation reaches 95% consolidation degree.

For simplification, the time at the end of construction was taken as t_{95} in this paper.

A number of authors [15–17] have reported the significant reduction of secondary consolidation when the soil is over-consolidated even to a modest degree. Laboratory and field experiments results indicated a decreasing exponential relationship between over-consolidation ratio (OCR) and C_α. The uniform expression is [16]:

$$C_\alpha = 10^{(A + B \cdot OCR)} + C \tag{7}$$

where, A, B, and C are fitting parameters, as recommended in papers [14, 16].

Kosaka [17] proposed the formula to determine OCR as follow:

$$OCR = \frac{\sigma'_{z,0} + \left(\Delta\sigma'_1 + \Delta\sigma'_s\right) \times U}{\sigma_{z,0} + \Delta\sigma'} \tag{8}$$

where, $\sigma_{z,0}$ is initial effective stress; $\Delta\sigma'$ is effective stress induced by design load; $\Delta\sigma'_1$ is effective stress induced by preloading; $\Delta\sigma'_s$ is effective stress induced by vacuum pressure; U is the consolidation degree at the time of preloading and vacuum removal.

3 Reliability-Based Analysis and Design Approach

3.1 Monte-Carlo Simulation Based on Analytical Solution

The analytical solution of vacuum consolidation and preloading can be incorporated into reliability procedure by using Monte-Carlo simulation (MCS), as indicated within the dashed box in Fig. 2. Comparing to deterministic analysis, MCS computes PCS repeatedly (usually > 5000 times) based on randomly generated samples of soil parameters, and perform statistical analysis on output to extract its probabilistic characteristics. MCS has been widely used in probabilistic analysis of geotechnical engineering problems, such as slope stability analysis, retaining structures, and foundations [1]. However, no study was found in vacuum consolidation combined with preloading.

For soil consolidation, obvious correlations exist among soil parameters such as C_c, and C_r. Thus, these correlations need to be considered in random samples generation.

In this paper, MCS based on the analytical solution was programmed with GNU software, Octave 4.0.

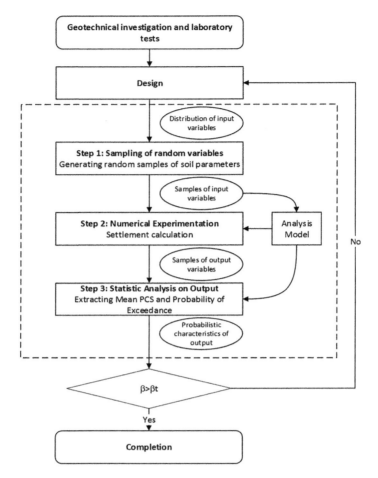

Fig. 2. Flowchart of RBD based on MCS

3.2   Reliability-Based Design Approach

Geotechnical design codes have been migrating towards RBD concepts for several decades [18]. The latest International Standard, ISO2394-2015 [19], has differentiated and related three levels of approach: risk-informed decision making, reliability-based design (RBD), and semi-probabilistic approaches. Comparing to semi-probabilistic approaches such as the load and resistance factor design (LRFD) approach in North American [20], and the characteristic values and partial factors used in the limit state design approach in Eurocode 7 [21] and AS 5100.3-2004 [22], RBD is based on a target reliability index that explicitly reflect the uncertainty of the parameters and their correlation structure, thus more suitable for large scale projects.

As shown in Fig. 2, the reliability index, β, need to be checked in RBD rather than overall factor of safety. EN1990-2002 [23] recommends the target reliability index, β_t.

For PCS, β_t for serviceability (irreversible) limit state in 50 years is 1.5. The Chinese Standard GB 50068-2001 [24] states that, β_t is between 0 to 1.5 for serviceability limit state, depending on the reversibility. The target reliability index of 0 and 1.5 are equivalent to failure probability of 50% and 6.7%, respectively.

In contrast to probability of failure for ultimate limit state, a low probability of exceedance needs to achieve for PCS. The concept of probability of exceedance and design consideration are illustrated in Fig. 3. Assume that settlement conforms reasonable well to a normal or log normal distribution. The probability density function (PDF) 1 has the same mean value μ_1, but smaller standard deviation than PDF 2, namely $\sigma_1 > \sigma_2$. The area under each PDF in the excessive settlement zone indicates probability of exceedance, P_e. Although PDF2 has the same mean value, which is usually the criterion in design, its P_e is obviously higher than that of PDF 1. This indicates that, the same design criterion does not necessarily mean the same level of P_e.

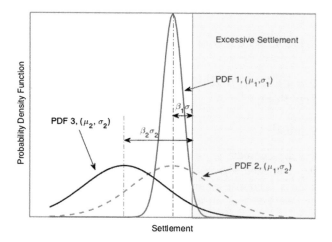

Fig. 3. The concept of P_e and design strategy

With the larger variation level (larger standard deviation) such as in PDF 2, the design needs to be offset to solution represented by PDF 3, which has the same standard deviation σ_2, but stringent design criterion (smaller mean value μ_2). Each P_e can be related to a reliability index, β, as indicated in Fig. 3.

Therefore, the variation of settlement can be considered in design criteria to achieve a specific level of P_e. When the settlement conforms to log normal distribution, Settlement Ratio (SR) can be applied in design according to the COV of settlement and a target exceedance probability from 6.7% to 25.0%, which is plotted in Fig. 4.

Given the maximum allowable settlement is S_{allow}, and the COV of settlement is 0.3, if the target P_e is 6.7%, then SR = 1.489 can be obtained from Fig. 4. This means the design criterion needs to be set at $S_{allow}/1.489$ to achieve the P_e of 6.7%, which has a reliability index of 1.5.

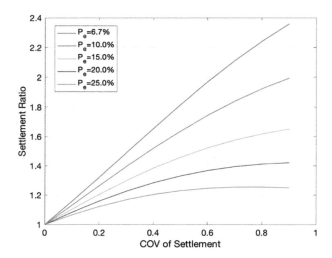

Fig. 4. Settlement ratio vs. COV for P_e of 6.7%–25.0%

4   Application in Design Review of the Reclamation of Kalibaru Port

4.1   Project Summary

The Kalibaru port development is located in the Jakarta bay, and will be constructed from dredged clay and sand materials, as indicated in Fig. 5. The proposed offshore development includes container terminals (CT2 and CT3), product terminals north of CT2 and CT3 and reserve area. In total, the area being constructed is approximately rectangular with dimensions of 2600 m by 900 m.

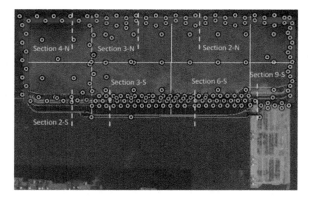

Fig. 5. Kalibaru port, Indonesia

The naturally occuring seabed is a soft Holocene clay material varying from 8 m up to 29 m thickness in some areas. This is followed by a stiff Pleistocene clay, and dense cemented sand. The reclamation works will involve placing grab-bucket dredged mud (GDM) and cutter-suction dredged mud (CSD) to RL 1.5 m. The water level is considered to be equal to the water level in the Jakarta Bay, which is RL 0.

Comprehensive geotechnical investigations have been performed prior to the design. The locations of boreholes and CPT tests are indicated in Fig. 5. Samples were then tested in laboratory to obtain soil parameters. In the original design, mean value of each parameter was adopted to assess the PCS. Generally, 90-day vacuum consolidation combined with preloading was proposed by the specialist contractor to target the criterion of 300 mm in 50 years after construction.

Geoinventions Consulting Services (GCS) was engaged by IPC to review the design, and RBD procedure was implemented to calibrate the design criterion of PCS. Deterministic analysis by using FEM was also carried out to consider the influence of construction stage.

4.2   Deterministic Results by Using FEM

The construction stage of the large-scale reclamation is complex. Preloading and vacuum pressure have to be applied section by section, due to limited volume of fill and quantity of pumps. The construction of filling for preloading and installation of PVDs are time-consuming which also elongate the construction period. To assess PCS under the real construction process, a 2D finite element software, OptumG2 (version 2017.05.20), was used in the deterministic analysis.

To simplify the modelling process, the analysis for each section starts from the stage when GDM/CSD have been built up to RL 1.5. The sections of wick drains/PVDs are modelled using fixed excess pressure lines. These allow the excess pore pressure to be fixed to any value. Pressure loss of 3 kPa/m was considered and average vacuum pressure was applied alone the fixed excess pressure lines. Main soil parameters adopted in the FEM analysis are listed in Table 1. The secondary consolidation coefficient, C_α, was not tested in the investigations. The test results performed at the Belawan Port which is approximately 200 km away from the site were used [25].

Table 1. Parameters adopted in Section 3-N analysis

Parameter	Upper Holocene Clay	Lower Holocene Clay	Stiff Clay	GDM/CDM
γ (kN/m^3)	14.3	14.6	15.8	12.0
C_c	0.96	0.85	0.59	1.08
C_r	0.148	0.108	0.046	0.18
C_v (m^2/year)	1.33	1.37	4.43	3.50
C_α	0.036	0.036	–	0.036
e_0	3.00	2.54	1.68	3.50
OCR	1.00	1.00	2.50	1.00

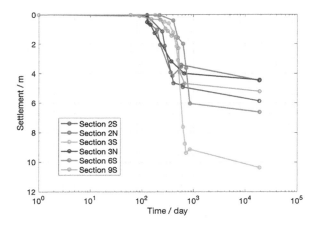

Fig. 6. Settlement development by FEM

A total of six cross sections have been considered across the project – Sections. 2-S, 2-N, 3-S, 3-N, 6-S, and 9-S. These sections are shown in Fig. 1. The settlement results for all six sections is illustrated in Fig. 6. These results indicate that the maximum total settlement occurs for Section 3-S, with a result of 10.4 m. Figure 6 also indicates that Section 2-N and 3-N exhibit very similar settlement profiles. Sections 6-S and 9-S are also similar, but exhibit a 1 m difference in total settlement due to varying soil profile and different construction process. All sections show a similar gradient to the secondary settlement line (after approx. 1000 days), except for Section 3-S which has almost doubled the thickness of soft clay.

4.3   Reliability-Based Design Calibration

According to the FEM analysis results, the soft soil and GDM/CDM are most critical for settlement analysis. GDM/CDM is the dredged layer which can be considered reasonably uniform. Thus, the thick soft soil layer dominates the variation of PCS. Statistical analysis was implemented on soft soil based on gathered test results from the principal geotechnical designer, LAPI. Log normal distribution function was adopted to fit each parameter. Six parameters: unit weight; compression index, C_c; recompression index, C_r; Coefficient of consolidation, C_v; secondary consolidation coefficient, C_α, and

Table 2. Mean Values and Variation of Parameters

Parameter	Mean Value	Standard Deviation	Range
γ (kN/m^3)	14.949	2.516	12.26–19.60
C_c	0.883	0.324	0.110–2.080
C_r	0.124	0.012	0.030–0.300
C_v (m^2/year)	4.068	1.666	0.370–4.248
C_α	0.037	0.003	0.001–0.499
e_0	2.574	2.515	0.040–3.780

initial void ratio, e_0, are considered in MCS. OCR in Table 1 was ignored in RBD because of its estimation in tests were rough thus less meaningful to taken into account.

Correlations exist among parameters listed in Table 2. The most important correlation is between C_c and C_r. Test results shows that the ratio of C_c/C_r is in a range of 2.81–27.60, which can be fitted by log normal PDF with log normal parameters $\mu_{LN} = 2.042$, $\sigma_{LN} = 0.629$, and it is in a range of 3–27. In order to obtain the correlated parameters C_c and C_r, C_c was generated randomly first with parameters listed in Table 3, then the ratio of C_c/C_r was generated and C_r was obtained by C_c and the ratio. The generated samples and test data were shown in Fig. 7. The generated samples conform to the range of test data.

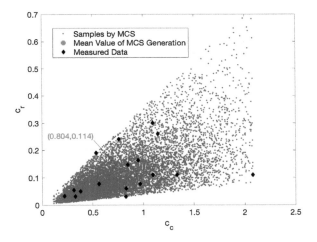

Fig. 7. Random generation of correlated parameters in MCS

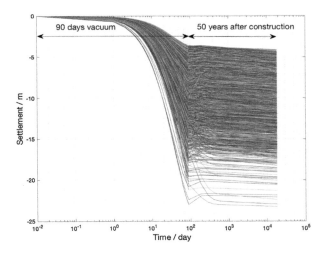

Fig. 8. Repeatedly settlement calculation in MCS

Figure 8 illustrated 10,000 times calculated settlement procession results. The settlement during vacuum and preloading is between 4–23 m depending on parameters generated in MCS. This is a reasonable result with a mean settlement close to FEM result. The settlement development shows different patterns due to combination of parameters.

In total, 10,000 PCS can be obtained in MCS. The absolute value of the output was statistically analysed with log normal PDF, and the result is shown by histogram in Fig. 9. The mean PCS is 917 mm, and the coefficient of variation is 1.164. This indicated a large variation exists because of the variation of soil properties. According to Fig. 4, the settlement ratio is 2.6, 2.1, 1.7, 1.4, 1.2 when target P_e is 6.7%, 10%, 15%, 20%, 25%, respectively. Taking the target P_e is 20%, GCS sets the design criterion at 300 mm/1.4 = 214 mm.

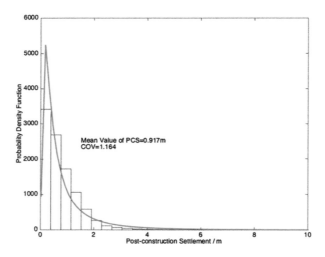

Fig. 9. Probabilistic analysis of MCS outputs

5   Conclusions

A reliability-based geotechnical design procedure with analysis approach was proposed in this paper. And this method was applied in the design review of Kalibaru port, Indonesia. Based on the practice, conclusions can be drawn as follow:

(1) A design criterion requires to be calibrated by using reliability-based design method in large scale projects to mitigate potential risk. This guarantees that a target probability of exceedance can be achieved for PCS in design;

(2) The settlement ratio diagram was developed to determine a settlement criterion based on the COV of PCS and a target exceedance probability, which is straightforward in the design process;

(3) The Monte-Carlo simulation based on analytical solution of vacuum consolidation and preloading was coded and incorporated in the RBD of Kalibaru port.

The COV was found to be 1.164, and the settlement ratios were recommended to target exceedance probabilities of 6.7%–25%. This RBD strategy can take into account the uncertainty induced by soil properties variation, and provide a reasonable criterion for engineering purpose.

References

1. Phoon, K.K., Ching, J.: Risk and Reliability in Geotechnical Engineering. CRC Press (2014)
2. Indraratna, B., Rujikiatamjorn, C., Geng, X.: Performance and prediction of surcharge and vacuum consolidation via prefabricated vertical drains with special reference to highways, railways and ports. In: International Symposium on Ground Improvement, 31 May–1 June 2012, Brussels, Belgium, pp. 145–168 (2012)
3. Indraratna, B., Kan, M.E., Potts, D., Rujikiatkamjorn, C., Sloan, S.W.: Analytical solution and numerical simulation of vacuum consolidation by vertical drains beneath circular embankments. Comput. Geotech. **80**, 83–96 (2016)
4. Chu, J., Yan, S.W., Yang, H.: Soil improvement by the vacuum preloading method for an oil storage station. Geotechnique **50**(6), 625–632 (2000)
5. Tang, T.Z., Dong, J.P., Huang, J.Q., Zhang, X.Z., Guan, Y.F.: Experimental research on hydraulic filled mud consolidated by vacuum preloading method combining long and short boards with thin sand cushions. Chin. J. Geotech. Eng. **34**(5), 899–905 (2012)
6. Holtan, G.W.: Vacuum stabilization of subsoil beneath runway extension at Philadelphia international airport. In: Proceedings of 6th ICSMFE, vol. 2 (1965)
7. Yee, K., Ooi, T.A.: Ground improvement – a green technology towards a sustainable housing, infrastructure and utilities developments in Malaysia. Geotech. Eng. J. SEAGS & AGSSEA **4**(3), 1–20 (2010)
8. Seah, T.H.: Design and construction of ground improvement works at Suvarnabhumi airport. Geotech. Eng. J. SE Asian Geotech. Soc. **37**, 171–188 (2006)
9. Ikeda, H., Kawano, M., Kiyoyama, T.: Performance and prediction of dredged clay reclaimed land by vacuum consolidation method. Int. J. GEOMATE Geotech. Constr. Mater. Environ. **8**(1), 1300–1307 (2015)
10. Indraratna, B.: Recent advances in the application of vertical drains and vacuum preloading in soft soil stabilisation. Aust. Geomech. J. **45**(2), 1–53 (2010)
11. Liqiang, S.: Theory and model test study of recently reclaimed soil foundation. Ph.D thesis, Tianjin University (2010)
12. Hansbo, S.: Consolidation of fine-grained soils by prefabricated drains. In: Proceedings of 10th International Conference SMFE, vol. 3, Stockholm, Sweden, pp. 677–682 (1981)
13. Indraratna, B., Sathananthan, I., Rujikiatkamjorn, C., Balasubramaniam, A.S.: Analytical and numerical modeling of soft soil stabilized by prefabricated vertical drains incorporating vacuum preloading. Int. J. Geomech. **5**(2), 114–124 (2005)
14. Alonso, E.E.: Precompression design for secondary settlement reduction. Géotechnique **51**(51), 822–826 (2001)
15. Mesri, G.: Coefficient of secondary compression. J. Soil Mech. Found. Div. ASCE **99**(1), 77–91 (1973)
16. Fukazawa, E., Yamada, K., Kurihashi, H.: Predicting long-term settlement of highly organic soil ground improved by preloading. J. Geotech. Eng. JSCE **493**(3–27), 59–68 (1994)
17. Kosaka, T., Hayashi, H., Kawaida, M., Teerachaikulpanich, N.: Performance of vacuum consolidation for reducing a long-term settlement. Japan. Geotech. Soc. Spec. Publ. **2**(59), 2015–2020 (2016)

18. Fenton, G.A., Naghibi, F.: Reliability-based geotechnical design code development. Vuluerability, Uncertainty, Risk ASCE, 2468–2477 (2014)
19. International Organization for Standardization. General principles on reliability of structures. ISO2394: 2015, Geneva, Switzerland (2015)
20. Federal Highway Administration (FHWA). Load and Resistance Factor Design (LRFD) for highway bridge substructures. Publication No. FHWA-HI-98-032 (2001)
21. European Committee for Standardization (CEN). Eurocode 7: geotechnical design — part 1: general rules. EN 1997–1:2004, Brussels, Belgium (2004)
22. AS 5100.3-2004. Bridge Design, Part 3: Foundations and Soil-Supporting Structures (2004)
23. European Committee for Standardization (CEN). Eurocode — basis of structural design. EN 1990:2002, Brussels, Belgium (2002)
24. GB 50068-2001. Unified Standard for Reliability Design of Building Structures, Beijing (2001)
25. Toha, F.X.: Secondary compression of belawan clay. In: International Symposium on Soft Soils in Construction and Deep Foundations, IGEA - JSSMFE, Jakarta (1987)

Numerical Investigation on Slope Stability of Deep Mixed Column-Supported Embankments Over Soft Clay Induced by Strength Reduction and Load Increase

Zhen Zhang[1(✉)], Yan Xiao[1], Guan-Bao Ye[1], Jie Han[2],
and Meng Wang[1]

[1] Department of Geotechnical Engineering, Tongji University,
Shanghai 200092, China
zhenzhang@tongji.edu.cn
[2] Civil, Environmental and Architectural Engineering (CEAE) Department,
The University of Kansas, Lawrence, KS 66045, USA

Abstract. The stability of the column-supported embankment may become one of the major concerns when constructed over soft clay. The numerical methods referred to as strength reduction method and the load increase method can be adopted to analyze the stability. However, limited studies have been conducted to examine the differences between load increase and strength reduction methods in the stability analysis of the column-supported embankments over soft soil. A three-dimensional (3-D) finite element method incorporated in the ABAQUS software was used in this study to investigate the contribution of deep mixed (DM) columns to the stability of the embankment over soft soil. The strength reduction method and the load increase method were implemented to obtain the factors of safety in stability. The maximum moments carried by the DM columns below the embankment crest were much less than those carried by the columns below the embankment slope. The failure modes of DM columns under embankment can be classified into four zones from the centerline to the toe, namely, compression zone, shear zone, a combination of compression and bending zone, and a combination of extension and bending zone. The factors of safety based on the strength reduction were equal to or higher than those based on the load increase.

Keywords: Column-supported embankment · Deep mixed column
Stability · Failure mode · Factor of safety

1 Introduction

When the embankments for railways and highways are constructed over soft soils, the slope stability of the embankment may become one of the major concerns. In the last decades, column-supported embankment has been increasingly used to increase the bearing capacity, reduce the total and differential settlement and enhance the stability around the world (Jamsawang et al. 2016). Among all kinds of columns, it is no doubt

© Springer Nature Singapore Pte Ltd. 2018
L. Li et al. (Eds.): GSIC 2018, *Proceedings of GeoShanghai 2018 International Conference: Ground Improvement and Geosynthetics*, pp. 89–96, 2018.
https://doi.org/10.1007/978-981-13-0122-3_10

that the deep mixed (DM) columns can be regarded as one of the most commonly used techniques (CDIT 2002).

When assessing the slope stability of DM column-supported embankment over soft soils, the reliability of calculation for factor of safety need to be verified. Navin and Filz (2006) indicated that the factor of safety by plane strain analysis was conservative as compared with that by three-dimensional (3-D) analyses. Han et al. (2004) and Zhang et al. (2014) presented that Bishop's method yielded a higher factor of safety than the numerical method. Besides, various failure modes of DM columns were identified, such as DM columns may have possible failure modes of shear, bending, sliding, tilting, or a combination of the above failure modes under an embankment loading (Broms 1999; Kitazume and Maruyama 2007). It is obvious that the DM column-supported embankment may fail by at least one of the above failure modes which yields the minimum factor of safety under certain conditions.

The objective of this paper is to assess the stability of DM column-supported embankment over soft soil based on the strength reduction and the load increase methods. The three-dimensional numerical analysis was carried out using the undrained parameters since the embankment load was assumed to be applied in a short period. The lateral displacements of DM columns, the failure modes of DM columns, the slip plane, and the factor of safety obtained based on the strength reduction and the load increase were compared and analyzed. The stress transfer mechanism during the procedure of embankment instability was investigated and the recommendations were proposed for engineering application.

2 Definition of Factor of Safety

In the shear strength reduction technique, a series of trial factor of safety values are used to adjust the cohesion and the friction angle of soil using the following equations until the system becomes unstable (Dawson et al. 1999):

$$c_{\text{trial}} = c/FS_{\text{trial}} \tag{1}$$

$$\phi_{\text{trial}} = \arctan(\tan \phi/FS_{\text{trial}}) \tag{2}$$

in which, c and c_{trial} are the initial cohesion and the trial cohesion reduced by the trial factor of safety FS_{trial}, respectively; and ϕ and ϕ_{trial} are the initial friction angle and the trial friction angle reduced by the trial factor of safety FS_{trial}, respectively. The factor of safety induced by the strength reduction FS_{SR} can be defined as being equal to the trial factor of safety when the deformation of the characteristic monitoring point (e.g., embankment toe) is varied rapidly.

The load increase method can be regarded as a reverse process as compared with the strength reduction method. In this method, the strength parameters are kept constant and a series of surcharge is applied on the embankment crest until the slope become unstable. The factor of safety induced by the load increase FS_{LI} can be defined as a ratio of the ultimate load when the slope becomes unstable to the initial embankment load:

$$FS_{\mathrm{LI}} = \frac{W + Q_{\mathrm{ult}}}{W} \qquad (3)$$

In which, W is the total weight of embankment load; Q_{ult} is the ultimate surcharge load applied on the embankment crest. Based on the definitions of the factors of safety induced by the strength reduction and the load increase, it can be noted that these two types of factors of safety are proposed based on different concepts. The strength reduction method is to reduce the sliding resistance force of the system, while the load increase is to increase the sliding driving force of the system.

3   3-D Finite Element Modeling

3.1   Numerical Model

The software ABAQUS was adopted to establish the numerical model of the DM column-supported embankment over soft soils. To simplify the mesh generation, the circular DM columns were approximated by square columns with equivalent cross-sectional area with a side length of 440 mm. Considering the symmetry of the cross section and the repetition of the span along the traffic direction, only a slice of one-half the embankment was simulated (see Fig. 1). The DM column was 500 mm in diameter and 10 m in length was installed in a square pattern with a spacing of 1.4 m, which is corresponding to an area replacement ratio of 10%. The DM columns starting from the center line were denoted in order from No. 1 to 14. The bottom boundary was fixed in both horizontal and vertical directions. The two side boundaries were fixed in the horizontal direction but allowed to move freely in the vertical direction.

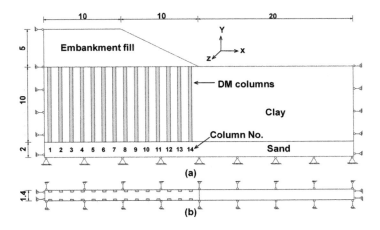

Fig. 1. Numerical model (unit: m): (a) cross-section; (b) plan view

3.2   Material Properties

The embankment fill, soft soil, dense sand and DM columns were modeled as linearly elastic to perfectly plastic materials with the Mohr-Coulomb (MC) failure criteria. The unconfined compressive strength of DM columns were in a wide range from 0.2 to 3 MPa depending on the cement content, mixing times, soil conditions, curing conditions and so on forth (Baker 2000). The unconfined compressive strength of DM was assumed as 0.3 MPa in analysis. The elastic modulus and the Poisson's ratio were taken as 100 MPa and 0.3 respectively. Table 1 tabulates the material properties in the numerical analysis.

Table 1. Material properties

Material	γ(kN/m³)	μ	E(MPa)	c(kPa)	ϕ(°)
Embankment	18	0.3	30	15	35
Soft soil	16	0.45	5	25	0
Sand	18	0.3	50	0	35
DM columns	20	0.3	100	150	0

Note: E = elastic modulus, γ = unit weight,
μ = Poisson's ratio, ϕ = friction angle, c = cohesion.

4   Results and Discussion

4.1   Lateral Displacements

Figure 2 shows the lateral displacements of the DM columns at different locations when the instability of the embankment occurred. It can be seen that the lateral displacements of DM columns obtained by the load increase method were larger than

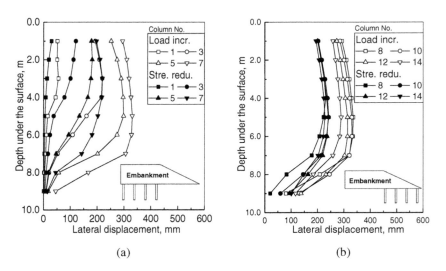

Fig. 2. Lateral displacement of DM columns in the model: (a) under the crest; (b) under the side slope

those by the strength reduction method but the distribution manners were similar. The lateral displacements of columns under the embankment crest were generally increased from column No. 1 to No. 7 and the depths of maximum lateral displacements were increased as well. However, the lateral displacements along the DM columns under the side slope had a similar manner irrespective to the column locations. The centrifugal test results also confirmed the similar findings (Kitazume and Maruyama 2007).

4.2 Failure Modes

Table 2 tabulates the maximum moments occurred on each DM column induced by the strength reduction method and load increase method when the embankment failed. Figure 3 illustrates the undeformed and deformed meshes of DM columns based on strength reduction and load increase. The deformed modes in the model by strength reduction were similar with the model by load increase, therefore, only the deformed mesh in the model by load increase were illustrated in this paper.

Table 2. Maximum bending moments of DM columns

Column no.	1	2	3	4	5	6	7
$M_{SR\text{-}max}$ (N · m)	7	4	9	6	12	21	10
$M_{LI\text{-}max}$ (N · m)	13	11	17	20	38	67	160
Column no.	8	9	10	11	12	13	14
$M_{SR\text{-}max}$ (N · m)	11	12	63	516	710	465	126
$M_{LI\text{-}max}$ (N · m)	681	1625	1538	1357	492	202	300

Note: $M_{SR\text{-}max}$ is the maximum absolute moment induced by strength reduction method; $M_{LI\text{-}max}$ is the maximum absolute moment induced by load increase method.

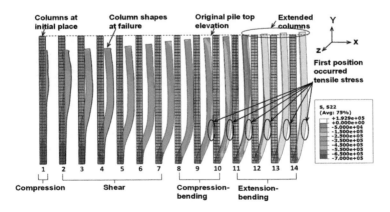

Fig. 3. Deformed and undeformed shapes of DM columns in the model by load increase

The column on the center line of embankment (i.e., column No. 1) was mainly subjected to compression deformation. The columns denoted from No. 2 to No. 7 in the model were subjected both compression and lateral displacement while the moments in these columns were small (see Table 2). It can be realized that these columns were failed due to shear. The columns from No. 8 to 10 in the models were subjected to significant moments while they also had compression deformation. The rest columns (i.e., the columns from No. 11 to 14) deformed with a combination of extension and bending, since the final length of column were larger than the initial length, the tensile stresses (gray areas in Fig. 3) were generated on the column shaft, and the significant moments were produced. Based on the above analysis, the failure modes of DM columns under embankment can be classified into four zones: from the centerline to the toe, namely, compression zone, shear zone, a combination of compression and bending zone, and a combination of extension and bending zone. The proposed classification of failure modes possess similarities and differences as compared with that indicated by Broms (1999). Broms (1999) explained the DM columns under embankment might fail under bending, while those close to and away from embankment might fail under tension.

4.3 Slip Surface

The slip surface can be obtained by investigating the contour of the plastic strain extracted from the numerical results as shown in Fig. 4. It can be seen that the numerical analyses did not have continuous slip surfaces and the slip surface had a certain thickness which is also called slip band hereafter. The upper and lower boundary of slip band agreed well with the curvature points of deformed DM columns. The slip bands determined by the strength reduction and the load increase were similar to circular slip surface but the slip bands were not consistent with each other.

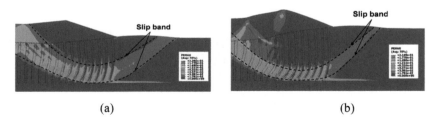

(a) (b)

Fig. 4. Contours of plastic strain magnitude: (a) by strength reduction; (b) by load increase

4.4 Factor of Safety

Based on the previous description, the factors of safety of the models by strength reduction and load increase can be obtained. The models with DM column-supported embankment with different surcharge on the crest (i.e., 0 kPa, 20 kPa, 40 kPa, 60 kPa, 80 kPa, 100 kPa, 110 kPa) were established and the factor of safety of each model was obtained based on the strength reduction and load increase. Figure 5 shows a comparison of the factors of safety of each model. It can be seen that the factors of safety based on the strength reduction were equal to or higher than those based on the load increase. This is opposite to the findings by Griffiths (2015). It might be the reason that Griffiths (2015) investigated the safety factor in bearing capacity of homogeneous soil slope which is not the case for the safety factor in stability on the DM column-supported embankment.

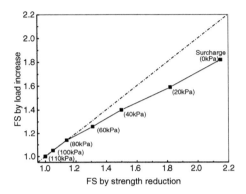

Fig. 5. Factors of safety based on strength reduction and load increase

5   Conclusions

The numerical analysis was conducted to assess the stability of DM column-supported embankment over soft soil based on the strength reduction method and the load increase method. Based on the results and discussion, the following conclusions can be drawn:

1. The lateral displacements of DM column subjected to the embankment load were depending on the locations of the columns under the embankment. The lateral displacements of the columns located under the embankment crest and the depths of maximum lateral displacements increased with an increase of distance from the embankment centerline. While the DM columns deformed almost consistently under the embankment side slope.

2. The DM columns failure modes under embankment load can be divided into four zones, from the centerline to the toe, namely, compression zone, shear zone, a combination of compression and bending zone, and a combination of extension and bending zone based on the numerical analysis.

3. The numerical analyses did not have continuous slip surfaces. The slip bands determined by the strength reduction and the load increase were similar to circular slip surface but the slip bands were not consistent with each other. When the stability of DM column-supported embankment moved towards to the safe state, the load increase method was more conservative than the strength reduction method.

Acknowledgments. The authors appreciate the financial support provided by the Natural Science Foundation of China (NSFC) (Grant No. 51508408 & No. 51478349) and the Pujiang Talents Scheme (No. 15PJ1408800) for this research.

References

Baker, S.: Deformation behavior of lime/cement column stabilized clay. Chalmers University of Technology, pp. 30–39 (2000)

Broms, B.B.: Can lime/cement columns be used in Singapore and Southeast Asia? In: 3rd GRC Lecture, 19 November, Nanyang Technological University and NTU-PWD Geotechnical research Centre, 214p (1999)

Coastal Development Institute of Technology (CDIT): The Deep Mixing Method: Principle, Design and Construction. A.A. Balkema Publishers, Tokyo (2002)

Dawson, E.M., Roth, W.H., Drescher, A.: Slope stability analysis by strength reduction. Geotechnique **49**(6), 835–840 (1999)

Griffiths, D.V.: Observations on load and strength factors in bearing capacity analysis. J. Geotech. Geoenviron. Eng. **141**(7), 06015004 (2015)

Han, J., Chai, J.C., Leshchinsky, D., et al.: Evaluation of deep-seated slope stability of embankments over deep mixed foundations. Anim. Reprod. Sci. **159**, 163–171 (2004)

Jamsawang, P., Yoobanpot, N., Thanasisathit, N., et al.: Three-dimensional numerical analysis of a DCM column-supported highway embankment. Comput. Geotech. **72**, 42–56 (2016)

Kitazume, M., Maruyama, K.: Centrifuge model tests on failure pattern of group column type deep mixing improved ground. Soils Found. **40**(4), 43–55 (2007)

Navin, M.P., Filz, G.M.: Numerical stability analyses of embankments supported on deep mixed columns. In: Geoshanghai International Conference, pp. 1–8 (2006)

Zhang, Z., Han, J., Ye, G.B.: Numerical analysis of failure modes of deep mixed column-supported embankments on soft soils. In: Geoshanghai International Conference, pp. 78–87 (2014)

Performance of Clay Bed with Natural and Lightweight Aggregate Stone Columns

Trudeep N. Dave[(⊠)] and Veerabhadrappa M. Rotte

Institute of Infrastructure Technology Research and Management,
Ahmedabad 380026, India
trudeepdave@iitram.ac.in

Abstract. Booming infrastructure activities demand suitable subsoil conditions to accomplish design requirements in terms of strength and serviceability. Soft soils encountered during such activities may pose problems such as very low strength, poor hydraulic conductivity, shrinkage and swelling with seasonal moisture variations etc. Utilization of sites with soft soils required the invention of various ground modification techniques. Stone columns is one of such techniques used in soft soils to improve its load bearing capacity, dissipate excess pore water pressure rapidly and reduce the total settlement effectively and economically. Stone columns require use of natural aggregates to be transported from a distant place to the construction site that causes an increase in construction cost, carbon footprint and diminish natural resources. This paper presents laboratory investigations of load test results on virgin clay bed and clay bed with stone columns. Two different conditions of stone columns viz. (1) un-encased stone columns with natural and lightweight aggregates (2) encased stone columns with natural and lightweight aggregates. Clay beds were prepared using slurry consolidation method by mixing dry soil with water content equal to 1.5 times liquid limit of the soil. Displacement controlled load tests were performed on clay beds with and without stone columns. Loads against applied displacements were continuously monitored for all the tests conducted. Clay bed with stone columns demonstrated higher load carrying capacity as compared to virgin clay beds. Un-encased stone columns indicated increase in bearing capacity by 31.7% for natural aggregates and 19.5% for light weight aggregates as compared to virgin clay bed. Further, provision of encasement to the stone columns increased load carrying capacity to 36.6% for natural aggregate stone columns and 31.7% for light weight aggregate stone columns as compared to virgin clay bed.

Keywords: Clay-bed · Stone-column · Aggregate · Encasement
Performance

1 Introduction

The rapid boom in infrastructural activities all over the world requires availability of land with desirable soil properties. Lack of availability of suitable land demands modification in pre-existing soil conditions as per the project requirements, particularly in coastal regions and on soft soils. Construction in coastal regions and on soft soils

© Springer Nature Singapore Pte Ltd. 2018
L. Li et al. (Eds.): GSIC 2018, *Proceedings of GeoShanghai 2018 International Conference: Ground Improvement and Geosynthetics*, pp. 97–104, 2018.
https://doi.org/10.1007/978-981-13-0122-3_11

areas will always be having excessive settlement problems at post-constructions stage. Therefore, appropriate ground treatment needs to be done before the construction works commence to avoid high maintenance costs after construction. In order to reduce consolidation time requirement, many techniques such as use of stone column, sand drains and prefabricated vertical drains are popularly used. Stone columns are found effective, feasible and economical to improving the soft and loose layered soil. Stone columns increase the unit weight and the bearing capacity of soil. It can densify the surrounding soil during construction. The stone columns not only act as reinforcing material to increase the overall strength and stiffness of the compressible soft soil, but also they promote consolidation through effective drainage [2].

The past researchers proved that the encasement materials provide a greater lateral support to the stone columns and enhances its load carrying capacity. The existence of geo-synthetic around granular column causes the possibility of an enormous settlement reduction, acceleration of the settlement rate, increase in shear strength of the surrounding soft soil and bearing capacity of the whole system [3, 4]. The encasement, besides increasing strength and stiffness of the stone column, prevents the lateral deformation of stone columns and thus enabling quicker and more economical installation [1]. Encasement materials also prevent the mixing of fine grained soil with stone material which has a negative effect on the stone column drainage efficiency during the consolidation process [5, 6]. Sivakumar et al. [7] performed a series of triaxial compression test on sand columns with and without geogrid sleeves having various sleeve lengths. It was reported that load carrying capacity of sand column was observed to increase with an increase in sleeve length. However, it could not be continued for the sand columns longer than approximately five times the diameter of sand column.

Review of previous studies highlighted that stone columns are cost effective and practical solution for improvement of saturated clayey soils. As stone columns involves use of natural aggregates, which are transported from a distant place to the construction site, causes increase in construction cost, carbon footprint and use of natural aggregates. Use of light weight (manufactured) aggregates during stone column installation may prove beneficial for improving load-settlement behavior of soft clays and one of the promising alternative to the natural aggregates. As lightweight aggregates are lightweight from materials such as clays, fly ash, etc. reduce the carbon footprint of the project and their use may be alternative solution for disposal of fly ash. Studies in this direction are limited and hardly addressed.

Keeping the above in view, this study is aimed at observing the settlement behaviour of the virgin clay beds and clay beds with stone columns prepared of natural and light weight aggregates with and without encasement through strain controlled load tests using small scale experimental setup.

2   Loading Tests on Stone Columns

2.1   Test Materials

The soil used in this study was collected from Surat riverfront project area on the banks of Tapi River. The properties of the above mentioned soil are summarized in Table 1.

In this study crushed basalt (10 mm down) was used as natural aggregates whereas, commercially available shell aggregates 'Nodullar' was used as lightweight aggregates. Properties natural and lightweight aggregates are presented in Table 2.

Table 1. Properties of model soil used in the present study

Parameters	Unit	Value
Specific gravity	–	2.68
Liquid limit (w_L)	%	59.5
Plastic limit (w_P)	%	31.3
Shrinkage limit (w_S)	%	18.4
Plasticity index (I_P)	%	28.2
Flow index (I_F)	–	6.84
Permeability	cm/sec	5.89×10^{-9}
Water content of prepared slurry (w)	%	89.25
Pre- consolidation pressure	kPa	15.41

Table 2. Properties of natural and light weight aggregates

Parameters	Natural aggregates	Light weight aggregate
Material	Crushed basalt	Siliceous materials
Size	5–10 mm	4–12 mm
Shape	Angular to sub-angular	Normal round nodules
Density	1500 kg/m^3	630 kg/m^3
Water absorption	1.5%	37.8%

2.2   Test Setup and Procedure

All the experiments were performed in the laboratory environment using the small-scale consolidation setup (Fig. 1a-c). It consisted of mild steel cylindrical containers of 280 mm diameter, 350 mm height and 2 mm thickness. Brass nozzle (of inside diameter 5 mm) was provided 10 mm above the bottom of the container so as to allow and monitor drainage of water. Slurry was prepared from the procured soil by adding water content equivalent to 150% of liquid limit. The slurry container was covered with polyethylene sheet for 24 h. After that it was stirred manually to get uniform consistency. To prepare uniform clay bed, the slurry was consolidated in cylindrical container by providing 50 mm medium coarse sand layers both at the top and bottom of the container. Also, a layer of non-woven geotextile was laid at the periphery of the container to facilitate uniform radial consolidation. Additionally, geotextile layer equivalent to the diameter of the container was placed at interface of slurry and sand layer to facilitate separation and in plane drainage. Approximate thickness of clay layer achieved through this arrangement was about 250 mm. In order to facilitate application of uniform surcharge on the clay layer, a 5 mm thick mild steel plate of 250 mm diameter was placed above the top sand layer.

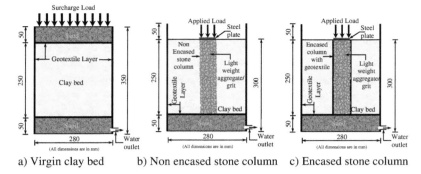

a) Virgin clay bed   b) Non encased stone column   c) Encased stone column

Fig. 1. Schematic diagram of preloading virgin clay and prepared stone columns

The slurry was consolidated at pressure equivalent to pre-consolidation pressure. A series of sequential photographs of the preparation of test sample for consolidation is depicted in Fig. 2. Starting from a pressure of 5 kPa, the pressure increments of 5 kPa was applied by placing dead weights above the mild steel plate. After application of surcharge, settlement and water discharge through the container were monitored. Consolidation was assumed to be completed when settlement was seized for about 24 h accompanied by no discharge of water. Surcharge pressure of 15 kPa was applied on all five soil containers before performing load test on the clay bed.

Fig. 2. Sequential photographs of preparation of test sample for consolidation (a-f)

Once a clay bed was fully consolidated under the applied surcharge pressure of 15 kPa, the surcharge and top sand layer was removed carefully from the clay bed. Centre position was marked on the top surface of the clay bed. In order to prepare stone column in virgin clay bed, thin wall sampling tube of 50 mm was penetrated in the clay bed up to its bottom and then carefully pulled out. The cylindrical space created was filled by aggregate (natural or lightweight) up to the top surface of the clay bed. Schematic diagrams of the experimental setup of stone column with and without encasement are illustrated in Figs. 1b and c. To perform load test on virgin clay bed, a mild steel plate of 50 mm diameter and 8 mm thickness was placed at the center of clay bed. The assembly was placed on the reaction type loading frame and test was performed at a displacement rate of 0.25 mm/min. Settlement behavior of all the five container with applied load was monitored. Figure 3 shows a sequential view of preparation of stone column for loading test.

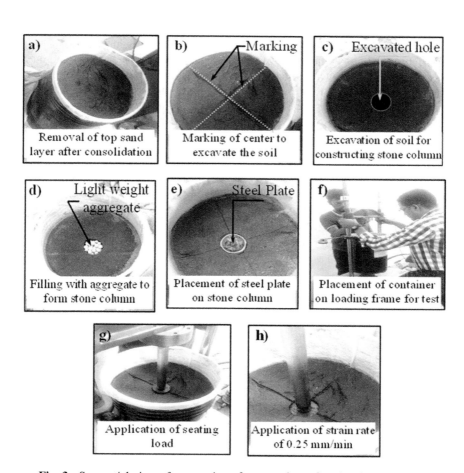

Fig. 3. Sequential view of preparation of stone column for plate load test (a-h)

2.3   Test Programme

A series of small scale experiments have been conducted to observe the performance of encased and non-encased stone column and to evaluate the efficiency of light weight aggregate for its use in stone column to enhance the bearing capacity of the soft soil. Load and settlements were monitored throughout the experiments. Five different types of load tests were performed: (1) on virgin clay bed (2) clay bed with natural aggregate stone column (3) clay bed with lightweight aggregate stone column (4) clay bed with encased stone column of natural aggregate (5) clay bed with encased stone column of lightweight aggregate. All the tests were performed up to settlement equivalent to $1/10^{th}$ of bearing plate diameter.

3   Results and Discussion

Load - settlement graphs were obtained for all the five containers (virgin clay bed, stone column with light weight aggregate, stone column with natural aggregate, encased stone column with light weight aggregate and encased stone column with natural aggregate). Considering plate diameter as 50 mm, the test results corresponding to load were converted into stress to obtain effect of stone column inclusion and to observe effect of providing encasement to the stone columns.

3.1   Effect of Provision of Stone Columns

Influence of providing stone columns is shown in Fig. 4a. Load test results on virgin clay bed revealed bearing capacity of 208 kPa. While, load tests on clay bed with stone columns prepared with natural aggregate and light weight aggregate showed bearing capacities as 275 kPa and 250 kPa, respectively. The improvement in bearing capacity due to provision of stone columns was observed as 31.7% for natural aggregate and 19.5% for light weight aggregates.

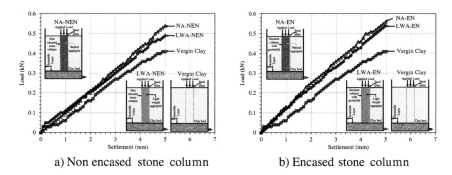

a) Non encased stone column   b) Encased stone column

Fig. 4. Load vs settlement curve for stone columns

3.2   Effect of Encasement to Stone Columns

Figure 4b depicts effect of providing encasement to the stone columns. Provision of encasement to stone columns prepared with natural aggregate and light weight aggregate showed bearing capacities as 285 kPa and 275 kPa, respectively. The improvement in bearing capacity due to provision of encasement to stone columns was observed as 36.6% for natural aggregate and 31.7% for light weight aggregates as compared to virgin clay bed.

3.3   Effect of Stone Column Materials on Bearing Capacity

Comparison between stone columns constructed with natural with and without encasement is shown in Fig. 5a. Results pointed out that compared to virgin clay bed, provision of natural aggregate stone column increased bearing capacity by 31.7%. Whereas, the encased natural aggregate stone column increased bearing capacity by 36.6%. It depicts provision of encasement increased the bearing capacity by about 5%.

Figure 5b presents the evaluation between stone columns prepared by using light weight aggregate with and without encasement. Compared to virgin clay bed, the provision of light weight aggregate stone column increased bearing capacity by 19.5%. Whilst, encased light weight aggregate stone column improved bearing capacity by 31.7%. It shows provision of encasement increased the bearing capacity by about 12%. It can be concluded that provision of encasement to lightweight aggregate stone columns was more beneficial as compared to natural aggregate stone columns.

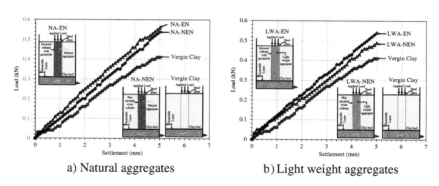

a) Natural aggregates   b) Light weight aggregates

Fig. 5. Load vs settlement curve of stone columns with and without encasement

4   Conclusions

Following are the salient conclusions from the present study:

- Bearing capacity of virgin clay bed increased by 31.7% and 19.5% due to provision of natural aggregate stone column and light weight aggregate stone column, respectively.

- Encasement of stone column increased the bearing capacity by 36.6% and 31.7% for natural aggregate stone column and light weight aggregate stone column, respectively.
- Encasement increased bearing capacity by 5% and 12% for stone columns with natural aggregates and light weight aggregates, respectively.
- The provision of encasement is more beneficial in light weight aggregate stone column as compared to natural aggregate stone column.

Acknowledgement. Authors are thankful to B. Tech. Students of batch 2011-15 of Department of Civil Engineering of Pandit Deendayal Petroleum University, Gandhinagar, India.

References

1. Bauer, G.E., Al-Joulani, N.: Laboratory and analytical investigation of sleeve reinforce stone columns. In: Geosynthetics: Application, Design and Construction, De Groot, pp. 463–466 (1996)
2. Bergado, D.T., Anderson, L.R., Miura, N., Balasubramaniam, A.S.: Soft Ground Improvement in Lowland and Other Environments. ASCE press, New York, p. 427 (1996)
3. Geduhn, M., Raithel, M., Kempfert, H.G.: Practical aspects of the design of deep geotextile coated sand columns for the foundation of a dike on very soft soils. In: Ochiai, H., Omine, K., Otani, J., Yasufuku, N. (eds.) Proceedings of the International Symposium Earth Reinforcement, Kyushu, Fukuoka, Japan, pp. 545–548 (2001)
4. Malarvizhi, S.N., Ilamparuthi, K.: Load versus settlement of clay-bed stabilized with stone and reinforced stone columns. In: Proceeding of the 3rd Asian Regional Conference on Geosynthetics, GEOASIA, Seoul, Korea, pp. 322–329 (2004)
5. Murugesan, S., Rajagopal, K.: Performance of encased stone columns and design guidelines for construction on soft clay soils. In: Proceedings of the 4th Asian Regional Conference on Geosynthetics, Shanghai, China, pp. 729–734 (2008)
6. Murugesan, S., Rajagopal, K.: Investigations on the behavior of geosynthetic encased stone columns. In: Proceedings of the 17th ICSMGE, Alexandrina, Egypt (2009)
7. Sivakumar, V., McKelvey, D., Graham, J., Hughes, D.: Triaxial tests on model sand columns in clay. Can. Geotech. J. **41**, 299–312 (2004)

Permeability Comparison of MgO-carboanted Soils and Cement-Treated Soils

Guang-Hua Cai[1,2(⊠)], Song-Yu Liu[2], Guang-Yin Du[2], Liang Wang[2], and Chuan Qin[2]

[1] School of Civil Engineering, Nanjing Forestry University, Nanjing 210037, China
caiguanghua@seu.edu.cn
[2] Institute of Geotechnical Engineering, Southeast University, Nanjing 210096, China

Abstract. Carbonation of reactive magnesia (MgO) is employed for treating soft soils, which has received attention in the ground improvement as an innovative technology. However, no literature on the permeability of reactive MgO-carbonated soils has been studied. Based on the previous research, this paper focuses on the permeability coefficient of reactive MgO-carbonated soils and PC-treated soils. Through the laboratory permeability tests, the influence of reactive MgO content, carbonation time, initial water content and CO_2 ventilation pressure on the permeability coefficient of carbonated silt and silty clay was systematically studied. Moreover, the permeability coefficient of PC-treated soils was used for comparison under the same conditions of MgO content and initial water content. The results show that: the permeability coefficient of reactive MgO-carbonated soils reduces with the MgO content increasing, it is the same magnitude with that of PC-treated soils at the same dosage; and the permeability coefficient of carbonated silt is obviously larger than that of carbonated silty clay. When both the MgO-stabilized silt and silty clay are carbonated for 6.0 h, the corresponding permeability coefficient could reach the minimum (10^{-6} m/s). Ventilation pressure has little effect on the permeability coefficient of reactive MgO-carbonated soils, which is slightly smaller when the ventilation pressure is 200 kPa. Therefore, the reactive MgO-carbonated soils have similar impermeability with PC-treated soils, and have a good prospect of popularization and application.

Keywords: Magnesia · Carbonation · Cement · Silt · Silty clay
Permeability

1 Introduction

In foundation reinforcement, the cementitious materials of Portland cement (PC) and lime have been extensively utilized for treating soft soils in order to improve the strength and decrease the permeability [1]. However, the PC production involves intensive energy consumption and severe environmental impacts (~ 0.85 to 0.95 t CO_2/t PC and $\sim 5\%$ to 8% of global anthropogenic CO_2 emissions). Thus, considerable efforts have been made to explore alternative low-carbon materials to completely or partially

© Springer Nature Singapore Pte Ltd. 2018
L. Li et al. (Eds.): GSIC 2018, *Proceedings of GeoShanghai 2018 International Conference: Ground Improvement and Geosynthetics*, pp. 105–113, 2018.
https://doi.org/10.1007/978-981-13-0122-3_12

eliminate the use of PC, including supplementary materials such as fly ash and slag [2], geopolymers [3] and calcium carbide residues [4]. In recent years, reactive MgO cements were also put forward to be as one of alternative materials owing to their higher hydration rate and greater potential in absorbing CO_2 [5–7]. Therefore, a prospective technology could generate the rapid and significant enhancement in soil strength as well as absorb lots of CO_2 [8, 9].

More studies of carbonated reactive MgO-stabilized soils are mainly reflected in the following aspects: (*a*) the reactive MgO-treated sandy soil with 5% dosage could complete the carbonation in 3–6 h, its unconfined compressive strength was more than two times that of PC-stabilized soil of 28d with the same dosage, and the CO_2 absorption reached 90% of the theoretical value [10]; (*b*) the strength of carbonated silt was studied, and the strength prediction formula with regard to MgO content and carbonation time were proposed [8]; (*c*) the strength, electrical resistivity and microstructure characteristics of carbonated silt were studied, and the predicted models of strength and electrical resistivity the ratio of initial water content to MgO content (w_0/c) were established [9, 11]; (*d*) the effect of MgO activity on the carbonation of muddy clay was studied through the modified tri-axial apparatus, indicating that the muddy clay could finish carbonation during 24 h and the higher the MgO activity was, the higher the strength and carbonation degree were [12]. In addition, the durability of carbonated soils was investigated through the freezing-thawing cycles, wetting- drying cycles and sulfate attack tests, and it was proved that the carbonated soils had better durability than cement-solidified soils [13–15].

The permeability coefficient is one of the important indexes in evaluating the treatment efficiency of soft soil in engineering application. However, there is no in-depth investigation about the permeability characteristics of carbonated soils. The paper adopts flexible-wall penetration test and aims at the permeability properties of MgO-carbonated silt and silty clay through the comparison of PC-treated soils.

2 Materials and Methods

2.1 Materials

The silt and silty clay used in this study were respectively collected from the highway construction site of Suqian city and Nanjing city, China. The Atterberg limits were determined according to ASTM D4318, and the soil pH was measured by employing a portable D-54 pH meter as per ASTM D4972. The physicochemical properties of the two soils were summarized in Table 1. The reactive MgO with a light-burned powder, was purchased from Xingtai, China; and Portland cement (PC, 32.5#) obtained from Nanjing, China was used for comparison. The CO_2 gas used with concentration of 99.9% was supplied by Nanjing Third Bridge Industrial Gases Co, Ltd. Based on ASTM D854, the specific gravity of reactive MgO and PC was 2.25 and 3.1 respectively. The particle analyses were tested by a laser diffractometry analyzer (Mastersizer 2000, Malvern), and the particle-size curves of materials were shown in Fig. 1. The compaction test of the soils was conducted as per ASTM D698. It is noted that the disperse medium used in the tests of particle-size analysis and specific gravity is water-free kerosene to avoid the hydration of reactive MgO and PC. Moreover, the chemical compositions of soils, MgO and PC, determined by X-ray fluorescence spectrometer, were shown in Table 2.

Table 1. Physicochemical properties of the materials.

Index	Value	
	Silt	Silty clay
Natural water content, w_n (%)	26.1	36.7
Specific gravity, G_s	2.71	2.72
Density, ρ (g/cm^3)	1.96	1.92
Dry density, ρ_d (g/cm^3)	1.55	1.45
Void ratio, e_0	0.745	0.939
Liquid limit, w_L(%)	33.8	37.2
Plastic limit, w_P(%)	23.9	20.9
pH (water/soil = 1)^a	8.78	8.33

^{a}Measured as per ASTM D4972 (2008).

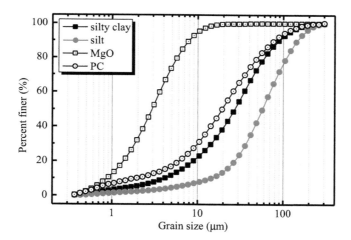

Fig. 1. Particle-size distribution curves of the materials.

Table 2. Chemical compositions of the materials (%)^a.

Soil	SiO$_2$	Al$_2$O$_3$	Fe$_2$O$_3$	CaO	K$_2$O	MgO	TiO$_2$	Na$_2$O	P$_2$O$_5$	SO$_3$
Silty clay	65.1	13.8	6.60	6.65	2.76	1.97	0.90	1.00	0.87	0.24
Silt	71.8	10.2	3.57	6.41	2.16	1.22	0.65	3.10	0.51	0.27
MgO	3.91	0.30	0.30	1.26	ND	91.8	ND	0.023	0.31	0.40
PC	27.4	11.5	3.43	48.8	1.31	1.16	0.48	0.14	0.13	3.28

Note: "ND" represents not detected;
^{a}Measured by using a X-ray fluorescence spectrometry.

2.2   Methods

Taking into account the curing agent content range of the in-situ mixing pile construction, the curing agent contents (c, the weight ratio of curing agent to the dry soil) were chosen as 10%, 15%, 20% and 25%. The water content of stabilized specimens was determined by the liquid limit, and the gravimetric water content (w_0, the weight ratio of deionized water to dry soil) was 0.75 times of liquid limit, therefore, the water content of the stabilized silt and silty clay was about 25% and 30%, respectively. When analyzing the influence of initial water content, the curing agent content was 20%, and the initial water content of silt and silty clay was chosen as 15%–30% and 20%–35%, respectively.

For the specimen's preparation, the natural soil was firstly dried in the oven at 105 °C, and was then ground into powder and finally passed through 2-mm sieves. After the pretreatment of soils, all raw materials (i.e., dry soil, reactive MgO or PC and distilled water) were calculated and weighed. The dry components including dry soil and reactive MgO or PC were initially mixed and homogenized for about 5 min in a laboratory mixer, and the mixtures continued to mixing for 5 min again after which distilled water was added. Next, the homogenous mixtures were placed into a steel cylindrical mold ($D50$ mm x $H100$ mm) until achieving the same compaction degree of $\sim$89%. Then, the specimens were immediately extruded from the molds by using the hydraulic jack and their weight, diameter and height were measured immediately. Finally, the MgO specimens were put into the sealed organic glass cylinders contained 99.9% CO_2 of 200 kPa (see Fig. 2), and were subjected to 12 hour's carbonation. The PC-treated specimens were cured for 28 d under the standard conditions (20 ± 2 °C, humidity of 94 ± 3%).

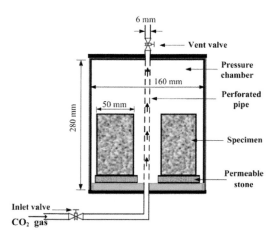

Fig. 2. Carbonation model diagram of MgO-stabilized specimens.

After the carbonation or curing of MgO/PC-stabilized specimens, the specimens were firstly conducted to vacumize for saturation and then carried out the permeability tests, and the triaxial flexible-wall permeameter used was redesigned and manufactured by Nanjing Soil Instrument Co. Ltd., China (see Fig. 3). During the permeation tests, the de-aired tap water flow was applied from the lower base to the upper side of the specimen under the seepage pressure in order to avoid any air entrapment. The confining pressure and the seepage pressure were respectively set as 200 kPa and 150 kPa. During the permeating process, the ambient temperature was strictly controlled at 22 ± 2 °C, and the weight of leachate collected was measured per 2 h.

Fig. 3. The permeability test instrument.

3 Results and Discussion

Figure 4 shows the permeability results of MgO-carbonated silt and PC-treated silt specimens under different curing agent contents, and Fig. 5 shows the permeability results of MgO-carbonated silty clay and PC-treated silty clay specimens. It can be found from Figs. 4 and 5 that the permeability coefficient of all carbonated or stabilized specimens decreases with an increase of MgO or PC, the permeability coefficient of MgO-carbonated specimens is relatively higher than that of PC-treated specimens. The pre-existing reference has shown that there is the threshold content for PC-treated soils and the limited value is 10% [16]. When the PC content exceeds this limited value, the PC-stabilized soils enter into the inert region, in which the PC content has little influence on the anti-permeability of PC-treated soils. Since the PC content is larger than 10%, the change ratio of permeability coefficient is not obvious. Compared with PC-treated soils, the permeability coefficient of MgO-carbonated soils has an obviously decreasing ratio when the MgO content is less than 25%, indicating that MgO-carbonated soils do not enter into the inert region.

Moreover, compared with Figs. 4 and 5, the permeability coefficient of MgO-carbonated or PC-stabilized silt is higher than that of MgO-carbonated/PC-stabilized silty clay. There is much larger cohesion between particles of silty clay, which would hinder the infiltration of CO_2 gas and carbonation of $Mg(OH)_2$, weakening the anti-permeability of carbonated silty clay to a certain extent. While there is larger silt particle and smaller cohesion, and carbonation products can rapidly fill the pores, causing the larger change of permeability coefficient. According to the above results, when MgO and PC have the same content ($\sim 10\%$ to 25%), the permeability coefficient of carbonated/stabilized soils for the same soil type would lie on the same magnitude order, indicating that the MgO-carbonated soil has a similar anti-permeability with PC-treated soil.

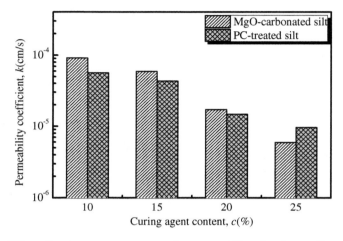

Fig. 4. Effect of curing agent content on the permeability of carbonated/treated silt.

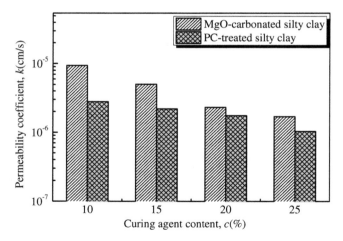

Fig. 5. Effect of curing agent content on the permeability of carbonated/treated silty clay.

Figure 6 describes the relationship between permeability coefficient and initial water content for MgO-carbonated silt and PC-treated silt, and Fig. 7 shows the relationship between permeability coefficient and initial water content for MgO-carbonated silty clay and PC-treated silty clay. It can be seen from Fig. 6 that the permeability coefficient of MgO-carbonated silt and PC-treated silt reduces with the initial water content increasing, and the permeability coefficient has no obvious change for the initial water content of 15% and 20%. It can be found from Fig. 7 that the permeability coefficient of PC-treated silty clay decreases with an increase of initial water content, while the permeability of MgO-carbonated silty clay increases with initial water content increasing. The permeability of MgO-carbonated silty clay is basically same as that of PC-treated silty clay under the initial water content of 30% and 35%. In addition, it can be seen from Figs. 6 and 7 that the permeability coefficient of MgO-carbonated soils is slightly less than that of PC-treated soils, which is opposite to the permeability under the influence of curing agent content.

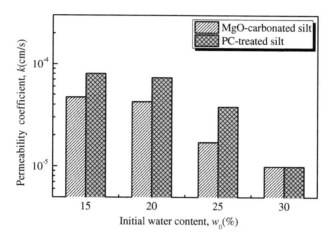

Fig. 6. Effect of initial water content on the permeability of carbonated/treated silt.

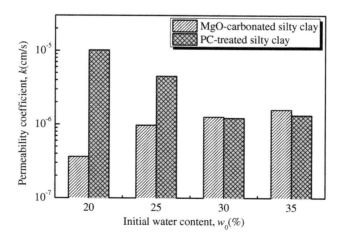

Fig. 7. Effect of initial water content on the permeability of carbonated/treated silty clay.

These abovementioned results attribute to the following aspects:

(1) The hydration of MgO needs a certain amount of water, and higher initial water content is more beneficial to the full hydration of MgO and the carbonation of Mg $(OH)_2$. However, the excessive water not only hinders the infiltration of CO_2 but also affects the carbonation reaction of $Mg(OH)_2$.

(2) When the initial water content is beyond to the required water of MgO hydration, and water will occupy the pore of samples, increasing the porosity and weakening the anti-permeability.

(3) The increasing initial water content facilitates the hydration of PC, and the cementing products can fill the pores of the stabilized soils and decrease the permeability.

4 Conclusions

This paper studies the permeability properties of MgO-carbonated silt and silty clay, and gives the comparison with those of PC-treated silt and silty clay. The following conclusions could be obtained:

(1) The permeability coefficient of MgO-carbonated soils decreases with an increase of MgO content. In the limited curing agent content of 25%, MgO-carbonated soils have similar permeability properties with PC-treated soils.

(2) The permeability coefficient of MgO-carbonated silt and PC-treated silt reduces with the initial water content increasing; the permeability coefficient of PC-treated silty clay decreases with an increase of initial water content, while the permeability of MgO-carbonated silty clay increases with initial water content increasing.

(3) The reactive MgO-carbonated soil has similar impermeability to PC-treated soil, and has a good prospect of popularization and application.

Acknowledgments. The authors appreciate the financial support of NSFC (41330641, 51279032), National key research and development projects (2016YFC0800201).

References

1. Liu, S.Y., Du, Y.J., Yi, Y.L., Puppala, A.J.: Field investigations on performance of T-Shaped deep mixed soil cement column–supported embankments over soft ground. J. Geotech. Geoenviron. Eng. **138**(6), 718–727 (2012)
2. Disfani, M., Arulrajah, A., Haghighi, H., Mohammadinia, A., Horpibulsuk, S.: Flexural beam fatigue strength evaluation of crushed brick as a supplementary material in cement stabilized recycled concrete aggregates. Constr. Build. Mater. **68**, 667–676 (2014)
3. Sukmak, P., Horpibulsuk, S., Shen, S.L., Chindaprasirt, P., Suksiripattanapong, C.: Factors influencing strength development in clay-fly ash geopolymer. Constr. Build. Mater. **47**, 1125–1136 (2013)
4. Horpibulsuk, S., Phetchuay, C., Chinkulkijniwat, A., Cholaphatsorn, A.: Strength development in silty clay stabilized with calcium carbide residue and fly ash. Soils Found. **53**, 477–486 (2013)
5. Harrison, A.J.W.: Reactive magnesium oxide cements. United States Patent 7347896 (2008)

6. Liska, M., Al-Tabbaa, A.: Ultra-green construction: reactive magnesia masonry products. Proc. ICE-Waste Res. Manag. **162**(4), 185–196 (2009)
7. Liska, M., Al-Tabbaa, A., Carter, K., Fifield, J.: Scaled-up commercial production of reactive magnesium cement pressed masonry units. Part I: Prod. Proc. ICE-Constr. Mater. **165**(4), 211–223 (2012)
8. Cai, G.H., Liu, S.Y., Du, Y.J., Zhang, D.W., Zheng, X.: Strength and deformation characteristics of carbonated reactive magnesia treated silt soil. J. Cent. South Univ. **22**(5), 1859–1868 (2015)
9. Cai, G.H., Du, Y.J., Liu, S.Y., Singh, D.N.: Physical properties, electrical resistivity and strength characteristics of carbonated silty soil admixed with reactive magnesia. Can. Geotech. J. **52**(11), 1699–1713 (2015)
10. Yi, Y.L., Liska, M., Unluer, C., Al-Tabbaa, A.: Carbonating magnesia for soil stabilization. Can. Geotech. J. **50**(8), 899–905 (2013)
11. Cai, G.H., Liu, S.Y., Cao, J.J.: Influence of initial water content on strength and electrical resistivity of MgO-carbonated silt. China J. Highw. Transp. **11**(30), 18–26 (2017)
12. Liu, S.Y., Li, C.: Influence of MgO activity on the stabilization efficiency of carbonated mixing method. Chin. J. Geotech. Eng. **37**(1), 148–155 (2015)
13. Zheng, X., Liu, S.Y., Cai, G.H., Cao, J.J.: Experimental study on freeze-thaw properties of carbonated reactive MgO-stabilised soils. J. Southeast Univ. (Natural Sci. Ed.) **45**(3), 595–600 (2015)
14. Zheng, X., Liu, S.Y., Cai, G.H., Cao, J.J.: Experimental study on drying-wetting properties of carbonated reactive MgO-stabilised soils. Chin. J. Geotech. Eng. **38**(2), 297–304 (2016)
15. Liu, S.Y., Zheng, X., Cai, G.H., Cao, J.J.: Study on resistance to sulfate attack of carbonated reactive MgO-stabilised soils. Rock Soil Mech. **37**(11), 3057–3064 (2016)
16. Hou, Y.F., Gong, X.N.: Permeability properties of cement soils. J. Zhejiang Univ. (Eng. Sci.) **34**(2), 189–193 (2000)

Resilient Modulus of Liquid Chemical-Treated Expansive Soils

Shi He, Xinbao Yu[(⊠)], Sandesh Gautam, Anand J. Puppala, and Ujwalkumar D. Patil

University of Texas at Arlington, Texas, USA
xinbao@edu.com

Abstract. Diluted acids are used as chemical stabilizers in Texas to treat expansive soils for residential projects via deep injection. Due to the proprietary nature of the chemical stabilizers, there are very limited studies on the resilient modulus (M_R) of chemically-treated expansive soils. This paper evaluates the effect of a liquid chemical stabilizer on the treatment of expansive soils collected from Texas and Colorado. The chemical solution, called ionic soil stabilizer (ISS) which contain sulfuric acid, phosphoric acid, citric acid, and water was used as an additive and tests were carried out on untreated and treated bulk soil samples in accordance with AASHTO T-307. The treated soil specimens were prepared by hand mixing the dry soils with the chemical stabilizer at three application ratios and two curing periods (7 and 28 days). The experiment results show that the value of resilient modulus increases with the increase of chemical application ratio. The resilient modulus of the treated sample cured for 28 days is much higher that of the untreated sample. Also, M_R test results were found to be highly dependent on the compaction, moisture content, chemical ratio and curing time. Finally, M_R test results are compared with compressive strength obtained from UCS test to find out the optimum treatment chemical dosage for field application.

Keywords: Resilient modulus · Unconfined compressive strength
Liquid chemical stabilizer · Ionic soil stabilizer

1 Introduction and Background

Resilient modulus (M_R) is defined as the ratio of cyclic deviator stress to the recoverable or resilient strain and is considered as one of the important parameters to design flexible pavement (Banerjee 2017; Buchanan 2007; Han and Vanapalli 2016; Rahman and Tarefder 2015; Sun et al. 2016). It is a stiffness measurement that is profoundly influenced by the stress state and moisture content (Rahman and Tarefder 2015).

In general, the M_R value of clay soil decreases when the moisture content increases (Buchanan 2007). Expansive soil is the kind of clay that tends to swell or shrink when the moisture content changes (Jones and Jefferson 2012). At least \$1 billion per year is spent on rehabilitating U.S. residential homes and pavements (Jones and Jones 1987). To prevent and mitigate the loss, a variety of treatment methods have been developed in the past decades. Essentially traditional chemical stabilizers such as lime, cement and

© Springer Nature Singapore Pte Ltd. 2018
L. Li et al. (Eds.): GSIC 2018, *Proceedings of GeoShanghai 2018 International Conference: Ground Improvement and Geosynthetics*, pp. 114–120, 2018.
https://doi.org/10.1007/978-981-13-0122-3_13

fly ash are utilized to control the swelling and enhance the soil stiffness (Katz et al. 2001; Rauch et al. 2002). Among these stabilizers, lime and fly ash are the most common stabilizer utilized in the U.S. Although the lime treatment increases the optimum water content as compared to the value of the control sample, M_R and UCS values of lime treated sample are much higher than those of control samples (Cokca 2001; Kumar et al. 2007; Little 1987; Punthutaecha et al. 2006; Rahman and Tarefder 2015; Sweeney et al. 1988). There is a lack of research on the resilient modulus (M_R) testing of liquid chemical-treated expansive soils.

In this study, an ionic soil stabilizer (ISS) is used to treat expansive soil collected from Texas and Colorado. The ISS is composed of sulfuric acid, phosphoric acid, citric acid, water, and surfactant. In the field, engineers dilute the ISS concentrate with water, and then deep inject it into the sublayer. According to the provider, this ISS is environmentally friendly, non-toxic, and efficient to treat expansive soil. Moreover, the transportation fee of the liquid chemical stabilizer is much less than that of traditional soil stabilizer (Katz et al. 2001).

Despite the several benefits as mentioned above, engineers are reluctant to implement the chemical treatment of the expansive soil in general practice. This is primarily due to the lack of literature explaining the mechanisms involved in treating the expansive soils with the chemical, especially the resilient modulus of the soil before and after treatment. In this research, the soil collected from Texas and Colorado were treated with ISS in the laboratory. Furthermore, a series of lab testing including M_R and UCS tests on expansive soils before and after treatment with different application ratio was carried out. Finally, the effect of ISS content is evaluated through analysis of M_R and UCS test results, and an optimum ISS ratio is recommended.

2 Material Properties

In this study, Texas soils were sampled from Caddo Mills in Dallas area, and the Colorado soils were collected from the state of Colorado. The Dallas soil for laboratory testing was collected at 3 feet below ground surface, and the topsoil above this depth was neglected to avoid the contamination. These soils were excavated in large chunks and transported to the laboratory. Before the specimen preparation for Atterberg Limits and Standard compaction test, the soil sample was oven dried and pulverized through No. 40 sieve. Plasticity index (PI) for Colorado and Texas soil was found to be 42 and 58, respectively.

In the construction site, the suppliers injected ISS provided by TX Prochem via high pressure. The ratio recommended by the supplier was to mix 8 gal of the liquid chemical concentration and 12 oz of surfactant with 6000 gal of water. To simulate the recommended field application ratio, 5 ml of the chemical concentration and 0.057 g of the surfactant were diluted into 1 gallon of water. Apart from this ratio, two other ratios were designed to evaluate the best possible ratio for soil treatment with ISS, and the tested ratios are shown in Table 1.

Table 1. Three liquid stabilizer dosage designs for soil treatment

ISS content	First ratio	Second ratio	Third ratio
Chemical concentrate (ml)	5	5	10
Surfactant (g)	0.057	0.057	0.114
Water (gallon)	1	2	1

2.1 Standard Compaction Test

Standard compaction test was conducted according to ASTM D698 to determine the relationship between maximum dry density (MDD) and optimum moisture content (OMC). For treated soils, around 1.8 kg of pulverized dry soils were mixed with various proportions of ISS in separate containers. As liquid chemical may continue to react with the soil after initial mixing, the prepared soil samples were stored in the moisture room at least for 24 h before proctor test. Figure 1 shows the standard compaction test results for Texas and Colorado soils with various ratios of treatment. Unlike control samples, There is an increase in OMC and decrease in MDD with adding ISS. Such behavior could be explained that the soil chemical reaction may not finish without sufficient water, which resulted in OMC increasing.

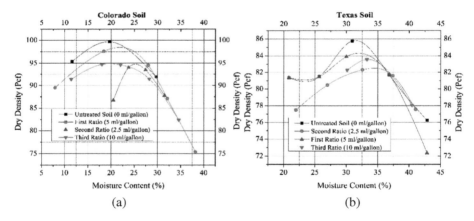

Fig. 1. Moisture-density relationship before and after treatment: (a) Colorado soil and (b) Texas soil.

3 Specimen Preparation

Before UCS and M_R tests, both untreated and treated samples were compacted in a split mold to reach the target density of 95% MDD and 100% OMC to simulate the field condition. The sample was 2.8 inches in diameter and 5.8 inches in height. For treated specimens, three different application ratios of ISS were utilized in this study. Dry soil was hand-mixed with ISS uniformly and then put inside a plastic bag and stored in a 100% humidity-controlled moisture room for overnight to ensure that there is sufficient time to allow soil-chemical reaction after initial mixing. After compaction, soil samples were cured in moisture room for 7 and 28 days respectively. At least two samples were prepared for each test for repeatability check.

4 Unconfined Compressive Strength Test

The UCS test was carried out both on untreated, and ISS stabilized specimens. ASTM D2166 is the standard to determine the relationship between unconfined compressive strength and axial strain.

5 Resilient Modulus Test

In this study, resilient modulus test of untreated and treated samples was conducted according to AASHTO T307 (Buchanan 2007; Rahman and Tarefder 2015). The test progress contained 15 stress sequences. Each sequence included a different combination of confining pressure and deviator stress. During the test, 0.1 s of load pulse was followed by 0.9 s of rest period. One loading cycle combined one load pulse and one rest period. Although each sequence includes 100 cycles, resilient modulus was only determined by averaging stress-strain responses of the last five cycles. Figure 2 shows the resilient modulus test equipment used in the lab.

Fig. 2. Resilient modulus test equipment

6 Results and Discussions

Figure 3 shows the UCS test results for untreated and treated samples with different ISS ratios. For the Texas soil, the sample treated with more ISS revealed higher unconfined compressive strength especially for soil treated by the third ratio. Furthermore, the strength of treated sample curing 28 days was greater than that of treated sample curing seven days. Perhaps the soil and chemical reaction were continuously happening after seven days. For specimen treated by the second ratio, the treated samples were even less than the control one. The decrease may be attributed to the decrease in maximum dry density and increase in optimum water content.

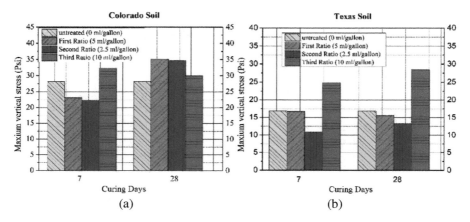

Fig. 3. Unconfined compressive strength test for treated soil with different dosages of ISS: (a) Colorado soil. (b) Texas soil.

Table 2 shows the resilient modulus of Texas and Colorado soil before and after treatment. The M_R value reduced with increasing deviator stress due to stress softening the effect of soils (Rahman and Tarefder 2015). Also, M_R value was found to be influenced by OMC. For instance, there was much more ISS content in the third ratio as compared with the first ratio. Texas soil treated by the third ratio after seven days of curing has M_R value in the range between 43 to 53.3 MPa. However, Texas soil treated by the first ratio after seven days of curing has M_R value between 66.3 and 77.6 MPa. The reduction of M_R value is attributed to the fact that OMC of soil treated by the third ratio is nearly 1.5% more than that of soil treated by the first ratio, which could be readily seen from Fig. 1. In sum, to some extent, M_R value of treated soil increased in comparison with control samples.

Table 2. Resilient modulus test results for treated soil with different dosages of ISS: (a) Texas soil. (b) Colorado soil.

Confining Pressure (kPa)	Deviator Stress (kPa)	Control M_R (MPa)	7 Days of Curing (Texas) M_R (MPa)			28 Days of Curing (Texas) M_R (MPa)		
			First Ratio	Second Ratio	Third Ratio	First Ratio	Second Ratio	Third Ratio
41.4	14	32	77.6	41.7	53.8	86.9	49	63.7
41.4	28	40.6	75.4	38.1	51.5	79.7	54.5	60.4
41.4	41	43.3	73.6	34.4	49	76.9	55.1	57.1
41.4	55	41.8	71.7	31.5	46.7	76	54.3	54.7
41.4	69	38.8	69.7	29.2	44.6	72.3	52.6	51.8
27.6	14	34.9	75.2	38.1	50.7	81.1	50.4	61.3
27.6	28	41.7	72.4	34.7	48.7	76.9	54.1	58.1
27.6	41	41.6	70.5	31.7	46.5	74.1	54.1	54.9
27.6	55	39.7	69.3	29.6	44.9	72	52.8	52.9
27.6	69	37.9	68.1	28.2	43.3	69.7	51.5	50.9
13.8	14	39	69.3	34.1	47.9	74.8	51.8	58.6
13.8	28	39.4	67.5	31.6	46.1	71.6	50.6	55.6
13.8	41	38.4	66.3	29.5	44.2	69.5	49.8	53.1
13.8	55	37.1	65.5	27.8	43	67.8	49.1	51.3
13.8	69	35.8	64.6	26.7	41.7	67.3	48.4	49.5

(a). Texas soil resilient modulus test before and after treatment

Confining Pressure (kPa)	Deviator Stress (kPa)	Control M_R (MPa)	Curing 7 days (Colorado) M_R (MPa)			Curing 28 days (Colorado) M_R (MPa)		
			First Ratio	Second Ratio	Third Ratio	First Ratio	Second Ratio	Third Ratio
41.4	14	61.6	83.4	59.3	65.6	101.4	63.8	37.7
41.4	28	56.9	76.1	65.5	62.8	94.6	73.6	43.8
41.4	41	54.7	69.7	71.4	60.8	87.5	77.8	48.4
41.4	55	59.3	64.6	76.7	60.8	82.2	83.5	53
41.4	69	63.7	60.1	80.8	60.3	77.5	87.2	56.8
27.6	14	40.7	77.8	52.8	63.6	99.5	60.6	35.8
27.6	28	58.7	71.2	60.3	58.9	90.1	66.4	41.4
27.6	41	61.3	65	64.8	56.3	83.2	70.9	45.1
27.6	55	62.4	61.1	69.5	55.9	78.5	74.6	48.9
27.6	69	62.4	58.5	74.2	55.9	75.6	79.5	52.5
13.8	14	55.3	74.8	50.7	56.3	96.5	59.7	36.8
13.8	28	58.9	67.8	57	52	87.3	64.3	40.8
13.8	41	56.4	62.9	60.5	49.5	80.8	66.7	43.9
13.8	55	56	58.9	64.9	49.1	76	69.2	46.6
13.8	69	55.5	56.4	69.4	49.9	73.2	71	49.6

(b). Colorado soil resilient modulus test before and after treatment

7   Conclusions

The conclusions obtained from the laboratory test before and after treatment are summarized as follows:

After treatment with ISS, soil strength displayed a significant increase, especially for soil treated by the third ratio. In general, the extended curing period for treated sample resulted in higher unconfined strength the sample would perform.

The M_R value increased due to the ISS application ratio for both the Texas and Colorado soil. Also, the moisture content in expansive soils has great influence on ISS treatment.

This paper summarizes the effect of ISS dosage on UCS and resilient modulus test results for Colorado and Texas soil. Considering the best fit results among UCS and resilient modulus, the first ratio is recommended for the use in the field.

Acknowledgements. The authors would like to appreciate TX Prochemical to provide the testing ISS and Mr. Ben Baker for the help of soil sample collection for this study.

References

Banerjee, A.: Response of unsaturated soils under monotonic and dynamic loading over moderate suction states. Doctoral Dissertation, University of Texas at Arlington, Arlington, Texas (2017)

Buchanan, S.: Resilient modulus: what, why and how? Technical report, p. 13 (2007)

Cokca, E.: Use of class C fly ashes for the stabilization of an expansive soil. J. Geotech. Geoenviron. Eng. **127**(7), 568–573 (2001)

Han, Z., Vanapalli, S.K.: Relationship between resilient modulus and suction for compacted subgrade soils. Eng. Geol. **211**, 85–97 (2016)

Jones, D.E., Jones, K.A.: Treating expansive soils. Civ. Eng.—ASCE **57**(8), 62–65 (1987)

Jones, L.D., Jefferson, I.: Expansive Soils. ICE Publishing, New York (2012)

Katz, L., et al.: Mechanisms of soil stabilization with liquid ionic stabilizer. Transp. Res. Rec. J. Transp. Res. Board **1757**, 50–57 (2001)

Kumar, A., Walia, B.S., Bajaj, A.: Influence of fly ash, lime, and polyester fibers on compaction and strength properties of expansive soil. J. Mater. Civ. Eng. **19**(3), 48–242 (2007)

Little, D.N.: Evaluation of structural properties of lime stabilized soils and aggregates (1987)

Punthutaecha, K., Puppala, A.J., Vanapalli, S.K., Inyang, H.: Volume change behaviors of expansive soils stabilized with recycled ashes and fibers. J. Mater. Civ. Eng. **18**(2), 295–306 (2006)

Rahman, M.T., Tarefder, R.A.: Assessment of molding moisture and suction on resilient modulus of lime stabilized clayey subgrade soils (2015)

Rauch, A., Harmon, J., Katz, L., Liljestrand, H.: Measured effects of liquid soil stabilizers on engineering properties of clay. Transp. Res. Rec. J. Transp. Res. Board **1787**, 33–41 (2002)

Sun, X., Han, J., Crippen, L., Corey, R.: Back-calculation of resilient modulus and prediction of permanent deformation for fine-grained subgrade under cyclic loading. J. Mater. Civ. Eng. **29**(5), 4016284 (2016)

Sweeney, D.A., Wong, D.K.H., Fredlund, D.G.: Effect of lime on a highly plastic clay with special emphasis on aging (1988)

Stabilization of Marine Soft Clay with Two Industry By-products

Yaolin Yi[(⊠)] and Pengpeng Ni

School of Civil and Environmental Engineering,
Nanyang Technological University, Singapore 639798, Singapore
yiyaolin@ntu.edu.sg

Abstract. Stabilization using Portland cement (PC) is one of the most widely used soft clay treatment methods. However, there are significant environmental impacts associated with the production of PC in terms of high energy consumption and non-renewable resources, as well as CO_2 emissions. Some industry by-products/wastes have potentials to be applied in soil stabilization. This paper presents an experimental study on the stabilization of a marine soft clay using the blend of two industry by-products: ground granulated blastfurnace slag (GGBS) activated by carbide slag (CS). The testing program involved unconfined compressive strength test, X-ray diffraction, and scanning electron microscopy. The results indicated that both CS-GGBS-stabilized and PC-stabilized clays had similar types of hydration products, including calcium silicate hydrates, calcium aluminates, hydrocalumite, and ettringite. However, the highest strength of CS-GGBS-stabilized clay was 1.5–3 times greater than that measured for PC-stabilized clay. The optimum CS/GGBS mass ratio varied from 0.2 to 0.3 for stabilized clay at different curing age and GGBS content.

Keywords: Marine soft clay · Stabilization · Industry by-products

1 Introduction

Soft clay with high water content usually has to be treated before construction of infrastructure, because it has a very low bearing capacity, and will induce a high settlement due to construction. The deep mixing method is commonly used for ground improvement, where cementitious binders, such as Portland cement (PC), are mixed in situ with clay to enable chemical reactions [1]. Chemical products will fill the pores within the soil, and soil particles will be bonded together, resulting in increased strength of stabilized clay. The environmental issues related to the production of PC have been pointed out by researchers. On average, producing 1 t of PC will consume approximately 5,000 MJ energy and 1.5 t non-renewable resources of limestone and clay, along with which about 0.95 t CO_2 will be emitted [2]. It is recommended to use industry by-products/wastes to replace PC for soil stabilization, which can minimize the detrimental effects of PC production, and save costs for waste disposal [3, 4].

Ground granulated blastfurnace slag (GGBS) is a by-product of the steel industry, which has a significant proportion of glassy phased Ca, Si, Al and Mg-based compounds. Higgins [2] indicated that a great reduction of energy consumption and CO_2

© Springer Nature Singapore Pte Ltd. 2018
L. Li et al. (Eds.): GSIC 2018, *Proceedings of GeoShanghai 2018 International Conference: Ground Improvement and Geosynthetics*, pp. 121–128, 2018.
https://doi.org/10.1007/978-981-13-0122-3_14

emissions can be expected, where manufacturing 1 t GGBS uses approximately 1,300 MJ energy and produces 0.07 t CO_2. GGBS is a latent material and needs to be activated by additives. Most investigations are focused on the beneficial effect of lime (CaO or $Ca(OH)_2$) activated GGBS binders for soil stabilization with emphasis on expansion issues [4–6]. For soft clay improvement, lime-GGBS mixture is also proved to be an effective option [7, 8]. However, the production of lime from calcination of limestone ($CaCO_3$) also results in unfavorable environmental impacts.

An alternative by-product of the calcium carbide industry, carbide slag (CS), shows its potential for use to replace lime as the activator for GGBS. Cardoso et al. [9] claimed that the major component of CS is $Ca(OH)_2$, which can be as high as 85%–90%. China produces the highest amount of calcium carbide (900,000–1,140,000 t) every year, which accounts for 90%–95% of the global market's total supply [10]. However, most of dry CS are finally disposed in landfills near chlor-alkali plants [11]. It should be emphasized that CS is not categorized as a dangerous-hazardous material, and landfill disposal is a waste of calcium resources [10]. Yi et al. [12] has demonstrated that CS can be used to activate GGBS for soft clay stabilization.

However, the study of Yi et al. [12] was focused only on one type of soft clay, and further investigations on the CS-GGBS treatment approach should be carried out. Hence, in this paper, a series of laboratory tests was conducted to evaluate the efficacy of the CS-GGBS treatment technique for another marine soft clay. The changes in properties of CS-GGBS-stabilized clay were examined by comparing the results from unconfined compressive strength test, X-ray diffraction, and scanning electron microscopy with those measured for PC-stabilized clay.

2 Experimental Program

2.1 Marine Soft Clays and Binders

The marine soft clay was taken from Dongshugang, Lianyungang, China. The Dongshugang clay has a very similar plastic limit compared to the Ganyu clay as previously studied by Yi et al. [12], but a different liquid limit. The natural water contents of the two clays were near their liquid limits. Details of physical properties of Dongshugang clay can be seen in Table 1. A higher bound of water content (80%) was chosen in the laboratory.

Table 1. Physical properties of the marine soft clay.

Plastic limit (%)	Liquid limit (%)	In situ water content (%)	Specific gravity	Bulk density (g/cm³)	Void ratio
33	74	75–81	2.62	1.6	1.9

Three binder materials of CS, GGBS and PC were all obtained from Nanjing, China. The primary chemical properties of three binders are summarized in Table 2. Two GGBS contents were selected as 20% and 30% to define the weight of GGBS over the weight of dry soil. From previous study [12], it has been found that the quantity of hydration products in stabilized soil is simply governed by the addition of GGBS, and

Table 2. Primary chemical properties (by % weight) of CS, GGBS, and PC.

Binder	CaO	SiO$_2$	Al$_2$O$_3$	SO$_3$	Fe$_2$O$_3$	MgO	K$_2$O	TiO$_2$	Loss on ignition
CS	67.98	4.01	2.30	0.32	0.13	0.27	<0.01	0.05	24.8
GGBS	34.00	34.30	17.90	1.64	1.02	6.02	0.64	1.17	2.67
PC	48.80	27.40	11.50	3.28	3.43	1.16	1.31	0.48	2.00

the activator can only promote the hydration rate. Five CS/GGBS ratios of 0.05, 0.10, 0.20, 0.30 and 0.40 were tested to determine the optimum mass ratio that enables the stabilized soil to develop the highest strength. PC contents of 20% and 30% were also used for comparison purpose.

2.2   Sample Preparation and Testing

The clay was initially oven dried at a temperature of 105 °C and ground into powders. All particles that passed a 2 mm sieve were collected. Similarly, the drying treatment was conducted on CS, after which CS became small and brittle particles. Binders were added in dry form to simulate the dry jet mixing method. The calculated amount of dry clay and binders was mixed and homogenized using a laboratory mixer for a period of 10 min. Water was then added for a continuing mixing of 10 min. The mixture was subsequently placed into cylindrical molds with a diameter of 50 mm and a height of 100 mm. The mold was filled in three consequent steps, between which manual compaction was performed using a steel rod. This was to minimize the possibility of forming air pockets within the specimen. The molds were put in a curing room with a controlled relative humidity of 95% ± 3% and a temperature of 20 °C ± 2 °C. Unmolding was carried out at 7, 28 and 90 days to allow testing for strength.

Three types of tests were performed to evaluate the properties of stabilized clay from both macro- and micro-perspectives. The unconfined compressive strength (UCS) tests were conducted in triplicate to check the repeatability of the specimen preparation. A displacement-controlled manner was used for the applied axial load with a rate of 1 mm/min. Selected specimens were tested using X-ray diffraction (XRD) and scanning electron microscopy (SEM). An ethanol bath of 7 days was conducted for all specimens to stop the hydration process, and a freeze drying by liquid nitrogen was performed. A separate vacuum sublimation was carried out for more than 48 h. SEM testing was performed on sample pieces with a size less than 10 mm, and XRD was conducted on ground sample powder with a size less than 75 μm.

3   Results and Interpretation

3.1   Unconfined Compressive Strength

At two binder contents of 20% and 30%, variations of UCS with CS/GGBS ratios for the stabilized clay are plotted in Fig. 1a for a curing period of 7 days, Fig. 1b for 28 days and Fig. 1c for 90 days, respectively. The UCS values for the Ganyu clay [13] are also included. In the figures, error bars are used to indicate the margin of error for three repetitive tests. From Fig. 1a, it can be seen that there is an optimum CS/GGBS ratio,

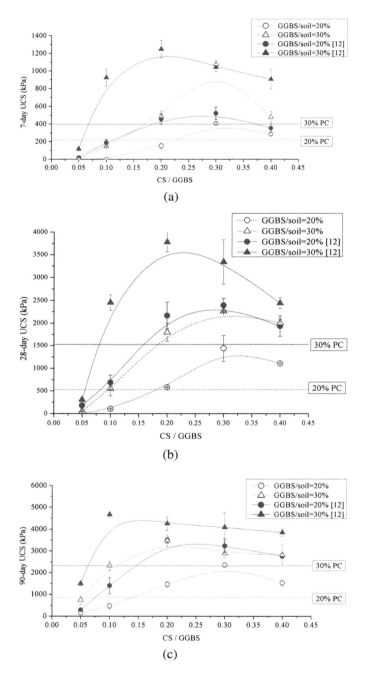

Fig. 1. UCS of stabilized clays at: (a) 7 days, (b) 28 days, and (c) 90 days.

where CS-GGBS-stabilized clay has a maximum strength that is much greater than PC-stabilized clay (e.g., maximum strength gain of 2.3 times). Basically, with the increase of CS/GGBS ratio, the strength of CS-GGBS-stabilized clay is raised from a value that is less than that of the PC-stabilized clay, until it reaches a maximum value, after which the strength is dropped with the CS/GGBS ratio. The optimum CS/GGBS ratio is defined as 0.3 and 0.2 for Ganyu clay with a GGBS/dry soil ratio of 20% and 30%, respectively [12]. Similarly, for Dongshugang clay, the optimum CS/GGBS ratio is determined as 0.3 for both GGBS contents.

Similarly, a concave down profile of UCS as a function of CS/GGBS ratio can be observed for stabilized clay at 28 days and 90 days in Fig. 1b and Fig. 1c as well. At 28 days, the maximum strength of Ganyu clay occurs at the optimum CS/GGBS ratios of 0.3 and 0.2 for 20% mixed GGBS and 30% mixed GGBS, respectively [12], and the corresponding CS/GGBS ratio is 0.3 for two conditions of Dongshugang clay. At 90 days, the optimum CS/GGBS ratio changes to 0.2 and 0.1 for 20% and 30% GGBS-stabilized Ganyu clay [12], and 0.3 and 0.2 for 20% and 30% GGBS-stabilized Dongshugang clay. This indicates that the optimum CS/GGBS ratio varies with the GGBS content, curing age, and clay; nevertheless, the variation is relatively small. For both soft clays, the advantage of the CS-GGBS stabilization is apparent, which produces stabilized clay with a strength of 1.5–3 times higher than that measured for PC-stabilized clay. However, an over-dosage of CS can result in a lower UCS value for GGBS-stabilized clay.

3.2   X-Ray Diffraction

X-ray diffraction analysis provided the crystalline phases for unstabilized and stabilized clays as shown in Fig. 2, where the intensities of crystalline phases for stabilized soils with 30% PC and 30% GGBS with a CS/GGBS ratio of 0.2 at 90 days are presented.

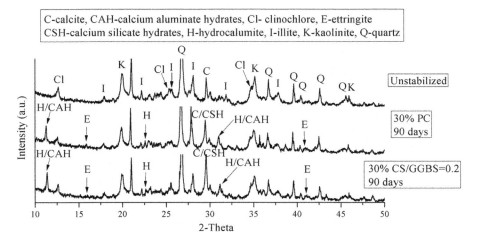

Fig. 2. X-ray diffraction diffractograms of unstabilized clay, PC-stabilized clay and CS-GGBS-stabilized clay at 90 days.

The unstabilized clay is primarily composed of quartz, calcite, kaolinite, illite, and clinochlore. Hydration products, including calcium silicate hydrates (CSH), calcium aluminate hydrates (CAH), and hydrocalumite, are detected in both PC- and CS-GGBS-stabilized clays. Hydrocalumite belongs to a family of hydrated calcium aluminates namely alumino-ferrite monosulfates (AFm) [13]. Additionally, weak peaks of ettringite has been observed in the XRD patterns of both stabilized clays. It is interesting that $Ca(OH)_2$ is not detected in CS-GGBS-stabilized clays, which should be consumed during the activation of GGBS.

(a)

(b)

Fig. 3. SEM images of stabilized clay: (a) 30% PC at 90 days, and (b) 30% GGBS with a CS/GGBS ratio of 0.2 at 90 days.

3.3   Scanning Electron Microscopy

Typical SEM micrographs at a magnification of 1000 times for 30% PC- and CS-GGBS-stabilized clays (CS/GGBS = 0.20) at 90 days are depicted in Fig. 3. Amorphous CSH is shown in both stabilized clays. Additionally, small needle-like crystals are also displayed in the SEM images for the two stabilized clays, which demonstrate the presence of ettringite. The production of both CSH and ettringite has been indicated by the XRD result as mentioned before. It can be seen that the pores between clay particles for CS-GGBS-stabilized clay are better filled by the hydration products compared to PC-stabilized clay, resulting in a relatively denser microstructure. This may explain why the 30% CS-GGBS-stabilized clay (CS/GGBS = 0.20) yielded a higher 90-day UCS compared to that obtained for the 30% PC-stabilized clay (Fig. 1), although they had similar hydration products (Fig. 2).

4   Conclusions

In this investigation, the efficacy of using carbide slag (CS) to activate ground granulated blastfurnace slag (GGBS) as the binder for soft clay stabilization is evaluated. Experiments have been conducted on a marine soft clay, including unconfined compressive strength test, X-ray diffraction, and scanning electron microscopy. An optimum CS/GGBS ratio is found to produce the best stabilization effect, which varies with the GGBS content, curing age, and clay. Nevertheless, the variation is relatively small, and the range is between 0.2 and 0.3 for the Dushugang marine soft clay used in this study. The UCS value measured for CS-GGBS-stabilized clay with the optimum CS content could be 1.5–3 times higher than the corresponding value determined for PC-stabilized clay. It should be noticed that an over-dosage of CS can lead to a reduction of UCS value. The main hydration products of CS-GGBS-stabilized clay are similar to those of PC-stabilized clay, including CSH, CAH, AFm, and ettringite.

Acknowledgments. The authors would like to thank Liyang Gu, Pengfei Guo, Zhao Wang, and Sheng Chang for their assistance in laboratory. The financial support from Singapore MOE AcRF Tier 1 grant (RG184/17) is appreciated.

References

1. Kitazume, M., Terashi, M.: The Deep Mixing Method. CRC Press/Balkema, Leiden (2013)
2. Higgins, D.D.: GGBS and sustainability. Constr. Mater. **160**(3), 99–101 (2007)
3. Jegandan, S., Liska, M., Osman, A.A.M., Al-Tabbaa, A.: Sustainable binders for soil stabilisation. Proc. ICE Ground Improv. **163**(1), 53–61 (2010)
4. Nidzam, R.M., Kinuthia, J.M.: Sustainable soil stabilisation with blastfurnace slag - a review. Proc. ICE Constr. Mater. **163**(3), 157–165 (2010)
5. Tasong, W.A., Wild, S., Tilley, R.J.D.: Mechanisms by which ground granulated blastfurnace slag prevents sulphate attack of lime-stabilised kaolinite. Cem. Concr. Res. **29**(7), 975–982 (1999)

6. Wild, S., Kinuthia, J.M., Jones, G.I., Higgins, D.D.: Suppression of swelling associated with ettringite formation in lime stabilized sulphate bearing clay soils by partial substitution of lime with ground granulated blastfurnace slag. Eng. Geol. **51**(4), 257–277 (1999)
7. James, R., Kamruzzaman, A.H.M., Haqueand, A., Wilkinson, A.: Behaviour of lime-slag-treated clay. Proc. ICE Ground Improv. **161**(4), 207–216 (2008)
8. Yi, Y., Gu, L., Liu, S.: Microstructural and mechanical properties of marine soft clay stabilized by lime-activated ground granulated blastfurnace slag. Appl. Clay Sci. **103**, 71–76 (2015)
9. Cardoso, F.A., Fernandes, H.C., Pileggi, R.G., Cincotto, M.A., John, V.M.: Carbide lime and industrial hydrated lime characterization. Powder Technol. **195**(2), 143–149 (2009)
10. Davis, S., Funada, C.: Calcium carbide. SRI Consulting, Calif (2011)
11. Li, Y., Sun, R., Liu, C., Liu, H., Lu, C.: CO2 capture by carbide slag from chlor-alkali plant in calcination/carbonation cycles. Int. J. Greenhouse Gas Control **9**, 117–123 (2012)
12. Yi, Y., Gu, L., Liu, S., Puppala, A.J.: Carbide slag-activated ground granulated blastfurnace slag for soft clay stabilization. Can. Geotech. J. **52**(5), 656–663 (2015)
13. Matschei, T., Lothenbach, B., Glasser, F.P.: The AFm phase in Portland cement. Cem. Concr. Res. **37**(2), 118–130 (2007)

Study on Strength and Microscopic Properties of Stabilized Silt

Xiaobin Zhang[1(✉)], Zhiduo Zhu[1], and Renjie Wei[2]

[1] Southeast University, Nanjing 210096, Jiangsu, China
zxb0615@163.com
[2] China Shipbuilding Industry Institute of the Engineering
Investigation & Design Co. Ltd., Shanghai 200063, China

Abstract. Stabilization is an effective method to improve the engineering properties of silt. In order to investigate the effect and mechanism of lime, cement + lime and a new curing agent stabilized silt, the unconfined compressive strength testes and scanning electron microscopy (SEM) testes are carried out. After the unconfined compressive strength testes, the impact of curing agent and admixture dosage on strength are analyzed. The strength increases significantly with the increase of admixture dosage and curing age, and the stabilizing effect of the new curing agent is better than that of the other two curing agents. Then microscopic testes of stabilized silt are carried out, and the scanning electron microscopy (SEM) images are obtained. The silt microstructure changes after stabilizing: particle size increases; particles are embedded more closely; the number of pores reduces. Based on the microstructure properties of stabilized silt, the mechanism of stabilizing silt is discussed, which has certain reference value for guiding the silt stabilizing design.

Keywords: Silt · Stabilization · Unconfined compressive strength
Microcosmic · Mechanism

1 Introduction

Silt has poor engineering properties, and it is widely distributed in China. Solidification/stabilization is an effective way to improve the engineering properties of silt. Solidification/stabilization technology is simplicity, efficiency, economy, and therefore has been widely used in geotechnical engineering. Many scholars have carried out extensive and detailed study on the solidification/stabilization technology in engineering applications. Stabilized object include silt, clay, expansive soil, loess, saline soil, organic soil and other soft soils. Curing agents include traditional calcium-based curing agents such as cement, lime and fly ash, as well as relatively new curing agents such as polymer materials, curing solutions, ionic curing agents, petroleum emulsions and ground polymers. In addition to traditional geotechnical engineering,

Foundation: National Natural Science Foundation of China (40872173); Jiangsu Province Transportation Science and Technology Project (2010Y36-3).

L. Li et al. (Eds.): GSIC 2018, *Proceedings of GeoShanghai 2018 International Conference: Ground Improvement and Geosynthetics*, pp. 129–136, 2018.
https://doi.org/10.1007/978-981-13-0122-3_15

solidification/stabilization technology also has a wide range of applications in environmental geotechnical engineering [1], such as remediation of contaminated sites and treatment of solid wastes.

Compared to other types of soil, study on the solidification/stabilization of silt is relatively few. Lo et al. studied the strength and dilatancy of cement and fly ash solidified silt [2], which indicated that the shear strength and dilatancy increased after curing, and the contribution of the curing agent to the strength could be divided into two parts: gelation and dilatation rate increase. Zhu et al. invented a new curing agent for silt, and the new curing agent can effectively improve the strength and deformation characteristics of silt [3]. Yao et al. studied the compaction characteristics of lime-fly ash solidified silt, which indicated that the content of lime-fly ash had a significant effect on the maximum dry density and the optimal water content, and with the increase of compaction degree the strength increased [4]. Cai et al. used activated magnesia carbonization to stabilize silt, which achieved good effect [5]. It can be seen from the current research that the studies mainly focuses on the macroscopic mechanical properties, and rarely involves the microscopic characteristics of solidified silt. In addition, the stabilizing mechanism of silt, the interaction of silt-curing agents and the long-term performance of stabilized silt still need to be studied in depth.

In this paper, lime, cement + lime and a new curing agent were chosen to stabilize silt, and the testes of unconfined compressive strength and scanning electron microscopy (SEM) were carried out. The effects of the dosage and the curing age on the unconfined compressive strength of stabilized silt were analyzed and the stabilizing effect of the three curing agents was compared. Based on the results of scanning electron microscopy (SEM), the mechanism of silt solidification was discussed from the perspective of microstructure.

2   Test Materials and Scheme

2.1   Test Materials

Silt. The soil used in the test was taken from a highway construction site in Dafeng City, Jiangsu Province. The particle size distribution of the soil is shown in Fig. 1. The contents of clay particles (<5 μm), silt particles (5 ~ 75 μm) and sand (>75 μm) are 10.7%, 81.5% and 7.8% respectively. The basic physical properties of soil are shown in Table 1. The chemical composition of the soil measured by X-ray fluorescence spectrometer is shown in Table 2. It can be seen from Table 2 that the main chemical constituents of the soil samples are SiO_2, Al_2O_3 and Fe_2O_3, which account for 91.37% of the total amount.

Curing Agents. The lime was purchased from a lime production plant in Nanjing. It is gray and massive, in which the active ingredient CaO + MgO accounts for 71.87%. The cement was common Portland cement P.O32.5 purchased from a cement production

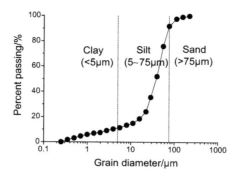

Fig. 1. Grain size distribution curve of soil

Table 1. Basic physical properties of soil

Water content (%)	Liquid limit (%)	Plastic limit (%)	Plasticity index (%)	Maximum dry density (g/cm³)	Optimum moisture content (%)
30.5	33.5	24.6	8.9	1.72	16.45

Table 2. Chemical composition of soil

Composition	SiO_2	Al_2O_3	Fe_2O_3	K_2O	MgO	CaO	Na_2O	TiO_2	P_2O_3	SO_3	Other
Content/%	74.64	10.20	6.53	1.44	0.74	0.63	0.57	0.56	0.084	0.036	4.57

plant in Nanjing. The main chemical components of cement are: CaO (55.37%), SiO_2 (25.41%) and Al_2O_3 (10.09%). Main components of the new curing agent are cement, fly ash and mineral powder, and the secondary components are alkaline excitation components, swelling components and surfactants.

2.2 Test Scheme

In order to fully consider the effects of content and curing age on the properties of the stabilized silt, the design scheme is shown in Table 3. The content of each curing agent is 4%, 6%, 8%, and the curing age of the sample is 7d, 28d, 60d and 90d. For the sake of simplicity, in this paper, the symbol "An" is used to represent the stabilized silt, in which the letter "A" represents the type of curing agent, and the number "n" represents the content (%). For example, in Table 3, L4 represents a sample containing 4% lime, and C2L4 represents a sample containing 2% cement + 4% lime.

Table 3. Test scheme

Numbering	Admixture	Content (%)	Curing age (d)	Tests
L4	Lime	4		
L6		6		
L8		8		Unconfined compressive strength
C2L2	Cement + lime	2 + 2	7, 28, 60, 90	Scanning electron microscopy (SEM)
C2L4		2 + 4		
C4L4		4 + 4		
S4	New curing agent	4		
S6		6		
S8		8		

3   Results and Analysis

3.1   Unconfined Compressive Strength Test

The results of unconfined compressive strength of stabilized silt are shown in Fig. 2.

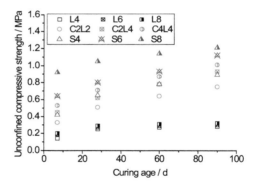

Fig. 2. Relationship between unconfined compressive strength and curing age of stabilized silt

It can be seen from Fig. 2 that the unconfined compressive strength of new curing agent stabilized silt is the highest, followed by cement + lime stabilized silt and lime stabilized silt. For lime-stabilized silt, the unconfined compressive strength increases slightly in the first 28 days, while with further increase of curing age, the unconfined compressive strength is almost unchanged. In addition, as the lime content increases, the unconfined compressive strength is only slightly improved. For new curing agent and cement + lime stabilized silt, the unconfined compressive strength increases significantly with the increase of curing agent content and curing age.

There are many scholars have studied the correlation between strength and curing age of solidified soil. Horpibulsuk et al. proposed the relationship between unconfined compressive strength and curing age of cemented soil [6], and the relationship is shown in the following equation.

$$UCS_t/UCS_{28d} = a + b \ln t \tag{1}$$

In the equation, UCS_t is the unconfined compressive strength at the curing age of t, and a, b is the correlation coefficient.

According to Horpibulsuk's research method, the experimental results of unconfined compressive strength of stabilized silt are normalized, and the normalization parameter is UCS_{28d}. The normalized results are shown in Fig. 3. It can be seen that the normalized unconfined compressive strength of stabilized silt has a good linear correlation with curing age in logarithmic coordinates. The normalized unconfined compressive strength of the lime, cement + lime and new curing agent stabilized silt are fitted as follows:

$$UCS_t/UCS_{28d} = 0.421 \lg t + 0.326 \tag{2}$$

$$UCS_t/UCS_{28d} = 0.656 \lg t + 0.116 \tag{3}$$

$$UCS_t/UCS_{28d} = 0.458 \lg t + 0.368 \tag{4}$$

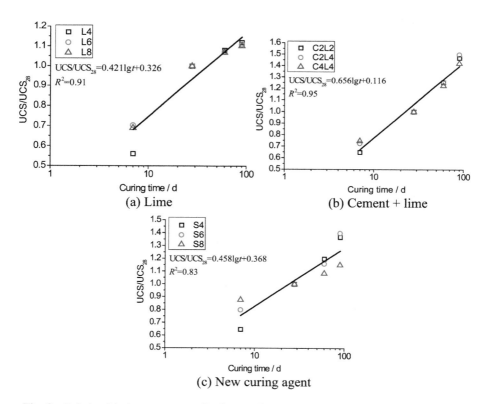

(a) Lime

(b) Cement + lime

(c) New curing agent

Fig. 3. Relationship between normalized unconfined compressive strength and curing age

3.2   Microscopic Test

Scanning electron microscopy (SEM) images of silt and stabilized silt are obtained by scanning electron microscopy (SEM) testes, as shown in Fig. 4 (taking 8% admixture content, curing age 28d as an example).

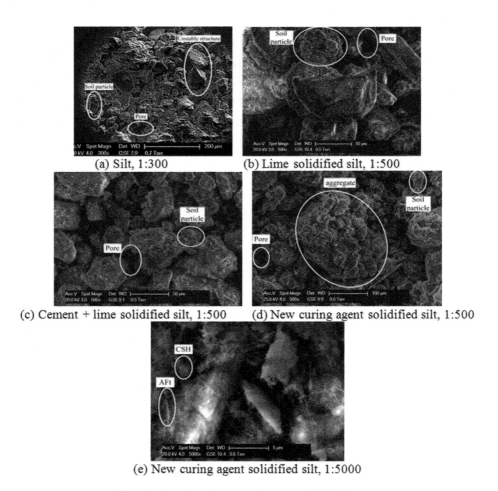

(a) Silt, 1:300

(b) Lime solidified silt, 1:500

(c) Cement + lime solidified silt, 1:500

(d) New curing agent solidified silt, 1:500

(e) New curing agent solidified silt, 1:5000

Fig. 4. Scanning electron microscopy (SEM) images

It can be seen from Fig. 4(a) that the particle size of silt is uniform and the size is mainly in the range of $5 \sim 75$ μm, which is consistent with the results of the granulation gradation in the previous text. Due to the lack of particles of other sizes to fill the skeleton pores, it is difficult to form an embedded and dense structure for the particles, which explains the compaction difficulty of silt. The shape of soil particles can affect the macro-mechanical properties of soil: if the soil particle shape is uniform and smooth spherical, the soil is isotropic material, and its strength is low; while if the soil particles shape is irregular, the soil is anisotropic material, and its strength and rigidity is high. It can be seen that the shape of silt particle is quite regular, mostly close to spherical, and lack of irregular shape like needle, column, and so on, which is one reason for low strength and large deformation of silt. From the arrangement of silt particles it can be seen that the particles arrange loosely, the boundaries between the particles are clear, and there is no cementing material between particles. In addition, the contact between particles is mostly edge-edge and edge-surface, and there are many overhead structures between particles. According to the principle of structural mechanics this type of structure is unstable, so the particles are very easy to rotate and slide. Due to the loose arrangement of particles, there are more pores between particles, and the size of pore is almost the same with particles.

It can be seen from Figs. 4(b), (c) and (d) that the particle size of stabilized silt is significantly larger than that of silt, which is due to the gelling product of hydration reaction and pozzolanic reaction of curing agent. The gelling product encapsulates the soil particles or cements the small particles together to form larger aggregates (Fig. 4(d)), increasing the particle size. The gelling products of hydration reaction and pozzolanic reaction include needle-like hydrated calcium sulphoaluminate (ettringite, AFt) and flocculent hydrated calcium silicate (CSH) [7], as shown in Fig. 4(e). The increase of AFt and CSH in soil can directly improve the strength of soil [8]. The increase in particle size of stabilized silt changes the uniformity of the silt particle, making it easier to form an embedded and stable structure. Compared with silt, solidified silt particles arrange more closely, and the boundaries between particles are less obvious, which is because AFt and CSH cement particles together and fill the pores between particles. In addition, due to dense arrangement of solidified silt particles, the number of pores is less than that of silt. By contrast, it can be found that the pore sizes of lime stabilized silt are obviously bigger than those of the other two curing agents, and the pore sizes of new curing agent stabilized silt are the smallest, which explains higher strength of new curing agent stabilized silt and the low strength of lime stabilized silt.

In summary, after stabilizing the silt microstructure changes include the increase of particle size (formation of aggregates), greater cementation of particles and the reduction of pore number. Huang et al. [9] proposed a solidified soil structure formation model, he believes that the formation of solidified silt structure includes two aspects: one is cemented hydrate fully wraps soil particles, the other is cemented hydrate or expansive hydrate fills the pores between particles. However, it can be seen from the above experimental results that the cemented product not only encapsulates the particles, but also binds the small particles together to form large aggregates (Fig. 4 (d)). In addition, Huang et al. [9] and Li et al. [10] all think that the filling of pores plays an important role in improving the strength of solidified silt, which indirectly illustrates the reason of low strength of lime solidified silt with more pores.

4 Conclusion

The strength of lime stabilized silt increases slightly in the first 28 days, and then almost stays the same with the increase of curing time, and the contribution of increasing lime content to strength is not obvious. On the contrary, for cement + lime and new curing agent stabilized silt, the strength is significantly improved with the increase of curing agent content and curing time. The stabilizing effect for silt of the new curing agent is better than the other two curing agents.

The unconfined compressive strength at 28 days is taken as the normalization parameter, and the relationship between normalized unconfined compressive strength and curing time is established, indicating that normalized unconfined compressive strength and curing time have a good linear relationship in logarithmic coordinates.

The reasons for the poor engineering properties of silt include: particle size is fairly uniform, particle shape is spherical, the contact of the particles is unstable and the number of pores is large. After hydration reaction and pozzolanic reaction of agent, AFt, CSH and other products change the microstructures of silt. The change includes: particle size increases (the formation of agglomerates), which improving the particle gradation, the particle gelation is more intense and the number of pores decreases.

References

1. Liu, S., Zhan, L., Hu, L., Du, Y.: Environmental geotechnics: state-of-the-art of theory, testing and application to practice. China Civil Eng. J. **49**(3), 6–30 (2016)
2. Lo, S.R., Wardani, S.: Strength and dilatancy of a silt stabilized by a cement and fly ash mixture. Can. Geotech. J. **39**(1), 77–89 (2002)
3. Zhu, Z., Liu, S., Shao, G., Hao, J.: Research on silts and silts treated with stabilizers by triaxial shear tests. Rock Soil Mech. **26**(12), 1967–1971 (2005)
4. Yao, Z., Lian, J., Ai, Y.: Compaction properties on Yellow River silty soil stabilized with lime-flyash. Chin. J. Geotech. Eng. **29**(5), 664–670 (2007)
5. Cai, G., Du, Y., Liu, S., Singh, D.N.: Physical properties, electrical resistivity, and strength characteristics of carbonated silty soil admixed with reactive magnesia. Can. Geotech. J. **52**(11), 1699–1713 (2015)
6. Horpibulsuk, S., Miura, N., Nagaraj, T.S.: Assessment of strength development in cement-admixed high water content clays with Abrams' law as a basis. Geotechnique **53**(4), 439–444 (2003)
7. Du, Y., Jiang, N., Wang, L., Wei, M.: Strength and microstructure characteristics of cement-based solidified/stabilized zinc-contaminated kaolin. Chin. J. Geotech. Eng. **34**(11), 2114–2120 (2012)
8. Hoshino, S., Yamada, K., Hirao, H.: XRD/Rietveld analysis of the hydration and strength development of slag and limestone blended cement. J. Adv. Concr. Technol. **4**(3), 357–367 (2006)
9. Huang, X., Ning, J., Xu, S., Lan, M.: Structure formation model of stabilized soil. Ind. Constr. **36**(7), 1–6 (2006)
10. Li, Y., Qian, C., Liu, S., Zhu, Z.: Mechanism of silt stabilization. Chin. J. Geotech. Eng. **26**(2), 268–271 (2004)

Stabilisation of Expansive Soil Against Alternate Wetting–Drying

An Deng[(✉)]

The University of Adelaide, Adelaide, SA 5000, Australia
an.deng@adelaide.edu.au

Abstract. Expansive soils widely deposit in major areas of Adelaide of South Australia. The soils tend to expand as they absorb a volume of moisture and shrink when a portion of the moisture escapes. The alternate changes in volume of the soil pose severe damage to foundations of residential and commercial buildings. A solution is to amend the expansive soils by stabilisation. The stabilisation becomes less effective over time, in particular, when the soils are exposed to cyclic wetting–drying impact due to seasonal changes which regularly occur in Adelaide. Laboratory investigations were conducted to examine the stabilisation durability. Expansive soil was stabilised by three separate additives, i.e. cement, lime and fly ash. These additives were included at different doses, aiming at determining the choice of suitable additive and optimising the feed. Stabilised soil samples were exposed to cyclic wetting–drying conditions. Swelling and shrinkage of the samples were measured in the course of the cyclic processes. Some interesting results were presented.

Keywords: Expansive soil · Wetting–drying · Stabilisation · Swelling

1 Introduction

Expansive soils outcrop over a wide range of areas in Adelaide South Australia [1]. The soils are prone to substantial volume changes when in contact with water—expand as soils absorb a volume of water and shrink when parched [2]. This moisture-sensitive nature leads expansive soils problematic as foundation in Adelaide which has a seasonal climate of alternate rainfall and drought. The foundation rises in rainfall seasons and goes down when the soils dry out in drought. The undesired foundation shifting easily causes damage to buildings mainly double-brick houses which undergo walls cracking or even splitting. Each year in Australia there is a good number of houses reported of cracking problems which have accounted for a large portion of all housing insurance claims. It is thus important to assess solutions developed to resolve the swelling problem and examine solutions when applied to the local expansive soils.

Stabilisation is usually a preferred solution to cope with the swelling problem of expansive soils [3, 4]. The stabilisation is a process conducted by incorporating a certain mass of cementitious materials into the expansive soils. When in contact with water the cementitious materials solidify and grow strength to counteract swelling potential of expansive soil. The stabilisation effect, however, is time dependent, in particular, when exposed to severe conditions such as the alternate moisture environment. In this context,

© Springer Nature Singapore Pte Ltd. 2018
L. Li et al. (Eds.): GSIC 2018, *Proceedings of GeoShanghai 2018 International Conference: Ground Improvement and Geosynthetics*, pp. 137–144, 2018.
https://doi.org/10.1007/978-981-13-0122-3_16

it is of interest to examine response of stabilised expansive soils when subjected to cyclic wetting-drying. This cyclic impact is applied replicating Adelaide's climate; mild winters with moderate rainfall and hot, dry summers. Similar studies such as Estabragh et al. [4] have been conducted. The past studies however were conducted on soils with compositions different from soils present in Adelaide. Use of a soil representing local expansive soil profile helps develop guideline suitable for stabilisation practice in Adelaide. Another research value lies in the use of multiple cementitious materials in one study and to the same soil, enabling a comparative study.

2   Materials and Method

2.1   Materials

The Adelaide region has different types of soils which can cause soil swelling [1]. To represent the varying soil types, an artificial soil was developed based on geological data in Adelaide. The composition of the artificial soil was determined in terms of soil samples recovered within 0–5 m depth from various Adelaide areas which are accessible from Bowman and Sheard [1]. As a result, the soil used in this study was composed of 45% clay, 35% coarse sand, and 20% silt. The clay content was further refined based on three typical minerals: montmorillonite, illite and kaolinite. This clay content however was not accessible from the geological data; and had to be perceived by knowledge. To pick up swelling potential, a major content was assigned to montmorillonite and a minor for the rest. The clay contents were 80% montmorillonite and 20% kaolinite. Illite, an inert mineral, was not included to simplify clay contents. The soil ingredients became 36% montmorillonite, 9% kaolinite, 35% coarse sand and 20% silt. Compositions of montmorillonite and kaolin are provided in Table 1.

The soils then were subjected to Atterberg limits test in accordance with Australian

Table 1. Chemical composition of montmorillonite and kaolin.

Oxide	Montmorillonite (%)	Kaolin (%)
SiO_2	63.8	64.9
Al_2O_3	13.6	22.4
Fe_2O_3	2.8	1.0
CaO	0.2	0.1
Na_2O	2.3	0.2
MgO	2.0	0.7
K_2O	0.2	2.8
TiO_2	0.3	1.4

Standards 1289.3.9 and 1289.3.2.1, and standard Proctor compaction test in accordance with AS 1289.5.1.1. Test results were used to classify the soils and develop compaction curve, respectively.

Cementitious materials used for expansive soil stabilisation included Portland cement, lime and fly ash. These materials are widely used additives as a solution to cope with soil swelling problems. The chemical compositions of the lime and fly ash, both from the materials supplier, are provided in Table 2. As shown in the table, the fly ash was classified as Class F in terms of ASTM C618 which means an essence of low percent weight of CaO, SiO_2, Al_2O_3 and Fe_2O_3. Class F is the sole fly ash product generated and accessible in Australia due to a lack on lignite and sub-bituminous coal in the area. Although being less preferable due to a low CaO content, it is worth assessing capacity of this type of fly ash in promoting pozzolanic reactions when mixed with expansive soils and the likelihood of stabilising the soils.

Table 2. Chemical composition of additives.

Oxide	Lime (%)	Fly ash (%)
SiO_2	0.3–0.6	49.01
Al_2O_3	0–2	30.96
Fe_2O_3	0–0.7	2.81
CaO	85–95	5.35
Na_2O	N/A	3.76
MgO	0.5–1.5	2.52
MnO	N/A	0.05
K_2O	N/A	1.17
TiO_2	N/A	2.10
P_2O_5	N/A	0.90
SO_3	N/A	0.25

The three additives were added to the expansive soils in terms of mixture dosages provided in Table 3. This table outlines six sample sets. Each set includes three samples stabilised separately by the three additives but at the same dosage. The dosage varied from 5% to 20% which improved ranges versus past studies. The increased percent ranges added two values: (i) to assess effect of a single additive on stabilisation durability when increasing the additive content; (ii) to cross check between the additives where the dosage varies. This was to develop dosage suitable for local practice. The sample sets also accounted for sample curing time, enabling comparison between 'pre-matured' and 'matured' samples. Curing was conducted at room temperature. Control samples were also prepared to cross check stabilisation.

Table 3. Percent weight of additives.

Sample set	Cement	Lime	Fly ash	Curing
A	5%	5%	5%	16 h
B	10%	10%	10%	16 h
C	5%	5%	5%	7 day
D	10%	10%	10%	7 day
E	15%	15%	15%	7 day
F	20%	20%	20%	7 day

2.2   Methods

Blend proportioned dry ingredients (clay, silt and sand) into expansive soil. Split the soil into seven portions and add corresponding additives into each portion. Transfer the portion into an air-proof bag, wrap up the end and shake it for about one minute. This encapsulated air-shaking is one of the best approaches to attaining a uniform mixture. Pile the mixture onto a mixing tray and grow its water content to its optimum moisture content (OMC). The OMC which was attained in the standard Proctor test was selected in order to compact all samples into a consistent dry unit weight. Make use of a spatula to toss the mixture until water appears uniformly distributed.

Place sample into a consolidation ring. The ring was made of polyvinyl chloride pipe of 70 mm internal diameter and 40 mm high. The ring acted in a way similar in concept to the oedometer ring when used to measure swelling and shrinkage of samples resulted from cyclic wetting and drying processes. Due to the small size of the ring, standard Proctor compaction was not applicable. An alternate solution was to apply static compaction, targeting a dry unit weight consistent with that obtained in standard Proctor test. In the process of the static compaction, determine the mass required to achieve the desired dry unit weight; split the mass into three equal portions; transfer each portion at a time into the ring and compact the portion into one third of intended volume. Scarify the interface to improve soil homogeneity. The intended volume was worked out based on the dry unit weight desired for individual samples. The compaction was static and completed by a metal rod of 30 mm diameter. Force down the rod by using a moment-driven plunger. Conduct compactions in a rotary motion enabling a uniform energy input. In the meantime, encase the consolidation ring with a rigid metal ring to improve confinement and prevent possible damage to the consolidation ring. Upon completion of compactions, a soil sample occupied the bottom half of the consolidation ring, leaving the top 20 mm for soil to travel when having swelled. That is, the ring was able to accommodate a 100% swelling potential. Record start height of individual samples before moving into the next testing phase.

Dock the consolidation rings onto a wetting board and then place the board into a wetting container as shown in Fig. 1. On the board there were nine spots to dock the rings. Each spot included a grid of 1 mm holes perforated through the 10 mm wetting board enabling access to water underneath. Handles were designed on the sides of the board to allow for easier transport. Attach a filter paper above and below the sample. Seal with silicone gel the bottom edge of the rings to eliminate water access from the sideways. Fasten each ring to its corresponding spot by adhesive tapes to prevent the ring from possible relocations during subsequent tests. In addition place a porous disc on top of a sample to prevent possible escape of fine particles from the sample surface. The porous disc also helped record deformation consistently; record deformation repeatedly against the cross marked out at the centre of a disc.

Apply wetting and drying phases in an alternating fashion. In a wetting phase fill up with tap water the wetting container to a level approximately 20–30 mm above the top of the rings. Keep recording swelling potential of the control sample on a daily base until a criterion of less than 1 mm deformation per 24 h has been obtained. This criterion is usually taken to decide the completion of the wetting phase. Benchmark the stabilised samples in wetting time against the control, as the stabilised samples were

Fig. 1. Wetting apparatus: (a) container; (b) wetting board.

supposed to be quicker than the control to come to a completion of swelling. Start a drying phase by transporting the samples together with the board into an oven. Set the oven to 45 °C in accordance with ASTM D559. Record samples height once a day against the same criterion used in the wetting phase. Upon completion of the drying phase, subject the samples to a new wetting phase, and so forth.

3   Results and Discussion

3.1   Atterberg Limits and Compaction

Atterberg limit results were 103.4% for liquid limit, 25.55% for plastic limit and 77.85 for plasticity index. The soil was sandy silty clay, high plasticity, with a USCS code CH. Compaction results for sample sets A to D and controls are provided in Table 4. As both the OMC and the maximum dry unit weight were consistent between samples, it was assumed that compaction was independent of additives content. As a result, compaction tests were not conducted on sample sets E and F, which were supposed to give rise to similar compaction results. Meanwhile compact all sample sets at a moisture content of 18%.

Table 4.   Standard Proctor compaction test results.

Soil	Optimum moisture content (%)	Maximum dry unit weight (kN/m^3)
Control	18.0	16.5
5% cement	18.0	16.2
5% lime	18.2	16.1
5% fly ash	18.0	16.6
10% cement	20.5	15.8
10% lime	18.9	16.1
10% fly ash	18.0	16.2

3.2 Swelling and Shrinkage

The swelling and shrinkage values of the stabilised and control samples are illustrated in Fig. 2. The stabilised samples were subjected to up to five wetting–drying cycles. As expected, swelling and shrinkage were clearly dependent on the choice of additives, and then curing time offered prior to the testing. Additive percentage however marginally influenced the swelling and shrinkage results throughout the cycles. The cement and lime were able to stabilise the soils noticeably from the beginning and the fly ash barely stabilised the swelling. This ranking agreed with the one reported in Estabragh et al. [4]. The other agreement was the residual swelling potential occurred to controls. Both studies found that control samples, where having swelled, hardly shrank to its original thickness. A residual swelling potential remained which was in the range of 20% to 30% in this study and 10% to 20% in Estabragh et al. [4]. The higher residual swelling potential observed in this study was likely caused by the higher bentonite content which was 80% in relation to 60% in Estabragh et al. [4]. The higher the bentonite content, the more water retention likely entrapped in the bentonite crystal structure upon the completion of a drying phase. This mechanism was supportive of a similar shifting occurred to swelling potential amplitude. The swelling potential amplitude was 60% on average in this study and 35% on average in Estabragh et al. [4] which cross checked significance of the bentonite. Another cross check related to swelling potential oscillations. Both studies demonstrated consistent curve oscillations for the controls throughout the cycles. The difference lied in oscillation gaps: about 55% in this study and 45% in Estabragh et al. [4]. Overall, bentonite content played an active role in swelling potential curve development where no additives were applied.

Where additives were applied, cement demonstrated the best outcome of counteracting swelling. Cement-stabilised samples developed consistently flat swelling-shrinkage curves where less than 1% deformation had occurred throughout the five cycles. These flat trendlines excluded a couple of incidental 'jumps' or 'drops' which were possibly caused by inappropriate operation of recording, but these outliers were not carried forward nor influenced assessment results. It is seen that the flat curve development was independent on the cement content and the curing time examined in this study. That is, increasing cement content (from 5% to 20%) seemed to give rise to marginal stabilisation effect; a 5% cement content suffices. This finding disagrees with results in Estabragh et al. [4] which suggested that increasing cement content (from 10% to 20%) further reduces soil swelling potential (from about 7% to 3%). To ensure the 5% cement content was not overdosed, it is worth exploring a threshold or optimal dosage that equally satisfies the stabilisation need. Therefore a next step to this study will be examining capacity of less than 5% cement samples in counteracting the soil swelling problem.

Lime-stabilised expansive soils demonstrated curing-dependent results. If having cured the samples for a short time (i.e. 16 h), lime was able to reduce swelling potential to about 20% (Fig. 2(a, b)); otherwise down to less than 5% where having cured the samples for 7 days (Fig. 2(c to f)). That is, lime is sensitive to curing and the sensitivity is independent of the lime dosage. In Fig. 2(a, b) samples demonstrated flat swelling potential (i.e. no oscillations) upon the completion of the first cyclic impact. This agrees with the argue of lime building up strength at a late stage. Where a curing time is 7 days

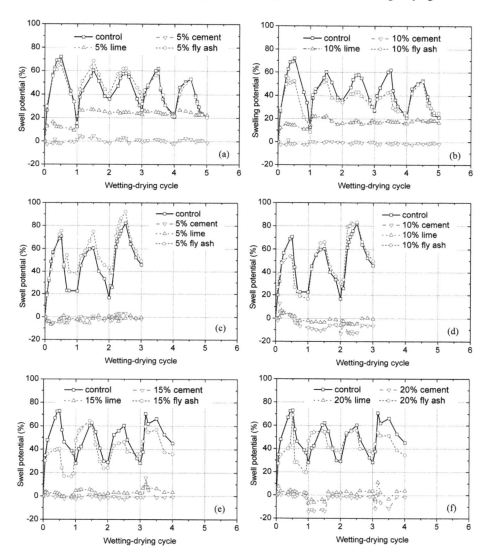

Fig. 2. Swelling potentials of (a) sample set *A*, (b) sample set *B*, (c) sample set *C*, (d) sample set *D*, (e) sample set *E* and (f) sample set *F*.

or more, a 5% lime content satisfactorily counteracted the swelling. The curing dependency agrees with results presented in Estabragh *et al.* [4] and Phanikumar [5]. So do the swelling potential changes (from around 20% to less than 5%). Similarly the next step study will be aiming at optimising the curing time and lime content.

Fly ash was not effective in stabilising expansive soil. Swelling potential curves largely replicated those of controls. The two set of curves (fly ash and control) were similar in magnitude and developed in a way independent of curing period, dosage and cyclic number. That is, fly ash demonstrated the least favourable capacity to stabilising the expansive soils. This outcome was associated with the composition of the fly ash. As aforementioned, it was Class F fly ash that was explored in this study. The Class F fly ash contained low quantity of calcium oxide and thus was inactive in giving rise to cement hydration reaction, nor to develop material strength. Many studies however validated Class C fly ash as an important additive for expansive soil. The argue was that this category of fly ash contains calcium oxide at a level in favour of cement hydration and strength development.

4   Conclusions

This study presented test results from an Honours project undertaken by K Al-Rashid, N Koirala, P Munoz and D Nguyen. Tests were conducted by subjecting stabilised expansive soil samples to wetting-drying cyclic impact. Three additives were assessed: cement, lime and fly ash. Cement outperformed the other additives in respect to counteracting soil swelling potential. A 5% cement content successfully reduced the swelling to within 1%. The success was independent of curing. If being cured, lime was also a solution to stabilising the same expansive soil. Due to low calcium oxide content, fly ash was not as effective to stabilise the soil.

References

1. Bowman, G.M., Sheard, M.J.: Soils, Stratigraphy and Engineering Geology of Near Surface Materials of The Adelaide Plains, Department of Mines and Energy South Australia, Parkside, South Australia (1996)
2. Basma, A.A., Al-Homoud, A.S., Malkawi, A.I.H., Al-Bashabsheh, M.A.: Swelling-shrinkage behavior of natural expansive clays. Appl. Clay Sci. **11**(2–4), 211–227 (1996)
3. Graber, E.R., Fine, P., Levy, G.J.: Soil stabilization in semiarid and arid land agriculture. J. Mater. Civ. Eng. **18**(2), 190–205 (2006)
4. Estabragh, A.R., Pereshkafti, M.R.S., Parsaei, B., Javadi, A.A.: Stabilised expansive soil behaviour during wetting and drying. Int. J. Pavement Eng. **14**(4), 418–427 (2013)
5. Phanikumar, B.R.: Effect of lime and fly ash on swell, consolidation and shear strength characteristics of expansive clays: a comparative study. Geomech. Geoeng. Int. J. **4**(2), 175–181 (2009)

Application of Cone Penetration Test Technology in Whole Process Inspection of Reinforcing Hydraulic Fill Sand Foundation

De-yong Wang[1,2,3(✉)], Sheng Chen[1,2], Xiao-cong Liang[1,2], and Hong-xing Zhou[1,2]

[1] CCCC Fourth Harbor Engineering Institute Co., Ltd.,
Guangzhou 510230, China
de_yong_wang@163.com
[2] Key Laboratory of Environmental Protection and Safety of Communication
Foundation Engineering, CCCC, Guangzhou 510230, China
[3] Civil and Transportation Institute, South China University of Technology,
Guangzhou 510641, China

Abstract. Cone Penetration Test (CPT) is one of the most widely adopted and accepted test means for determining geotechnical soil properties in ground investigation and inspection. Based on the whole process of ground improvement of a LNGI project in Middle East, the hydraulic fill material is mainly sand from medium to coarse, which contain some fine particles, and the soil was identified improved by vibroflotation. CPT technology is widely applied in different stages of foundation treatment, such as investigation stage, design stage, field trial stage, material monitoring stage and inspection & assessment stage, which were showed in the project case. These ideas and scheme shall provide meaning reference for similar project design and construction especially overseas projects.

Keywords: Cone penetration test · Hydraulic fill sand ground
Reinforcing · Inspection

1 Introduction

The cone penetration test (CPT) has been gaining in popularity for site investigations due to the cost-effective, rapid, continuous, and reliable measurements [1].

CPT technology is suitable for different types of soil, such as clay, silt and sand. CPT Datas are applied to soil profiling, material identification and evaluation of geotechnical parameters and design, where it is widely used to interpret soil strength, deformability and hydraulic properties. Moreover CPT is internationally regarded as one of the most widely used and accepted test methods for liquefaction assessment [2].

Formerly several geotechnical design parameters of the soil are associated with the SPT. In contrast CPT is becoming increasingly more popular for site investigation and geotechnical design, especially in deltaic areas, based on the soil type and testing method. Considering their respective advantages, the CPT-SPT correlations attracted wide attention and research [3].

© Springer Nature Singapore Pte Ltd. 2018
L. Li et al. (Eds.): GSIC 2018, *Proceedings of GeoShanghai 2018 International Conference: Ground Improvement and Geosynthetics*, pp. 145–156, 2018.
https://doi.org/10.1007/978-981-13-0122-3_17

CPT could be carried out by involving many researchers [4–7] for expansion, interpretation and developing many applications from this technology.

The paper will set forth different applications of the CPT technology in the whole process of reinforcing ground from dredging material.

2 Project Introduction

A new LNG regasification plant will be constructed to supply natural gas in the middleeast. The plant will also have the capability to re-export LNG. Due to land area requirements the LNG storage and process area will be located on an offshore reclaimed land, the layout of the project is shown in Fig. 1.

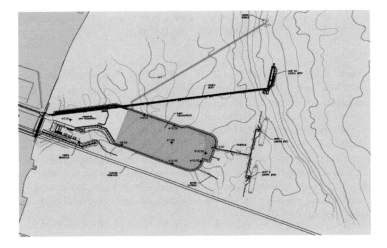

Fig. 1. Layout of the LNGI project

The reclamation material are mainly hydraulic fill sand obtained by cutter suction dredger and trailing suction hopper dredger. This fill materials are reinforced by vibro-floatation supplemented with roller compaction in subsurface layers, its requirements of ground improvement were shown on Table 1, which indicated its high standards and strict requirements.

Table 1. Ground improvement requirements

S.no.	Acceptance requirement
1	Bearing capacity 200 kPa
2	Anticipated long-term settlement < 25 mm at design bearing pressure of 200 kPa 10 years after FAC
3	MDD (Maximum Dry Density) 95% in dry fill and MDD 90% in wet fill
4	No Liquefaction susceptible fill and subsoils; LPI < 2

3 Application of CPT for Reinforcing Hydraulic Fill Sand

CPT is the main inspecting means of performance test for soil improvement by vibroflotation in the LNGI project. Take typical test of the project, to explain the application of CPT in the whole process of soil improvement.

3.1 Material Source Surveying in Investigation Stage

An important factor in preparing an appropriate insitu testing program is to meet the specification of the engineers designing an investment project.

The geological conditions or other special circumstances warrant was invested by CPT for trial dredging, the aim was to confirm whether or not directly dredging mix sand to reclamation area from the sea, as shown on Fig. 2. What's more, from the CPT data, we could know the sand-grain-size and compaction degree distribution related to the suitability of dredging and reclamation.

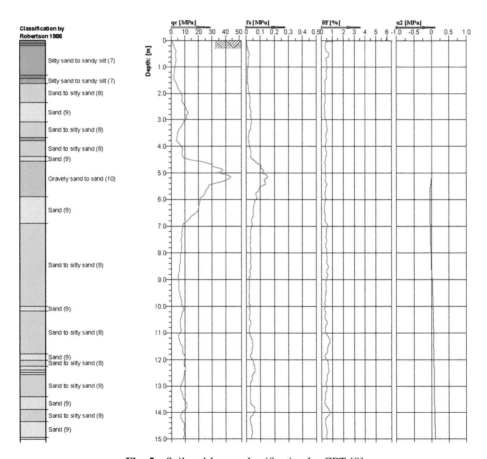

Fig. 2. Soil and layers classification by CPT [8]

3.2   Raising Target Curve in Design Stage

The target performance of improved soil will be derived in terms of CPT q_c since the CPT test is practical and specified performance management method.

The CPT-performance lines should meet the four requirements described in this Sect. 2, then the final CPT-performance line will be determined comprehensively.

3.2.1   Liquefaction Requirement

The liquefaction resistance (CRR) of the reclamation fill is calculated in accordance to Youd and Idriss [9] based on CPT. Three fill scenarios with fines content of 5%, 10%, and 15% have been considered to investigate the effect of grain characteristics of the reclamation fill. The clean-sand equivalent normalized CPT q_c, $(q_{c1N})_{cs}$, is calculated by Robertson and Wride formula [10] based on the soil type behavior Index, I_c, as follows:

$$(q_{c1N})_{cs} = K_c q_{c1N} \qquad (1)$$

Where K_c is the correction factor for grain characteristics

$$\text{for } I_c \leq 1.64, \, K_c = 1.0$$

$$\text{for } I_c > 1.64, \, K_c = -0.403I_c^4 + 5.581I3_c^3 - 21.63I_c^2 + 33.75I_c - 17.88 \qquad (2)$$

The calculated correction factor, K_c, is presented on Table 2.

Table 2. Correction factor for grain characteristics, K_c

Fill scenarios	Fines content	Soil behavior type, I_c	Correction factor, K_c
Best fill	5%	1.64	1.00
Good fill	10%	1.88	1.17
Acceptable fill	15%	2.07	1.41

The back-estimated CPT tip resistance, q_c, satisfying a safety factor of 1.25 is presented in Fig. 3 for ground level at +7.5 m ACD and Ground water table: +2.7 m ACD. In the assessment, the following assumptions are made: $f_s = 0.005q_c$.

As shown on Fig. 3, relatively good fill material requires higher cone penetra-tion resistance since the lower fine contents will result in higher liquefaction susceptibility.

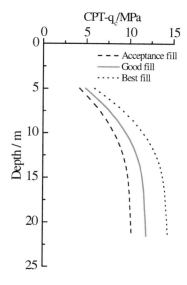

Fig. 3. Minimum CPT-*qc* against Liquefaction Criterion (SSE; Mw = 6.5, FOS = 1.0)

3.2.2 Compaction Requirement

The required compaction of reclamation fill of 95% MDD for Dry Fill and 90% MDD for Wet Fill had been interpreted to an equivalent relative density (D_r), as shown on Table 3.

Table 3. Equivalent relation of R_C and D_r

	Best fill		Good fill		Acceptable fill	
	Wet	Dry	Wet	Dry	Wet	Dry
$R_C(\%)$	90	95	90	95	90	95
$D_r(\%)$	52	75	60	80	68	85

The relative density then needs to be converted into CPT q_c by appropriate correlation equation. Baldi et al. [11] had conducted CPT tests by different density, and the relation between q_c and D_r was established: $D_r = ln[q_c/C_0/(\sigma'^{C_1})]/C_2$, where C_0, C_1, C_2 are coefficients and σ' is the effective self-weight stress. In consideration of possible carbonate nature of reclaimed fill sand, Jamiolkowski carried on related tests, to minimize creep settlement, it is recommended that the carbonate content of reclamation fill shall be lower than 30%. Three CPT q_c correlations equations were shown on Fig. 4.

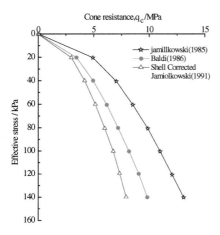

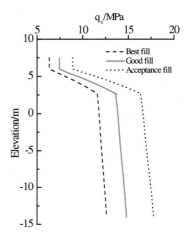

Fig. 4. Correlation equations between q_c and D_r

Fig. 5. Minimum CPT-q_c performance

Although the performance target has been set to that of the good fill, this performance target can be adjusted during construction in accordance with the fine content of actual fill. The target CPT tip resistance for RGLs of +7.50 m ACD was presented in Fig. 5.

3.2.3 Bearing Capacity Requirement

There are three different failure modes that are punching shear failure, local shear failure, and general shear failure. The failure modes are related with the embedment depth of a foundation and relative density of soil as shown on Fig. 6. From Fig. 6, the relative density of sand shall be greater than 0.7 to ensure the foundation shall not be failed by local shear mode.

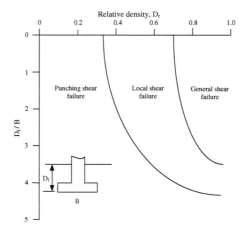

Fig. 6. Modes of foundation failure in sand [12]

The relative density of reclaimed fill mainly depends on the required degree of compaction, relative compaction. In this section, the achievable relative densities for given relative compaction are determined based on the laboratory tests. As per the field test and laboratory test, to good sand, the required Dr are 60% and 80% corresponding to the relative compaction of 90% and 95%, respectively. The design relative densities are higher than the required relative densities. Therefore the bearing capacity will be estimated only for the general shear failure condition.

Based on the study, noting that the bearing capacity is mainly governed by the angle of internal friction, the following performance requirement can be derived: characteristic angle of internal friction shall be greater than 33.5 for top 6 m, In terms of the bearing capacity, the performance requirement can be managed by: average q_c value $\geq$ 6.7 MPa. The result generally matches with the Robertson's value shown in Table 4.

Table 4. Summary of required q_c based on the bearing capacity requirement

Fill Scenarios	Soil behavior Type, Ic	q_c (MPa)				
		(1) Jefferies, Davies [13]	(2) Kulhaway, Mayne [14]	(3) Robertson [5]	EN 1997-2	Average (1, 2, 3)
Best fill	1.64	7.2	5.5	6.7	7.7	6.5
Good fill	1.88	6.6	5.3	7.8		6.7
Acceptable fill	2.07	6.2	5.1	9.3		6.9

3.2.4   Settlement Requirement

The performance requirement for the project settlement is allowable long term creep settlement of 25 mm. The long-term settlement of reclamation sand will happen in the following situations: (1) LS1: Subsoil due to the fill load; (2) LS2: Fill mass itself due to aging; (3) LS3: Fill due to the foundation load.

The allowable creep settlement of LS3 considering the creep settlement of LS1 during construction and LS3 of the optimistic scenario, the creep settlement of LS3 is estimated by Schmertmann method [15] as per EN 1997-2:2007. Where the formula can be expressed as:

$$S = C_1 \times C_2 \times (q - \sigma'_{v0}) \times \int_0^Z \frac{I_Z}{C_3 \times E} dz \qquad (3)$$

Where q is the net load at the bottom of foundation; I_Z is the influence coefficient of strain, approximately distributed as triangular; E is modulus of elasticity; C_1, C_2 are correction factor, $C_2 = 1.2 + 0.2 \times \log(t)$; σ'_{v0} is 'effective self-weight pressure' at the depth of foundation. Since soil settlement is governed by elastic properties of soil, based on the CPT data, the variation of the modulus of deformation E and its strain as a function of depth is shown in Fig. 7, and the minimum q_c at the top has been set as 7.2 MPa in consideration of the increasing trend of Young's modulus.

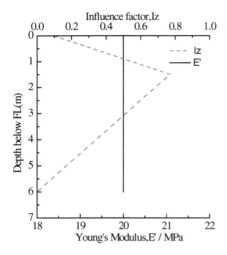

Fig. 7. Performance for creep settlement

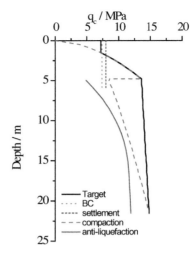

Fig. 8. CPT Performance targets for reclamation fill

Therefore Required Young's modulus shall be: Average Young's modulus for top 6 m: 20 MPa (equivalent CPT q_c of 8.0 MPa as per EN1997-2:2007)

After satisfying four requirements above, the target performance requirement of hydraulic fill can be summarized on Fig. 8. Figure 8 shows target performance lines for reclamation fill. It shall be noted that the minimum q_c for top 1.5 m has been set as 7.2 MPa since the value is just same q_c at the level and it satisfies the average q_c requirement.

3.3 CPT-SPT Establish Relation Established in Field Trial Stage

The CPT may refuse due to oversized pushing force, cone resistance, sleeve friction or vertical inclination. In order to fully evaluate and confirm the performance of the fill for the full depth, borehole/SPT was undertaken at the same location for comparison.

SPT-CPT correlation has been evaluated from available CPT q_c & Nspt log data from field trial test results. Lowest correlation equation considered to convert N60 blow counts to CPT profiles at lower elevations to complete the analysis for the full depth.

In order to determine the suitable vibrating spacing, three types of spacing d = 4.0 m, 3.75 m, 3.5 m are arranged. The comparison of actual CPT q_c and q_c derived from N-value based on proposed correlation equation are shown in Fig. 9.

It is most conservative approach from test results and cone resistances derived from SPT results have been added to the rolling mean q_c to complete the CPT profile.

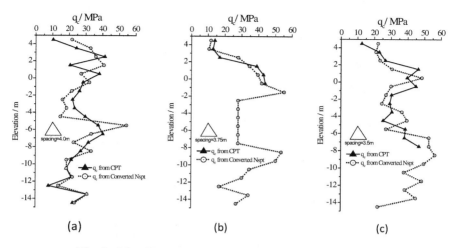

Fig. 9. Nspt Vs CPT q_c results at different vibrating spacing

3.4   Material Quality Control in Monitoring Stage

Vibroflotation without additional backfill treatment method is economical and practical and has been widely used in the treatment of foundations. While it only suitable to coarse grained soils with silt or clay content less than 10–15% [16]. Therefore the quality monitoring of gregding sand is very important. (Fig. 10)

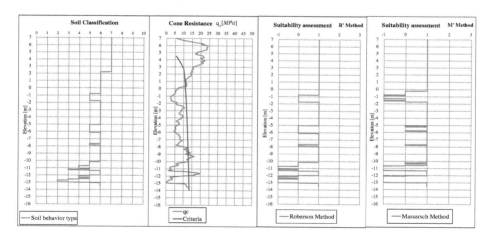

Fig. 10. Initial evaluation of hydraulic fill material

The most common CPT-based classification systems are based on behaviour characteristics and are often referred to as a soil behaviour type (SBT) classification [17]. Liang [18] thought the Robertson (1990) method can give the best result by comparison with the Robertson (1986), the Schmertmann [15] and Douglas [19] method.

Firstly the hydraulic material were classified into 9 categories by CPT data, from Robertson's method, 5th soil mean the corresponding fine content is from 35% to 50%, the experiment showed that 5th soil was barely suitable to vibroflotation, but the effect is not good. The SBT is bigger than 5, it is suitable to vibroflotation, while SBT smaller than 5, it is not.

Then the fill materials are assessed by Robertson's method and Massarsch's method [20], and the suitability was assigned values, 1 mean the soil is suitable for vibrolotation, 0 mean the soil belong to transition, −1 mean it is unqualified soil.

3.5   Quality Verification of Reinforcing in Assessment Stage

Vibro-flotation was carried out in 3.75 m triangular grid areas. Before vibro compaction, one CPT called Pre-CPT is carried out at the center of a box with 50 m × 50 m After vibro compaction, two CPT tests called Post-CPT shall be performed besides the Pre-CPT, where one CPT (a) shall be taken in the center of the dynamically reinforced column, the other CPT (b) at the one-third point between two diagonally adjacent columns as illustrated in Fig. 11 below.

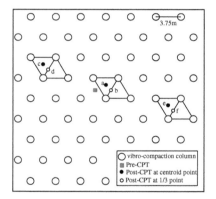

Fig. 11. Local layout of the vibroflotation and CPT

Rolling mean q_c shall be used to compare with the target performance curve and check compaction success. Rolling mean is the average taken over 0.5 m depth of two post CPTs (1/3rd and centroid). If first chance of Post CPTs (a, b) failed, two sets of additional CPTs (c, d, e, f) will be performed at both sides. The friction ratio (Fr) is also presented to allow a clear interpretation, which are described in Fig. 12.

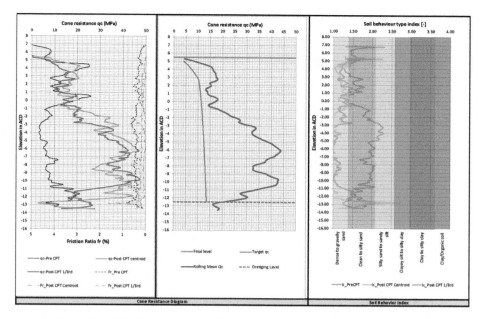

Fig. 12. CPT Assessment of reclamation sand reinforced by vibroflotation

4   Conclusions

The CPT technology was applied to different inspection and test stages of reinforcing hydraulic fill sand foundation. First CPT was used to investigate the reclamation area or suitable sand source for dredging; and the target curve for the whole ground treatment was raising by CPT; next the relation of SPT-CPT was established for the specific site; then the material obtained by dredging are assessed by CPT and corresponding suggestion was offered; finally the quality of soil improvement was inspected and tested by CPT technology. The CPT was applied throughout the process of ground treatment evaluation, which gives references to similar overseas projects.

References

1. Robertson, P.K.: Cone penetration test (CPT)-based soil behaviour type (SBT) classification system–an update. Can. Geotech. J. **53**, 1910–1927 (2016)
2. Robertson, P.K.: In-situ testing of soil with emphasis on its application to liquefaction assessment (1982)
3. Ahmed, S.M., Agaiby, S.W., Abdel-Rahman, A.H.: A unified CPT–SPT correlation for non-crushable and crushable cohesionless soils. Ain Shams Eng. J. **5**(1), 63–73 (2014)
4. Lunne, T., Robertson, P.K., Powell, J.J.M.: Cone-penetration testing in geotechnical practice. Soil Mech. Found. Eng. **46**(6), 237 (2009)
5. Robertson, P., Cabal, K.: Guide to Cone Penetration Testing, 6th edn. Gregg Drilling and Testing Inc., Signal Hill (2015)

6. Liu, S., Wu, Y.: On the strategy and development of CPT in China. Chin. J. Geotech. Eng. **26**(4), 553–556 (2004)
7. Hai-Bo, X.U., Song, X.J., Qian, C.F.: Research review on static cone penetration test. South-to-North Water Transf. Water Sci. Technol. **11**(5), 78–81 (2013)
8. Robertson, P.K., Campanella, R.G., Gillespie, D, et al.: Use of piezometer cone data. In: Use of In Situ Tests in Geotechnical Engineering, pp. 1263–1280. ASCE (1986)
9. Youd, T.L., Idriss, I.M.: Liquefaction resistance of soils: summary report from the 1996 NCEER and 1998 NCEER/NSF workshops on evaluation of liquefaction resistance of soils. J. Geotech. Geoenviron. Eng. **127**(4), 297–313 (2001)
10. Robertson, P.K., Wride, C.E.: Evaluating cyclic liquefaction potential using the cone penetration test. Can. Geotech. J. **35**(3), 442–459 (1998)
11. Baldi, G., Belloti, R., Ghionna, V., Jamiolkowski, M., Pasqualini, E.: Interpretation of CPTs and CPTUs. Part 2: drained penetration of sands. In: Proceedings of 4th International Geotechnical Seminar on Field Instrumentation and In Situ Measurements, pp: 143–156, Singapore (1986)
12. Vesic, A.S.: Analysis of ultimate loads of shallow foundations. Int. J. Rock Mech. Min. Sci. Geomech. Abs. **11**(11), 45–73 (1973)
13. Jefferies, M.G., Davies, M.P.: Use of CPTU to estimate equivalent SPT N60. Geotechn. Test. J. ASTM **16**(4), 458–468 (1993)
14. Kulhawy, F., Mayne, P.: Manual on estimating soil properties for foundation design Report no. EPRI-EL-6800, Electric Power Research Institute, EPRI (1990)
15. Schmertmann, J.H.: Guidelines for cone penetration test, performance and design, Report no. FHWA-TS-78-209, Washington D.C., US Department of Transportation, p. 145 (1978)
16. Zhou, J., Wang, G.Y., Jia, M.C.: Situation and latest technical progress of vibroflotation without additional backfill treatment. Yantu Lixue/Rock Soil Mech. **29**(1), 37–42 (2008)
17. Robertson, P.K.: Soil classification using the cone penetration test. J. Geotech. Geoenviron. Eng. **27**(1), 984–986 (1990)
18. Liang, X., Chen, S., Xie, X.: Comparative analysis of soil classifications based on CPT in ground improvement. Port Waterw. Eng. **501**(3), 41–46 (2015)
19. Douglas, J.B., Olsen, R.S.: Soil classification using electric cone penetrometer. In: Symposium on Cone Penetration Testing and Experience, Geotechnical Engineering Division, pp. 209–227. ASCE (1981)
20. Massarsch, K.R., Fellenius, B.H.: Deep vibratory compaction of granular soils. Elsevier Geo. Eng. Book Ser. **3**(05), 539–561 (2005)

Curing of Sand Stabilized with Alkali Lignin

Qingwen Yang[1], Chao Zheng[2], and Jie Huang[2(✉)]

[1] Chengdu University of Technology, Chengdu, China
[2] The University of Texas at San Antonio, San Antonio, TX, USA
jie.huang@utsa.edu

Abstract. Lignin and its derivatives from industrial wastes of paper pulp and biofuel production has been used as a soil stabilizer for various applications. This paper discussed the curing process of alkali lignin extracted from paper pulp residuals when it was used treated poorly graded sand (SP). Soil specimens treated by different dosages of lignin were cured in air and heat (i.e., an environment with elevated temperatures), respectively. It was found out that with higher dosage of alkali lignin the unconfined compressive strength (UCS) increased to 5,000 kPa at a lignin dosage of 8% if cured in heat, but increased monotonically with lignin dosage if cured in air. The UCS of air cured specimens showed strong correlation with moisture content, namely, higher initial moisture content led to low UCS. For heat curing, the ultimate UCS was not sensitive to temperature from 30 to 100 °C as it reached the same value given sufficient curing duration. This study overall proves that alkali lignin could be a good soil stabilizer.

1 Introduction

As one of the alternatives for soil improvement, soil stabilizers are often used to improve the engineering properties of the soil at the construction site when they do not meet the design requirements. The popularly used stabilizers generally can be categorized as traditional stabilizers and nontraditional admixtures (ARBA 1976; Das and Sobhan 2013).

Traditional stabilizers, primarily calcium-based chemicals such as lime, cement and fly ash, have been proven effective in improving soil strength and durability, decreasing shrink-swell potential or reducing the construction cost (Tingle et al. 2007). However, under various circumstances, the usage of calcium-based stabilizers becomes prohibitive. For example, due to the reaction between calcium ions and sulfate, these calciumbased stabilizers produce ettrigite (Ca6[Al(OH)6]2(SO4)3·26H20) and thaumasite ($Ca_3Si(OH)_6(CO_3)(SO_4)·12H_2O$) that lead to significant volume increase, generating heave and cracks on the surface of infrastructure (Binal 2016; Ingles and Metcalf 1972; Little 1995; Maaitah 2012; Sherwood 1993). In addition, the production of the cement and lime is energy consuming and releases enormous amount of CO_2 (Gilazghi 2014; Hunter 1988; Little 1995; Talluri et al. 2012).

Q. Yang—Visiting Ph.D. student at University of Texas at San Antonio.

© Springer Nature Singapore Pte Ltd. 2018
L. Li et al. (Eds.): GSIC 2018, *Proceedings of GeoShanghai 2018 International Conference: Ground Improvement and Geosynthetics*, pp. 157–168, 2018.
https://doi.org/10.1007/978-981-13-0122-3_18

To cope with deficiencies of calcium-based soil stabilizers, the possibility of using non-traditional soil stabilizers, for instance, enzymes, petroleum resins, ionic additives, liquid polymers and lignin, has been explored (Imbabi et al. 2012). Among them, lignin is getting popular due to its large quantity of availability and non-toxicity (Bin-Shafique et al. 2017; Katz et al. 2001; Newman et al. 2005; Rauch et al. 2002; Rezaeimalek et al. 2017; Scholen 1995; Tingle et al. 2007). Lignin is the second most abundant biopolymers after cellulose and before hemicellulose in plants, and can be found among the secondary cell wall and middle lamella (Stewart 2008; Yuan et al. 2008). It accounts for 15% to 40% weight of woody plants, and acts as binder which connects adjacent cells together, providing strength to the plant structure, making the plants resist to oxidizing and corrosions (Béguin and Aubert 1994; Lederberg and Joshua 2000; Pérez et al. 2002). The lignin and its derivatives are the major by-products from paper and biofuel industries (Christopher et al. 2014; Doherty et al. 2011; Tolbert et al. 2014). The annually yielded ligneous waste exceeds 30 million tons from both industries, which are usually burned or sent to landfills, adding additional cost to the production of paper and ethanol (Stewart 2008; Thielemans et al. 2002).

Many studies focused on finding valuable applications of lignin in construction. Depending on the chemical pathways, different ligneous wastes were produced during paper pulping and biofuel production, which provides diverse options for soil stabilization. Lignosulfonates in aqueous form has been used independently or together with other chemicals to improve soil shear strength, as it can form a thin layer of film at soil particle surfaces and bond the soil particles when dried (Ouyang et al. 2009). Puppala and Hanchanloet (1999) and Santoni et al. (2002) evaluated the stabilizing performance of both silty sand and clay soil with lignosulfonate and suggested the optimum additive content be 5% by dry weight. Palmer et al. (1995) treated three types of soils by mixture of sulfuric acid and lignosulfonate and found an increase of UCS by 30–130%. Tingle et al. (2007) reported satisfactory results when sulphite lignin was used to stabilize unpaved road. In addition, due to its high solubility in water lignosulfonate solution has used popularly as dust suppression agent in construction. However, the sulfur-free lignin has not been well studied even though it represents a large portion of the ligneous waste. Tingle and Santoni (2003) explored the feasibility of using sulfur-free lignin for soil stabilization and found out that it could yield comparable strength to fly ash treatment. Further study by Ceylan et al. (2010) and Gopalakrishnan et al. (2012a, b) showed that soil that was stabilized by sulfur-free lignin may provide excellent moisture resistance. Considering the missing information from literature on sulfur-free lignin, this study focused on the fundamental applications of alkali-lignin, which includes:

- A general guideline for the water lignin ratio design and curing procedure.
- The optimum quantity of additives under different curing conditions
- The UCS with different dosage of lignin

2   Materials

Alkali lignin
Lignin is a family of complex organic polymers that form the structural materials to support tissues of vascular plants, primarily existing in second wall of wood cells and middle lamella as shown Fig. 1(a). Lignin is made up of three basic precursor units known as p-coumaroyl alcohol, confineryl alcohol and sinapyl alcohol (Fig. 2) (Sadat 2013). These precursors linked through carbon-carbon and carbon-oxygen bonds to form longer molecular chains. Alkali lignin is the residual from paper pulping when NaOH is used, making pH value of the solubilized lignin higher than 7. As a result, the alkali lignin has a higher solubility than lignin and its average molecular weight ranges from 3600 to 30100 g/mol (Carrott and Carrott 2007). The alkaline lignin used in this work, having impurities less than 5%, was a dark brown powder that made directly from paper pulps of wheat straw plus various kinds of woods (Fig. 1(b)).

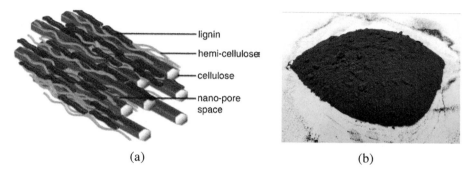

lignin
hemi-cellulose
cellulose
nano-pore space

(a)   (b)

Fig. 1. Lignin: (a) Lignin in wood structure and (b) alkali lignin powder used in this study

Sand
The selected sand (Fig. 3) is classified as poorly graded sand (SP) according to Unified Soil Classification System (USCS) and as A-1-a according to AASHTO classification. The result of sieve analysis is presented in Fig. 4. The sand has the coefficient of uniformity C_u as 1.78 and the coefficient of curvature $C_c = 1$. It has the maximum density of 1794 kg/m^3 and minimum density of 1570 kg/m^3, with an ultimate friction angle of 42°.

3   Methods and Procedures

The experimental study consisted of preparing, curing and testing soil specimens treated by alkali lignin. Since no specific dosage information was available for the alkali lignin as a soil stabilizer, this study varied alkali lignin from 5% to 15% as suggested by previous studies on lignosulfate (Tolbert et al. 2014). Two different

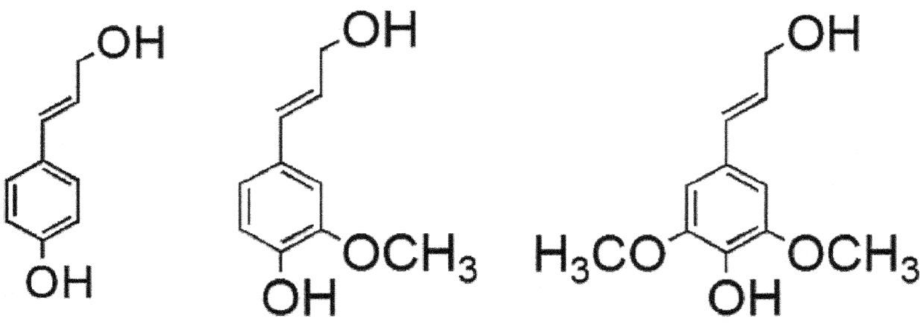

Fig. 2. Three precursors in lignin: (a) p-coumaroyl alcohol (b) confineryl alcohol (c) sinapyl alcohol

Fig. 3. Selected sand used in this study

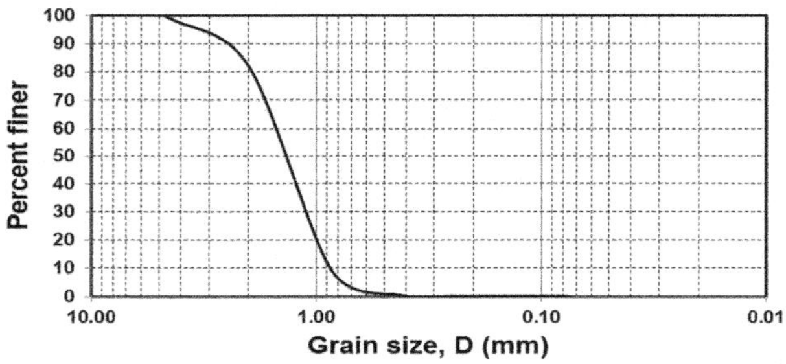

Fig. 4. Particle size distribution for the selected sand

curing environments, i.e., air and heat curing, were investigated in this study. For the air curing specimens, the water content was varied to achieve a water/lignin ratio from 0.3 to 0.9. As to heat curing specimens, water content was varied for specimens with 5% alkali lignin to achieve a water/lignin ration of 0.3, 0.6 and 0.8, respectively, and for the remaining of the specimens the water/lignin ratio was fixed at 0.6. For air curing, the specimens were placed in an environment where the temperature and humidity were regulated at 20 °C and humidity at 60% for up to 28 days, while for heat curing the specimens were placed in a temperature controlled oven. The unconfined compressive strength (UCS) was tested on the universal testing machine under a displacement-control mode. Table 1 summarizes the testing program.

Table 1. Lignin and water combination for specimen used in air and heat curing

Lignin content (%)	Air curing at T = 20 °C		Heating curing		
	Water content (%)	W:L Ratio	Water content (%)	W:L Ratio (water/lignin)	Temperature (°C)
5	1.5, 3.0, 4.0	0.3, 0.6, 0.8	1.5, 3.0, 4.0	0.3, 0.6, 0.8	30, 100
8	2.4, 4.8, 7.2	0.3, 0.6, 0.9	4.8	0.6	100
12	3.6, 6.0, 8.4	0.3, 0.5, 0.7	7.2	0.6	100
15	4.5, 6.8, 9.0	0.3, 0.45, 0.6	9.0	0.6	100

The specimens were compacted in cylindrical PVC molds with a 50 mm in diameter and 127 mm in length, so that the height-to-diameter ratio is 2.5 according to ASTM D2166 (Gopalakrishnan et al. 2012a, b; Santoni et al. 2005; Tingle and Santoni 2003). A thin layer of petroleum jelly was applied to the inner wall of the mold to make the specimen easy to extract. The amount of sand needed for each specimen was calculated by a relative density of 75% (17.32 kN/m^3), which was equivalent to a compaction greater than 90% according to the standard proctor test. The amount of lignin and water varied to achieve different target water/lignin ratios. The sand-lignin mixture was then weighted and compacted into the PVC mold by three equal layers to ensure the uniform density throughout the specimen. After compaction, the cylinders were wrapped by plastic sheets and set in a temperature controlled room for 24 h before extruded from the mold (Fig. 5(a)). Then the specimens were placed in appropriate curing environment for a certain time before tested for unconfined compressive strength (Fig. 5(b)). A set of three specimens were prepared for each test in order to assure repeatability of the results.

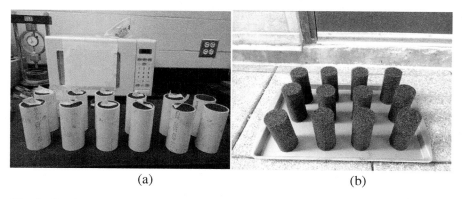

(a)   (b)

Fig. 5. Specimens for testing: (a) specimens prepared in PVC molds (b) cured specimens

4   Test Results and Discussion

Air Curing

The UCS test results for specimens with 5% lignin and 0.3, 0.6, 0.8 water/lignin ratio are presented in Fig. 6. It is obvious that all the specimen cured rapidly. Within 24 h, all the specimen reached more than 90% of their 28-day strength; consequently, it is not absurd to limit the curing of specimen to a day or so. The data also indicates that the UCS decreases with the increase of water/lignin ratio. This phenomenon can be analog to the effect of water-cement ratio on the concrete strength. The combined effect of water content and lignin content on the UCS can be clearly seen in Fig. 7, which shows the 28-day strength of specimens at 5%, 8%, 12% and 15% lignin content with different water contents. The general trend indicates that the strength will increase as the lignin content increases, and the optimum lignin content is plausibly at 12% as the increase of UCS becomes insignificant when the lignin content is great than 12%. Water content shows its effect at all lignin contents investigated: high water content leads to lower USC.

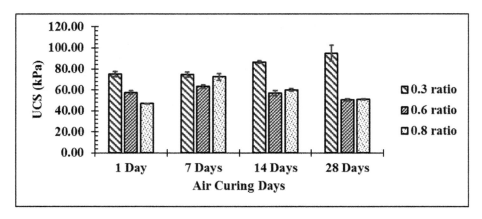

Fig. 6. Strength for 5% lignin treated specimens at various air curing time

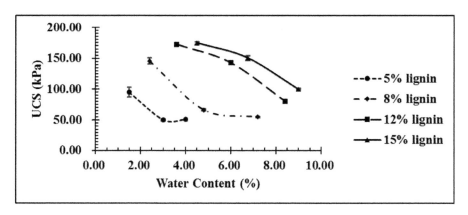

Fig. 7. 28 day strength for specimens treated with 5%, 8%, 12% and 15% lignin content at various water/lignin ratio

Figure 8 illustrates the strong linear correlation between water and lignin content of the cured specimens. This may be attributable to the absorption of lignin itself, which is a property of wood fiber.

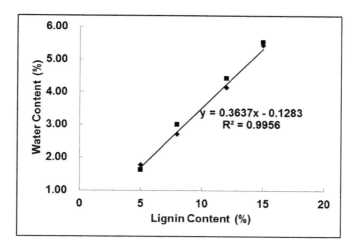

Fig. 8. Correlation between lignin and water content after curing

Heat Curing

The two curing temperatures, i.e., 30 and 100 °C, do not show salient effect on the UCS of the cured specimen. With 5% lignin, the UCS of the specimens for both temperatures was 2,400–2,500 kPa immediate after curing. However, further study disclosed that higher temperature made the curing faster. Under 100 °C, a peak UCS can be reached within 24 h, while it needed nearly three days under 30 °C to reach that level of UCS. However, the UCS decreased dramatically as the specimen cooled down. When the specimens were set in the 20 °C environment for a prolonged period, the UCS achieved a value that is about 30–40% higher than that achieved in air curing as indicated in Fig. 10, which compared the UCS after cooling with UCS of air cured specimens. This figure also manifests that 7-day cooling is needed to reach a constant UCS. Based on Figs. 9 and 10, it can be deduced that heat curing can improve the UCS up to 40%.

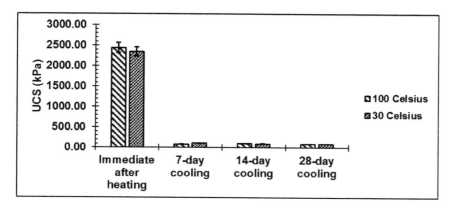

Fig. 9. Heat cured UCS strength for the specimens with 5% lignin

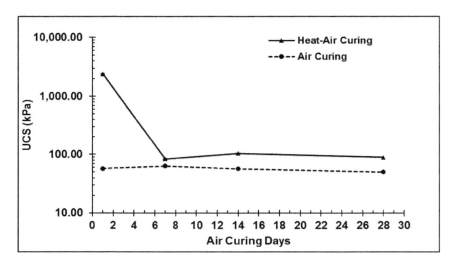

Fig. 10. UCS of heat cured specimen after cooling down vs. UCS air cured specimens (5% lignin)

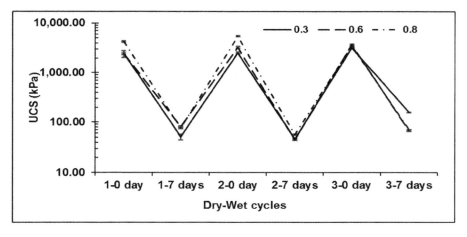

Fig. 11. Effect of heating and cooling cycle on UCS (5% lignin)

The effect of heating and cooling cycles on UCS was investigated by applying heating and cooling cycles. Each cycle included a one-day heat-curing, which was followed by a 7-day cooling. Figure 11 shows the USC at each heating and cooling. Generally, the results from heating-cooling cycles reveals that the UCS can be fully recovered by re-heating the specimen.

The UCS of the specimen cured in heat with 5%, 8%, 12% and 15% lignin content is compared in Fig. 12, which marks the UCS as a function of lignin content. The UCS reached maximum value of 5000 kPa at 8% lignin content and started to decline thereafter. With higher lignin content certainly, stronger bonds can be form between soil particles; however, when the lignin content becomes too high the pore fluid (i.e., lignin+water) is excessive and starts to bleed out of the soil skeleton as shown in Fig. 13. As a result of the bleeding, the UCS of the specimen with 15% lignin is less than 10% of that of the specimen with 8% lignin. The bleeding was not observed for the same specimen if cured in air. This discrepancy is attributable to the viscosity of aqueous lignin, which significantly decreases with the increase of temperature. Thus, it should be cautious to use high lignin content if the heat curing method be adopted.

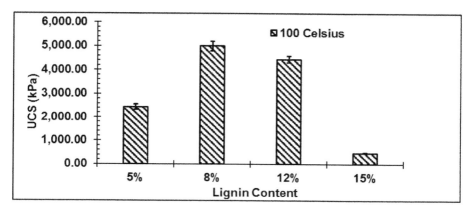

Fig. 12. UCS of specimens stabilized with 5%, 8%, 12% and 15% lignin content and 0.6 water/lignin ratio after 100 °C heat curing

Fig. 13. Lignin bleeding of 15% lignin specimen

5 Conclusions

This study used alkali lignin as soil stabilizer to treat the coarse sand. The performance was evaluated under different lignin contents, water contents, curing conditions and duration. The major findings are summarized as follow:

- Lignin stabilized sand cures quickly in air or heat. Generally, with various lignin contents the stabilized specimen can be cured within a week or so in the air, and within 24 h if cured in elevated temperature.
- For air curing, the UCS of the cured specimen increases with the increase of lignin content but decreases with the increase of water content.
- Heat curing can increase the stabilized soil strength significantly. The maximum UCS is 5,000 kPa if the specimen is tested immediately after heat curing. The UCS decreases gradually as the specimen cools down to room temperature and reach a steady value that is about 30–40% higher than what is achieved in air curing. Further study also indicates that the decrease of UCS due to cooling can be fully recovered if the specimen is reheated.
- Heat curing may lead to significant bleeding if high lignin content is used, which results in more than 90% UCS decrease. Thus, it should be cautious to use high lignin content if heat curing will be used.

In summary, the obtained results suggest that the heat curing not only increase the ultimate UCS but also shorten the curing duration.

References

ARBA: Materials for Stabilization. American Road Builders Association (ARBA), Washington, D.C. (1976)

Béguin, P., Aubert, J.: The biological degradation of cellulose. FEMS Microbiol. Rev. **13**(1), 25–58 (1994)

Bin-Shafique, S., Gupta, S.D., Huang, J., Rezaeimalek, S.: The effect of fiber type and size on the strength and ductility of fly ash and fiber stabilized fine-grained soil subbase. In: Geotechnical Frontiers, pp. 19–29 (2017)

Binal, A.: The effects of high alkaline fly ash on strength behaviour of a cohesive soil. In: Advances in Materials Science and Engineering (2016)

Carrott, P., Carrott, M.R.: Lignin–from natural adsorbent to activated carbon: a review. Bioresour. Technol. **98**(12), 2301–2312 (2007)

Ceylan, H., Gopalakrishnan, K., Kim, S.: Soil stabilization with bioenergy coproduct. Transp. Res. Rec. J. Transp. Res. Board **2186**, 130–137 (2010)

Christopher, L.P., Yao, B., Ji, Y.: Lignin biodegradation with laccase-mediator systems. Front. Energy Res. **2**, 12 (2014)

Das, B.M., Sobhan, K.: Principles of Geotechnical Engineering. Cengage Learning, Bostan (2013)

Doherty, W.O., Mousavioun, P., Fellows, C.M.: Value-adding to cellulosic ethanol: Lignin polymers. Ind. Crops Prod. **33**(2), 259–276 (2011)

Gilazghi, S.T.: Unconfined compressive strength and free swelling potential of polymer stabilized sulfate bearing soils (2014)

Gopalakrishnan, K., Ceylan, H., Kim, S.: Impact of bio-fuel co-product modified subgrade on flexible pavement performance, pp. 1505–1512 (2012a)

Gopalakrishnan, K., Ceylan, H., Kim, S.: Moisture susceptibility of subgrade soils stabilized by lignin-based renewable energy coproduct. J. Transp. Eng. **138**(11), 1283–1290 (2012b)

Hunter, D.: Lime-induced heave in sulfate-bearing clay soils. J. Geotechn. Eng. **114**(2), 150–167 (1988)

Imbabi, M.S., Carrigan, C., McKenna, S.: Trends and developments in green cement and concrete technology. Int. J. Sustain. Built Environ. **1**(2), 194–216 (2012)

Ingles, O.G., Metcalf, J.B.: Soil Stabilization Principles and Practice. Butterworth-Heinemann Ltd., Sydney (1972)

Katz, L., Rauch, A., Liljestrand, H., Harmon, J., Shaw, K., Albers, H.: Mechanisms of soil stabilization with liquid ionic stabilizer. Transp. Res. Rec. J. Transp. Res. Board **1757**, 50–57 (2001)

Lederberg, J.: Encyclopedia of Microbiology, Four-Volume Set. Academic Press, San Diego (2000)

Little, D.N.: Stabilization of Pavement Subgrades and Base Courses with Lime. Kendall Hunt Publishing, Dubuque (1995)

Maaitah, O.N.: Soil stabilization by chemical agent. Geotech. Geol. Eng. **30**(6), 1345–1356 (2012)

Newman, J.K., Tingle, J.S., Gill, C., McCaffrey, T.: Stabilization of silty sand using polymer emulsions. Int. J. Pavements **4**(1–2), 1–12 (2005)

Ouyang, X., Ke, L., Qiu, X., Guo, Y., Pang, Y.: Sulfonation of alkali lignin and its potential use in dispersant for cement. J. Dispers. Sci. Technol. **30**(1), 1–6 (2009)

Palmer, J.T., Edgar, T.V., Boresi, A.P.: Strength and density modification of unpaved road soils due to chemical additives (1995)

Pérez, J., Munoz-Dorado, J., de la Rubia, T., Martinez, J.: Biodegradation and biological treatments of cellulose, hemicellulose and lignin: an overview. Int. Microbiol. **5**(2), 53–63 (2002)

Puppala, A., Hanchanloet, S.: Evaluation of a new chemical treatment method on strength and resilient properties of a cohesive soil. Paper No. 990389. Transportation Research Board (1999)

Rauch, A., Harmon, J., Katz, L., Liljestrand, H.: Measured effects of liquid soil stabilizers on engineering properties of clay. Transp. Res. Rec. J. Transp. Res. Board **1787**, 33–41 (2002)

Rezaeimalek, S., Huang, J., Bin-Shafique, S.: Performance evaluation for polymer-stabilized soils. Transp. Res. Rec. J. Transp. Res. Board **2657**, 58–66 (2017)

Sadat, M.R.: The effect of upward and downward movement on the behavior of MSE wall, The University of Texas at San Antonio, ProQuest Dissertations & Theses Global (1442788104) (2013)

Santoni, R.L., Tingle, J.S., Nieves, M.: Accelerated strength improvement of silty sand with nontraditional additives. Transp. Res. Rec. J. Transp. Res. Board **1936**, 34–42 (2005)

Santoni, R.L., Tingle, J.S., Webster, S.L.: Stabilization of silty sand with nontraditional additives. Transp. Res. Rec. J. Transp. Res. Board **1787**, 61–70 (2002)

Scholen, D.E.: Stabilizer mechanisms in nonstandard stabilizers. In: Proceedings of Transportation Research Board Conference Proceedings. TRB, National Research Council, Washington, D.C., pp. 252–260 (1995)

Sherwood, P.: Soil Stabilization with Cement and Lime. H.M. Stationery Office, London (1993)

Stewart, D.: Lignin as a base material for materials applications: chemistry, application and economics. Ind. Crops Prod. **27**(2), 202–207 (2008)

Talluri, N., Gaily, A., Puppala, A.J., Chittoori, B.: A comparative study of soluble sulfate measurement techniques. In: GeoCongress 2012: State of the Art and Practice in Geotechnical Engineering, pp. 3372–3381 (2012)

Thielemans, W., Can, E., Morye, S.S., Wool, R.P.: Novel applications of lignin in composite materials. J. Appl. Polym. Sci. **83**(2), 323–331 (2002)

Tingle, J.S., Newman, J.K., Larson, S.L., Weiss, C.A., Rushing, J.F.: Stabilization mechanisms of nontraditional additives. Transp. Res. Rec. J. Transp. Res. Board **1989**, 59–67 (2007)

Tingle, J.S., Santoni, R.L.: Stabilization of clay soils with nontraditional additives. Transp. Res. Rec. J. Transp. Res. Board **1819**(2), 72–84 (2003)

Tolbert, A., Akinosho, H., Khunsupat, R., Naskar, A.K., Ragauskas, A.J.: Characterization and analysis of the molecular weight of lignin for biorefining studies. Biofuels, Bioprod. Biorefin. **8**(6), 836–856 (2014)

Yuan, J.S., Tiller, K.H., Al-Ahmad, H., Stewart, C.N., Stewart, N.R.: Plants to power: bioenergy to fuel the future. Trends Plant Sci. **13**(8), 421–429 (2008)

Combined Encased Stone Column and Vacuum Consolidation Technique for Soft Clay Improvement

Ganesh Kumar[1,2(✉)]

[1] CSIR-Central Building Research Institute, Roorkee 247667, India
`85sganesh@gmail.com`, `ganeshkumar@cbri.res.in`
[2] AcSIR, Geotechnical Engineering Division,
CSIR-Central Building Research Institute, Roorkee 247667, India

Abstract. Soft clay improvement using stone column is one of the popular ground improvement techniques adopted for soft clay deposits. The load carrying capacity can be further improved by encasing it with geosynthetic material around the column which offers effective lateral confinement, prevents column contamination with improved drainage and strength characteristics. The load carrying capacity can be further increased by improving the undrained shear strength of surrounding soil. Use of vacuum consolidation is a well-developed hydraulic modification technique which involves application of vacuum through prefabricated vertical drains in a sealed membrane system. The improvement in effective stress during vacuum application resulted in improved undrained shear strength of surrounding soil. In this research work, combined geosynthetic encased stone column and vacuum treatment method is examined for improving of extremely soft clay soils. Small scale unit cell tests were performed to evaluate the efficiency of encased stone column subjected vacuum loading under different pressure levels. The results were then compared with that of ordinary surcharge preloading conditions. It was observed that both time, rate of consolidation and undrained shear strength of soil increases significantly with vacuum application. With this reduction in time required for consolidation and improvement in undrained shear strength, the construction activity can be accelerated which makes the project economically beneficial.

Keywords: Encased stone columns · Vacuum consolidation
Ground improvement · Soft soils

1 Introduction

The rapid development and associated urbanization have compelled the engineers to carry out their projects, over soft clay deposits of low bearing capacity and high compressibility characteristics. In such conditions ground improvement technique plays a major role in improving the engineering characteristics of foundation soils. Ground improvement using stone columns is an effective soil reinforcement technique in the case of soft clay soils for improving bearing capacity and for reducing total and differential settlements [7, 11]. However, stone columns when installed in soft clay

© Springer Nature Singapore Pte Ltd. 2018
L. Li et al. (Eds.): GSIC 2018, *Proceedings of GeoShanghai 2018 International Conference: Ground Improvement and Geosynthetics*, pp. 169–177, 2018.
https://doi.org/10.1007/978-981-13-0122-3_19

deposits, there may be chances for inadequate lateral confinement from the surrounding soil and entry of finer clay materials inside the column which resulted in column contamination and failure in load carrying capacity [10]. To overcome these limitations, encasement using geosynthetic material was adopted around the column which offers better lateral confinement, prevents intrusion of finer clay particle inside the column and provides effective drainage path with improved load bearing capacity characteristics [2, 6, 12, 14, 15, 17]. The performance of geosynthetic encased stone columns can be further increased by improving the undrained shear strength of surrounding soil. Soft soil improvement using hydraulic modification is a type of treatment technique where consolidation of the soil is achieved by preloading the virgin soil. The applied surcharge removes excess pore water pressure inside the soil grains. The reduction in pore water pressure leads to the improvement in effective stress which increases the undrained shear strength of surrounding soil. Initially the surcharge preloading was performed with sand drains and later it was modified with prefabricated vertical drains. Provision of these vertical drains enhances the horizontal drainage and facilitates the dissipation of pore water pressure from the soft clay soils [1]. Consolidation using prefabricated vertical drains was then modified with vacuum consolidation [9]. In vacuum consolidation, vacuum pressure is applied through prefabricated vertical drains in a sealed membrane system. The applied vacuum creates negative pore pressure in the soil which increases the effective stresses without altering the total stresses. The increase in the effective stresses leads to compression of soil thus improving the shear strength and stiffness of the virgin soil [3–5, 8, 13, 16].

In this research work, soft clay improvement by combined geosynthetic encased stone column subjected to vacuum treatment was performed. The encased stone column installed inside the small scale unit cell was subjected to varying vacuum pressure level to evaluate the strength and time-rate of consolidation behaviour. For comparison, tests also performed under surcharge loading with similar pressure levels. Encased stone column subjected to vacuum loading achieved maximum settlement compared to conventional surcharge conditions. It was also observed that application of vacuum improves the shear strength of surrounding soil which additionally increases the load carrying capacity of encased stone columns.

2 Description of Experiments

2.1 Material Used and Sample Preparation

The clay soil used for the study was a fine grained silty clay soil obtained from the lake bed inside the campus of Indian Institute of Technology Madras, Chennai, India. The liquid limit and plastic limit of the soil were determined as 44% and 20%, respectively. The soil contains 20% of sand. The silt size and clay sized particles were 31% and 41%, respectively. The stone aggregates used to form the stone columns were angular granite chips of size 2 to 10 mm and having uniform gradation. The stone aggregate in all the tests was compacted to a dry unit weight of about 16 kN/m^3. The peak angle of internal friction of stone aggregates was determined as 42° using direct shear box of 300 mm × 300 mm size. Woven type geotextile was used for the study for encasing the stone columns. The tensile strength of the geotextile was determined by conducting Wide Width Tensile strength test. The tensile strength was found to be 56 kN/m at a strain level of 12%. The thickness of the textile and the Apparent Opening Size are 0.5 mm and 0.3 to 0.425 mm, respectively.

The clay soil collected from the lake bed was initially soaked in water and then mixed thoroughly using a stirrer at water content equal to 1.5 times the liquid limit of the soil. This was done to completely remove the previous stress history. The slurry was then poured inside the unit cell at equal intervals. Then the slurry was allowed to consolidate by applying loads in stages. Drainage was permitted at the top and bottom during the consolidation. The constant consolidation pressure of required magnitude was applied using pneumatic loading system. The initial consolidation pressure applied was 10 kPa (representing ultra-soft clay soil). Time–settlement data were recorded continuously in order to evaluate the completion of consolidation. After consolidation, soil samples were collected for performing vane shear strength tests and water content measurements. The water content and shear strength values obtained after slurry consolidation was found to be 48 ± 1% and 2.5 kPa respectively. The clay bed was prepared in the same manner for all the tests in order to maintain uniformity between the tests.

2.2 Unit Cell Arrangement and Installation Procedure for Encased Stone Column

Usually the stone columns are installed in triangular or square pattern in such a manner that each column influences certain area of soil, called unit cell [IS: 15284 (Part1):2003]. As unit cell is representative of an area treated with stone columns, it was decided to carry out unit cell experiments in the present study. The diameter and height of the unit cell used for the study are 150 mm and 160 mm, respectively, using the conventional CBR mould. The experiments were carried out to evaluate the efficiency of vacuum under different pressure levels through tests and compared with that of surcharge preloading conditions. An encased stone column of diameter 50 mm diameter was selected and installed at the centre of the unit cell, after consolidating the clay from slurry from a water content of 1.5 times the liquid limit water content.

The encased stone column was installed by displacement method using a casing pipe having an outer diameter equal to the diameter of the stone column and encasement was done by wrapping the geotextile around the casing pipe. The edges of the geotextile were stitched to have maximum seam strength and provided with small base plate, having the diameter equal to that of stone column, as an anchor at the bottom to hold the geotextile around the casing pipe. The casing pipe with the geotextile was then carefully pushed in to the consolidated clay soil bed through a guide plate in the unit cell vertically at the centre till it reaches the bottom. Only static force was used to carefully push the casing pipe at a very slow rate with minimal disturbance to the surrounding clay soil. The soil displaced during the installation of the tube was taken out of the tank at periodic intervals. The soil surface was leveled after the installation was completed. The quantity of the stone aggregate required to form the stone column was measured a priori and charged into the casing pipe in layers of 50 mm thickness. The stone aggregate was moistened prior to charging into the column to prevent it from absorbing moisture from the surrounding soil. After placing each layer of stone aggregate, the casing pipe was lifted up gently to a height such that a minimum overlap of 15 mm between the bottom of the casing pipe and the stone fill within the casing pipe was maintained. The stone aggregate was compacted with a tamping rod (10 mm diameter and 1 m long) with 25 blows falling from a height of 250 mm. Each layer of the aggregate was compacted using the same number of blows. The method of compaction gave a dry unit weight of aggregate of approximately 16 ± 0.2 kN/m^3 which corresponds to a relative density of about 73% representing field installation conditions. Details of soil sample preparation and schematic view of mould used for the experimental study is shown in Fig. 1.

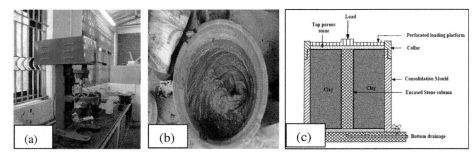

Fig. 1. Sample preparation and CBR mould used (a) Rotary motor used for slurry preparation (b) Clay slurry for consolidation (c) Schematic diagram of CBR mould used for the experiments

2.3   Vacuum Preloading

For vacuum preloading, the unit cell arrangement was modified with a special arrangement for applying vacuum through the bottom of the mould. After the installation of encased stone column, the top of the clay soil surface with the stone column was carefully leveled and a rubber sheet of 150 mm diameter was placed on it. Two settlement plates placed above the rubber sheet. The top of the clay surface was made air-tight by pouring bentonite slurry on the rubber sheet. The bottom of the stone column was connected to the vacuum pump through a vacuum chamber and a vacuum regulator. The arrangement adopted for applying is shown in Fig. 2(a). Required level of vacuum pressure was applied by the vacuum regulator to the soil through the encased stone column for the consolidation to take place. Three different pressure levels of 20 kPa, 40 kPa and 80 kPa were selected for determining the time-Settlement response and ultimate settlement.

2.4   Surcharge Preloading

In surcharge preloading, the unit cell with encased stone column was loaded with a loading plate having diameter equal to that of unit cell diameter of 150 mm. The mould was then placed in a modified consolidation loading frame and subjected to consolidation pressure levels of 20 kPa, 40 kPa and 80 kPa similar to that of vacuum preloading pressure conditions. During loading, drainage was allowed at both top and bottom. The photographic view of the loading arrangement is shown in Fig. 2(b). During loading, the time settlement data was continuously recorded.

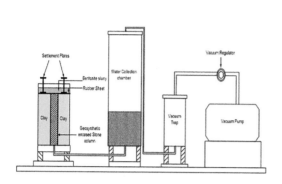 

Fig. 2. Loading conditions 2(a) Schematic diagram of vacuum consolidation test set-up 2(b) Photographic view of surcharge loading arrangement

3   Results and Discussions

3.1   Time-Settlement Response

The time-settlement response for 50 mm diameter stone column subjected to 20 kPa, 40 kPa and 80 kPa vacuum pressure level was presented in Fig. 3. For better comparison, unit cell subjected to corresponding pressure levels under surcharge loading conditions also presented. It can be understood that encased stone column subjected vacuum preloading found to achieve more settlements than surcharge loading for the same time intervals. The ratio of maximum consolidation settlement obtained in case of vacuum preloading was about 1.25 to 1.67 times higher than that of surcharge preloading in all three pressure levels.

Under vacuum preloading, the effective stress gets transferred to the soil mass and because of the negative pressure levels applied to the pore phase, higher settlements were achieved. Whereas in case of surcharge loading, the applied load was partly taken by the column and partly taken by surrounding soil which resulted in minimized settlements. The distribution of vacuum pressure was further verified by taking water content values at different locations i.e., top, middle and bottom after vacuum preloading and the same compared with that of surcharge loading conditions. For vacuum preloading, the water content values were found to be 37.8% to 38.4% from top to bottom and were approximately constant with depth. In case of surcharge loading, the obtained water content values were 40.6% for top and 42.4% for middle and 44.6% for bottom. This uniformity in water content values for vacuum preloading

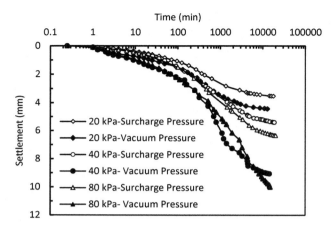

Fig. 3. Comparison of time-settlement graph for vacuum and surcharge preloading

proved that vacuum applied uniformly inside the system causing maximum settlements. In surcharge preloading, the water content increases with depth as the effect of surcharge pressure decreases with depth. Another reason could be the stone column carries some part of the applied surcharge thus preventing the clay soil from feeling the excess pressures.

3.2   Consolidation Response After Vacuum Treatment

To understand the consolidation behaviour of vacuum treated encased stone columns, consolidation tests were performed on 80 kPa vacuum treated sample. The diameter of the encased stone column adopted for this study was 50 mm. After installation of 50 mm diameter encased stone column, vacuum pressure of 80 kPa was applied at the bottom of the unit cell and time-settlement readings were recorded continuously. After completion of consolidation, the surface of the soil was leveled and placed inside the modified consolidation loading frame (Fig. 2(b)). Consolidation test was then performed with pressure levels of 25 kPa, 50 kPa, 100 kPa and 200 kPa. For comparison, control tests were also performed on clay soil without stone columns, with stone columns and clay with geosynthetic encased stone columns. The variation of vertical strain with consolidation pressure is shown in Fig. 4. Encased stone columns performed better than the corresponding ordinary stone column. The vacuum consolidated clay samples have undergone lower strains when compared with the clay soil consolidated under surcharge loading. The observed lower strains for vacuum consolidated soil showed that maximum consolidation settlement was achieved during vacuum application and confirms both strength and stiffness of the surrounding soil increased after vacuum application. The increase in undrained shear strength of surrounding soil due to vacuum application provides improved lateral confinement to stone column in addition to that of geosynthetic encasement which resulted in improved load carrying capacity of encased stone column.

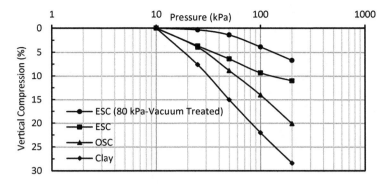

Fig. 4. Comparative graph for vertical compression - pressure response for different systems

3.3   Improvement in Shear Strength

To evaluate improvement in undrained shear strength after vacuum loading, cylindrical clay soil samples of 38 mm diameter and 76 mm height (3 numbers) were collected from the unit cell after 80 kPa vacuum loading for conducting unconfined compression test. The sample was collected by statically pushing a sampling tube of 38 mm diameter and 100 mm long without much disturbance. The compression test was conducted on the sample at a rate of 0.625 mm deformation per minute. Experimental test results shows that the cohesive strength of the clay soil after vacuum treatment is increased to 19.68 kPa from an initial strength of 2.5 kPa (obtained after initial con-solidation using surcharge loading of 10 kPa). In all the samples the value of shear strength almost constant and found that the strength increases about 0.25 times the applied pressure [6].

3.4   Predicted Improvement in Load Carrying Capacity of Stone Columns

The limiting axial stress for the case of ordinary stone column is given by (IS: 15284-Part-1:2003)

$$\sigma_v = \left(\sigma'_{r0} + 4c_u \right) \times K_{pcol} \tag{1}$$

Where, σ_v – limiting axial stress in column, σ'_{r0} – Initial effective stress (= $K_o \times \sigma_{vo}$ where K_o – coefficient of lateral pressure and $\sigma_{vo} = \gamma' \times 2 \times D$, ($\gamma'$ – Effective unit weight and D – Bulge depth, two times diameter of stone column), c_u - undisturbed undrained shear strength of surrounding soil, K_{pcol} – coefficient of passive earth pressure of stone columns, $\tan^2\left(45^o + \frac{\varphi_c}{2}\right)$ where φ_c angle of internal friction of granular material.

Considering diameter of stone column as 50 mm, angle of internal friction as 42° and assuming Ko - 0.6, γ' – 14 kN/m^3. The limiting axial stress in stone column can be predicted using Eq. 1, for the obtained undrained shear strength values.

When the undrained shear strength of soft clay before treatment is 2.5 kPa, the predicted limiting axial stress of stone column is 58.91 kN/m^2 and calculated yield load of stone column is 115 N. The same soft clay soil when treated with vacuum pressure of 80 kPa applied through encased stone column, the undrained shear strength of soft clay increases to 19.68 kN/m^2 and substituting the value in Eq. (1), the limiting axial stress of stone column is found to be 405.5 kN/m^2 and the yield load of stone column is 796 N. The improvement in load carrying capacity is about 6.9 times higher than that of untreated soil and maximum consolidation settlement is achieved within that shorter time intervals additionally. The above values were calculated without considering the effect of geosynthetic reinforcement which additionally contributing the improvement in load carrying capacity of encased stone columns. The predicted values proved that the strength and stiffness of the column can be improved when there is an improvement in the strength of surrounding soil. This will resulted in reduction in number of stone columns actually required for ground improvement and makes system cost effective and economically beneficial.

4   Conclusions

Based on the experiments carried out on encased stone columns subjected to vacuum loading, the following conclusions are drawn

1. The shear strength of the soil increases significantly with vacuum application. The value of shear strength increases about 0.25 times the applied vacuum pressure
2. The application of vacuum through encased stone columns is able to distribute the vacuum pressure evenly to the clay soil to cause rapid consolidation settlement.
3. The load carrying capacity of encased stone column increases with the improvement in undrained shear strength of surrounding soil along with the provision of geosynthetic encasement.
4. Application of vacuum through encased stone column can be viable ground improvement technique for improvement of soft clays. The c/c spacing between the columns can be increased due to the improvement in shear strength after vacuum treatment which resulted in reduction in number of stone columns actually required for treatment and makes the system cost effective.

Acknowledgement. This investigation was performed as part of a sponsored research project supported by the Department of Science and Technology of Government of India under sanction number SR/S3/MERC/065/2007, dated 19-03-2009.

The author sincerely acknowledges his guides **Prof. K. Rajagopal** and **Prof. R. G. Robinson**, Civil Engineering Department, Indian Institute of Technology Madras for their mentorship and continuous support during his research period at IIT Madras.

The author would like to thank Director, CSIR-Central Building Research Institute, Roorkee for giving permission to publish this research work.

References

1. Barron, R.A.: Consolidation of fine-grained soils by drain wells. Trans. Am. Soci. Civ. Eng. **113**(2346), 718–724 (1948)
2. Brokemper, D., Sobolewski, J., Alexiew, D., Brok, C.: Design and construction of geotextile encased columns supporting geogrid reinforced landscape embankments: Bastions Vijfwal Houten in the Netherlands. In: Proceedings of the 8th International Conference on Geosynthetics, Yokohama, Japan, pp. 889–892 (2006)
3. Chai, J.C., Carter, J.P., Hayashi, S.: Vacuum consolidation and its combination with embankment loading. Can. Geotech. J. **43**(10), 985–996 (2006)
4. Chu, J., Yan, S.W., Yang, H.: Soil improvement by vacuum preloading method for an oil storage station. Geotechnique **50**, 625–632 (2000)
5. Cognon, J.M., Juran, I., Thevanayagam, S.: Vacuum consolidation technology – principles and field experience. In: Yeung, A.T., Félio, G.Y. (eds.) Vertical and Horizontal Deformations of Foundations and Embankments, vol. 40, no. 2, pp. 1237–1248. Geotechnical Special Publications, ASCE (1994)
6. Ganesh Kumar, S., Robinson, R.G., Rajagopal, K.R.: Improvement of soft clays by combined vacuum consolidation and geosynthetic encased stone columns. Indian Geotech. J. **44**(1), 59–67 (2014)
7. Greenwood, D.A.: Mechanical improvement of soils below ground surface. In: Ground Engineering, Proceedings of the Conference by the Institution of Civil Engineers, London (1970)
8. Indraratna, B., Bamunawita, C., Khabbaz, H.: Numerical modeling of vacuum preloading and field applications. Can. Geotech. J. **41**(6), 1098–1110 (2004)
9. Kjellman, W.: Consolidation of clay soil by means of atmospheric pressure. In: Proceedings of Conference on Soil Stabilization, pp. 258–263. MIT, Cambridge (1952)
10. McKenna, J.M., Eyre, W.A., Wolstenholme, D.R.: Performance of an embankment supported by stone columns in soft ground. Geotechnique **25**(1), 51–59 (1975)
11. Mitchell, J.K., Huber, T.R.: Stone column foundations for a waste water treatment plant – a case history. Geotech. Eng. **14**, 165–186 (1983)
12. Murugesan, S., Rajagopal, K.: Studies on the behavior of single and group of geosynthetic encased stone columns. J. Geotech. Geoenviron. Eng. **136**(1), 129–139 (2010)
13. Qiu, Q.C., Mo, H.H., Dong, Z.L.: Vacuum pressure distribution and pore pressure variation in ground improved by vacuum preloading. Can. Geotech. J. **44**(12), 1433–1445 (2007)
14. Raithel, M., Kirchner, A., Schade, C., Leusink, E.: Foundation of constructions on very soft soils with geotextile encased columns - state of the art. Geotechnical Special Publication, No. 130-142, pp. 1867–1877 (2005)
15. Short, R.D, Prashar, Y., Metcalf, B.: Repairing railway spur roadbed failure using geotextile encased columns. In: Transportation Research Board. 84th Annual Meeting Compendium of Papers, Washington, D.C (2005)
16. Tang, M., Shang, J.Q.: Vacuum preloading consolidation of Yaoqiang airport runway. Geotechnique **50**(6), 613–623 (2000)
17. Van Impe, W., Silence, P.: Improving of the bearing capacity of weak hydraulic fills by means of geotextiles. In: Proceedings of the 3rd International Conference on Geotextiles, Vienna, Austria, pp. 1411–1416 (1986)

Numerical Simulation of Bearing Capacity and Consolidation Characteristics of PHC Pile Foundation

Shen Gong[1], Guojun Cai[1(✉)], Songyu Liu[1], and Anand J. Puppala[2]

[1] Institute of Geotechnical Engineering, Southeast University,
Nanjing, Jiangsu, China
SEUGS1992@163.com, focuscai@163.com
[2] Department of Civil Engineering, The University of Texas at Arlington,
Arlington, TX, USA

Abstract. This paper investigates the bearing capacity and consolidation characteristics of PHC pile foundation through numerical simulation method. The numerical simulation results greatly shows the process of field static load test and the soil deformation characteristic around the pile, and the ultimate bearing capacity calculated by finite element method has a good agreement with the measured value. The numerical simulation method is used to study the consolidation behavior of composite foundation in one year after bearing load. The excess pore pressure dissipation regular, displacement of pile side soil, and pile stress distribution are systematically studied by numerical simulation method. The result shows that the modified Cam-clay model can greatly simulate the mucky soil consolidation characteristic especially the creep properties. And the consolidation settlement also has great agreement with the real value. This study proves the feasibility of the numerical simulation method in the study of the bearing capacity and consolidation of the pile foundation.

Keywords: Numerical simulation · PHC pile · Bearing capacity
Consolidation characteristics

1 Ultimate Bearing Capacity of PHC Pile

PHC is an acronym for pre-stressed high strength concrete pipe piles. Numerical simulation is widely used in pile bearing capacity and consolidation analysis [1–8]. Two-dimensional model is used in numerical analysis in this research. A PHC pile with 0.4 m diameter and 17 m length penetrates two layers, the first layer is sand with the thickness of 2 m, and the second layer is marine clay with the thickness of 32 m. Finite element model is shown in Fig. 1. The pile side soil of the model is 10 times of the PHC in order to show the lateral deformation and stress status in different situation. In bearing capacity analysis process, the whole foundation soil is based on the Mohr Coulomb model. The parameters of Mohr Coulomb model are summarized in chart 1 (Table 1).

© Springer Nature Singapore Pte Ltd. 2018
L. Li et al. (Eds.): GSIC 2018, *Proceedings of GeoShanghai 2018 International Conference: Ground Improvement and Geosynthetics*, pp. 178–185, 2018.
https://doi.org/10.1007/978-981-13-0122-3_20

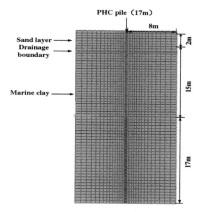

Fig. 1. Finite element model

Table 1. The parameters of Mohr Coulomb

Soil	E/kPa	υ	c/kPa	$\varphi/°$	$\Psi/°$
Sand	24500	0.35	0.1	34	0.1
Marine clay	8000	0.4	23	17	0.1

1.1 Research on Stress and Deformation of Pile and Soil in Static Load Test

The load values of the field static load test are: 50 kN; 350 kN; 650 kN; 950 kN and 1250 kN. The numerical simulation method is used to study the stress and soil deformation characteristics of the pile during static load test. Figure 2 shows the plastic deformation of soil at pile top and pile end in the loading process. When the final load is applied, soil around both of the pile top and end appears plastic deformation, which means the load exceed the ultimate bearing capacity and soil around the pile end appears plastic damage. S_{22} (Fig. 3) shows the vertical stress status of the soil in the final load, and greater stress concentration appears in soil around the end of the pile.

When the load is under the ultimate bearing capacity, the vertical stress of the soil changes linearly with the load and the soil is in elastic deformation stage. While the load exceed the ultimate bearing capacity, S_{22} increases rapidly while the pile occurs rapid piercing deformation and the soil around the pile end start to damage. It is shown in Fig. 4 that the Q-S curve of the numerical simulation is in good agreement with the Q-S curve of the static load test, which proves the accuracy of the numerical simulation. The ultimate bearing capacity value calculated by numerical simulation is 995 kN, which only has the error of 5.3% compared with the true value 1050 kN calculated by standard method. Figure 5 shows the change of axial force of the PHC pile along depth under different load conditions. The axial force of the pile is nonlinearly reduced along the depth, and the greater the load, the greater range of axial force changes, which is consistent with the actual result.

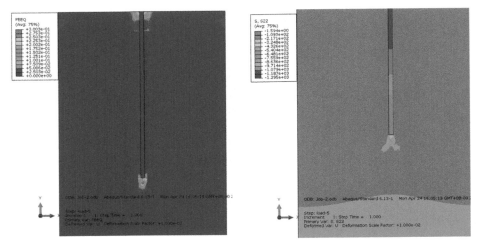

Fig. 2. Plastic deformation of soil   **Fig. 3.** S_{22} of the foundation

1.2   Research on Working Behavior of PHC Pile Under Design Load

Because of the marine clay in the site is very deep (the average thickness is over 80 m), the pile foundation can not penetrate the whole clay layer, and the PHC pile bearing process can be considered as undraining in a short time. Based on this condition, 1800 s time period was set to simulate the un-drained process and the excess pore pressure would not dissipate in this process.

Figure 6 is the lateral deformation of pile soil when it is under the design load 900 kN. Numerical simulation result shows that squeezing soil effect occurs when the design load is applied, and the squeezing effect zone is similar to spherical distribution which means that the closer to the pile end, the more obvious squeezing effect is.

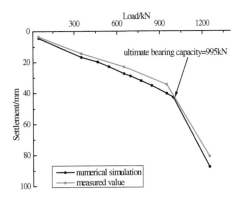

Fig. 4. Q-S curve

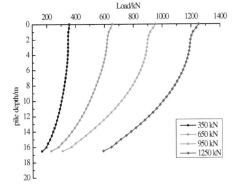

Fig. 5. The variation of pile axial force in different load

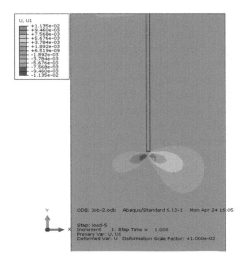

Fig. 6. Lateral deformation of pile soil

In fact, the squeezing effect is more obvious in the process of pile sinking and has effect on pile working process. Figure 7 is the differential settlement between the pile and soil, the differential settlement on pile-soil interface. Generally speaking, the pile-soil differential sedimentation gets smaller with depth,and the minimum value locates in the pile end. The results of finite element analysis show that the differential settlement at the pile ends suddenly becomes smaller. This is due to the plastic deformation of the soil around the pile end shows a tendency of release between pile and soil, which is realized through the separation among the element nodes in numerical simulation. In fact, this tendency is the reflection of the pile piercing deformation. On the other hand, the differential settlement in sand layer is bigger than the marine clay layer. This is due to the pile axial force in sand layer is greater than it in the marine clay layer, which is shown in Fig. 5.

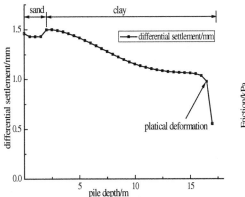

Fig. 7. Differential settlement

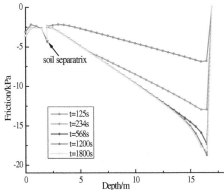

Fig. 8. Pile side resistance changes

The variation of pile side friction is consistent with the theory, which is generally liner distribution in different time (Fig. 8). The maximum friction is 19.6 kPa at t = 1200 s, and this value is almost constant in the subsequent loading process.

Figures 9 and 10 shows the excess pore pressure contour at pile end. This excess pore pressure is relative value, which is the pore pressure increment caused by the load. It shows that due to the soil compacting effect, positive excess pore pressure appears in the soil under the pile end and the pressure gradually increases to 39.5 kPa. Due to the soil around the pile side has a tendency of sliding upwards after compression, negative excess pore pressure appears and gradually increases to −33 kPa. The contact surface of pile and soil is un-drainage interface and therefore no excess pore pressure produced. Both of the positive and negative excess pore pressure increase simultaneously and remain stable. There is no doubt that the excess pore pressure would dissipate and the subgrade start consolidation.

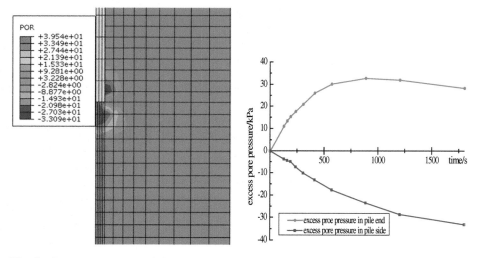

Fig. 9. Excess pore pressure field distribution   **Fig. 10.** Pore pressure variation

2   The Consolidation Characteristics of PHC Pile Foundation

Numerical simulation was used to research the consolidation characteristic in 1 year. Figures 11, 12, 13 and 14 show the excess pore pressure dissipation process of the marine clay in 1 year.

From the variation of excess pore pressure field, the dissipution process can be divided into two stages. The first is global expansion period. In this stage, the pore pressure around the pile end appears global expansion due to the pile pressure, at the same time the pressure around the pile side also appears global expansion due to the pile friction. This stage has a great effect on the consolidation settlement and can be

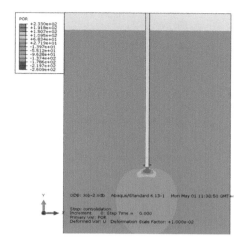

Fig. 11. Initial excess pore pressure field

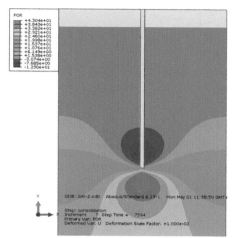

Fig. 12. Excess pore pressure field after 2 h

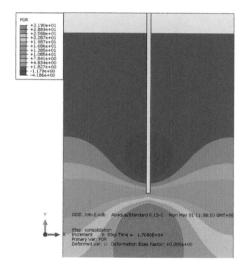

Fig. 13. Excess pore pressure field after 15d

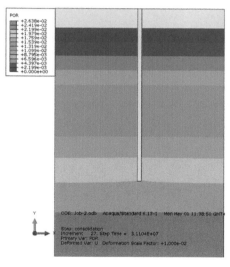

Fig. 14. Excess pore pressure field after 360d

regard as primary consolidation process. The second stage is vertical seepage. After the local expansion period, the pore pressure starts dissipute vertically until the end of the calculation, and this stage has little effect on consolidation settlement, which can be regard as secondary consolidation process.

Figures 15 and 16 show the whole consolidation settlement process, and indicate that after 78 days, the consolidation settlement has already accomplished, both of excess pore pressure and pile settlement have been stable. The two curves have same

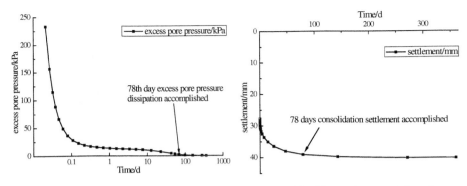

Fig. 15. Excess pore pressure time-history curve

Fig. 16. The settlement of the pile

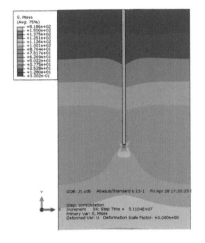

Fig. 17. Mises stress distribution

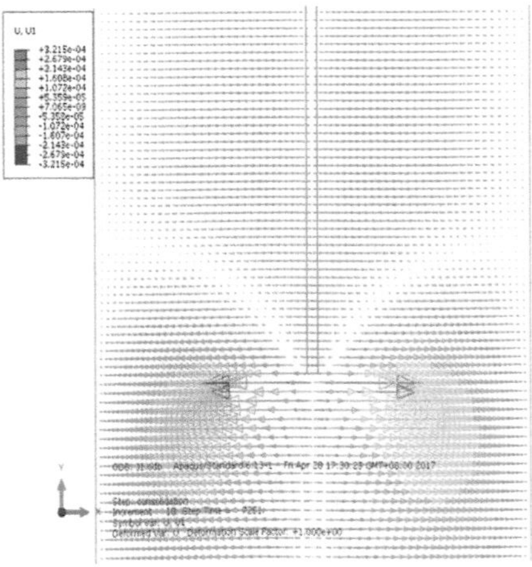

Fig. 18. Foundation horizontal displacement vector

variation features, and both of them changes rapidly at the beginning and remain stable until the end. From Fig. 16, total settlement is 40.25 mm after 1 year consolidation. The initial instantaneous settlement is 27.73 mm which is about 68.89% of the total value. Primary consolidation settlement is 11.44 mm which is about 28.42% of the total value. Secondary settlement is 2.68% of the total value.

When the consolidation accomplished, the mises stress distribution and foundation horizontal displacement vector are shown in Figs. 17 and 18. There is still stress concentration at the pile end and the stress contour of pile lateral soil has been flat. At the same time, the horizontal displacement vector shows that the soil around the pile side has a tendency to move towards the pile. This is due to the consolidation effect, excess pore pressure dissipated and the soil shrinks to the pile, which means that the consolidation effect strengthen the foundation soil.

3 Conclusion

This article mainly researched the characteristics of PHC pile foundation in bearing capacity and consolidation process through numerical simulation method. From this study the following general conclusions may be drawn.

(1) Characteristics of static load test: the results of numerical simulation have a great agreement with the real result, and the ultimate bearing capacity is 995 kN. In loading process, the axial force of the pile is nonlinearly reduced along the depth and the tendency is the variation in different load is consistent.
(2) Working behavior of PHC pile under design load: the spherical squeezing effect occurs due to the lateral settlement. The pile-soil differential sedimentation gets smaller with depth,and the minimum value locates in the pile end. The pile side friction is generally liner distribution in different time. Positive pore pressure appears in the pile end and negative pore pressure in the pile side.
(3) Consolidation characteristics: The variation of pore pressure distribution is consist with global expansion period and vertical seepage period. The initial instantaneous settlement is about 68.89% of the total settlement which means the instantaneous settlement accounts for most of the settlement, and the rest is consolidation settlement.

References

1. Baziar, M.H., Ghorbani, A., Katzenbach, R.: Small-scale model test and three dimensional analysis of pile-raft foundation on medium-dense sand. Int. J. Civ. Eng. **7**(3), 170–175 (2009)
2. Sheil, B.B., McCabe, B.A.: A finite element–based approach for predictions of rigid pile group stiffness efficiency in clays. Acta Geotech. **9**(3), 469–484 (2014)
3. Shao, Y., Wang, S.C., Guan, Y.: Numerical simulation of soil squeezing effects of a jacked pipe pile in soft foundation soil and in foundation soil with an underlying gravel layer. Geotech. Geol. Eng. **34**(2), 493–499 (2016)
4. Zhang, W.J.: Consolidated analysis of rigid pile composite foundation based on modified Cam-clay model, pp. 34–35. Guangzhou South China University of Technology (2013)
5. Lu, M.M., Xie, K.H., Zhou, G.Q., et al.: Analytical solution for consolidation of composite ground with impervious pile. Chin. J. Geotech. Eng. **33**(4), 574–579 (2011)
6. Sheng, D., Sloan, S.W., Yu, H.S.: Aspects of finite element implementation of critical state models. Comput. Mech. **26**(2), 185–196 (2000)
7. Yang, T., Li, G.W.: Consolidation analysis of composite ground with undrained penetrating piles under embankment load. Chin. J. Geotech. Eng. **29**(12), 1831–1836 (2007)
8. Xie, K.H., Lu, M.M., Hu, A.F., et al.: A general theoretical solution for the consolidation of a composite foundation. Comput. Geotech. **36**(1–2), 24–30 (2009)

Experimental Study on Electro-Osmosis Consolidation with Solar Power for Silt

Huiming Tan[1,2(✉)], Jianjun Huang[1,2], and Jiawei Wang[1,2]

[1] Key Laboratory of Coastal Disaster and Defense of Ministry of Education, Hohai University, Nanjing 210098, China
thming2008@163.com
[2] College of Harbour Coastal and Offshore Engineering, Hohai University, Nanjing 210098, China

Abstract. Electro-osmosis is an effective method for accelerating the consolidation of soft soils while it has not been widely used in practice because of its high electric energy consumption. With development of renewable solar energy, the solar energy is used in electro-osmosis method and the solar electro-osmosis technology has been proposed. This paper presents the performance of solar electro-osmosis method to improve the consolidation of soft clay based on the laboratory experiment, including the electric current and voltage, electric potential, water discharge, water content, shear strength. The energy consumption and carbon emission have also been discussed based on the measured data. The measured results show that both electric current and voltage in solar electro-osmosis are not stable and vary periodically in day time; the solar electro-osmosis can accelerate consolidation of soft clay effectively and the solar electro-osmosis technology has a well drainage efficiency and environmental benefits.

Keywords: Electro-osmosis · Solar energy · Drainage
Energy consumption · Carbon emission · Non-constant direct current

1 Introduction

The disposal of dredged soft clay (i.e. mud) generated from harbour, waterway and coastal engineering is a challenge in practice. Most dredged soft clay not only contains a high water content but also has a low permeability. Several methods are proposed to accelerate silt, such as the preloading method and the vacuum method. Compared to the improvement methods above, the electro-osmosis method has a better drainage behavior as well as a less time-consuming. The principle of this method is to drive the pore water from the anode to the cathode when an electrical field is applied. Consequently, an electro-osmosis water flow is formed by direct current.

Electro-osmosis technology has been used since 1930s for removing water from clay and silt soils [1]. Since then, several successful applications have been reported by researchers such as Casagrande [2], Bjerrum et al. [3], Chappell and Burton [4], Jones et al. [5] and so on. At the same time, a series of studies on electro-osmosis have been conducted by scholars, of which the energized form is one of the main factors

© Springer Nature Singapore Pte Ltd. 2018
L. Li et al. (Eds.): GSIC 2018, *Proceedings of GeoShanghai 2018 International Conference: Ground Improvement and Geosynthetics*, pp. 186–193, 2018.
https://doi.org/10.1007/978-981-13-0122-3_21

of electro-osmosis consolidation. Hamir [6] used electro-osmosis for soil stabilization with stable voltage and stable electric current correspondingly. The result showed that both of two energy controls have the same consolidation behavior. On the basis of Hamir's research, Yoshida et al. [7] reported that controlling voltage can get a better drainage than controlling electric current with the same energy consumption. Lockhart et al. [8] proved that the effect of electro-osmosis on soft soil can be enhanced by applying stepped voltage.

Moreover, other energized forms were further put forward, including current intermittence [9, 10] and electrode conversion [11, 12]. On the basis of effectiveness of electrode conversion, electro-osmosis with alternating current had been put forward. Estabragh et al. [13] showed a better drainage behavior of electro-osmosis with alternating current among a certain range of electrical frequency which had a weak relationship with waveform.

However, there are limited studies on the energized form effects on treatment of soils by electro-osmosis and most of them are only focused on the consolidation result of electro-osmosis [14–16]. In this work, electro-osmosis with solar energy has been put forward to research on the characteristics itself. This paper presents the unsteady voltage as well as non-constant electric current of solar energy and the performance of solar electro-osmosis to improve the consolidation of soft clay by laboratory experiment. Based on the measured data, the energy consumption and carbon emission have also been discussed.

2 Materials and Methods

2.1 Experimental Apparatus and Soil

The experimental apparatus is consisted of two main cells, one solar cell panel and three pairs of anode and cathode as shown in Fig. 1. One of the two main cells was connected to a solar cell panel while the left cell was unconnected to any power source to serve as a control which is called natural consolidation. The soft clay used for electro-osmosis experiments is from Nanjing City of China. The soft clay is saturated with the density of 1.54 g/cm^3 and the initial water content is 77.03%.

The main section of the apparatus is the cell that consists of a rectangular tank that holds the silt sample with 250 mm height. It was made of plexiglass plates with 1 cm thickness and had an inner dimension of 300 × 300 × 350 mm (length × width × height). Three pairs of anode and cathode were added on the two sides of the main cell, of which anode is made of graphite in order to decrease erosion and cathode is made of steel pipe with small holes 5 mm in diameter. Besides, the distance of homopolar is 100 mm while the distance of heteropolar is 200 mm. In order to prevent the small holes of cathode from jamming and measure the voltage in certain time intervals during the process of the test, cathode was wrapped up with wet geotextile as reversed filter and a number of voltage probes made of copper were installed at the surface of the cell at distances of 1, 7, 13 and 19 cm from the anode. The solar cell panel was used as power supply. The cell dimensions are 1580 × 808 mm. The open circuit voltage and the maximum current are 44 V, and 5.58 A, respectively.

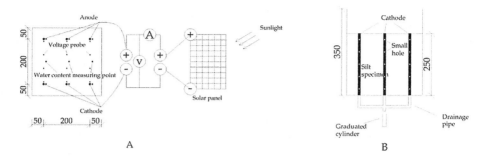

Fig. 1. (A) Apparatus diagram of Plan. (B) Apparatus diagram of Profile.

2.2 Experimental Procedures

Two types of test were performed in laboratory. The conditions of tests are summarized in Table 1. The procedure is described in 6 steps as follows.

(1) Before the test, the anode and the cathode are added on the two sides of the rectangular tank except the natural consolidation condition.
(2) The rectangular tank is filled with soft clay until 250 mm high.
(3) In order to monitor the water discharge, the graduated cylinder is put under the drainage pipe of the tank.
(4) The anode and the cathode are connected to power supply and then electro-osmosis starts working.
(5) The electric current, voltage and water discharge are measured and recorded in each hour.
(6) After the schedule working time, the experiment is stopped.

Table 1. Summary of experiments (a) Natural consolidation, (b) Solar electro-osmosis

Experiment code		Power supply	Voltage (V)	Treatment time (h)		Drained water (ml)
				Power-on time (h)	Power-off time (h)	
(a)	Natural consolidation	None	0	0	72	48
(b)	Solar electro-osmosis	Solar panel	0–40	36	36	605

3   Results and Discussion

3.1   Electric Current and Voltage

Figures 2 and 3 show the profile of applied voltage and electric current of solar electro-osmosis in 72 h. With the development of electro-osmosis consolidation, the resistivity of soft clay is improved gradually because of the increasing water discharge so that electric current decreases obviously. However, an intermittent variation is found in solar electro-osmosis, of which there are three stages every 24 h including rising section, steady section and descent section. Besides, an increase of current has been showed at the beginning of steady section in the next 24 h compared to the ending of steady section in the last 24 h. Moreover, the applied voltage of solar electro-osmosis varies during the daytime. As seen in the Fig. 3, the applied voltage rapidly increases after the sunrise (i.e. 6 am), and peaks during the day (i.e. between 10 am and 4 pm). Afterward, the voltage starts to decrease and diminish shortly after sunset (i.e. 6 pm).

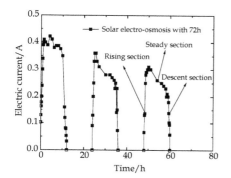

Fig. 2. Evolution of current with time

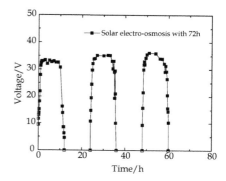

Fig. 3. Evolution of voltage with time

3.2   Water Discharge

Drainages of natural consolidation and solar electro-osmosis are showed in Fig. 4. It is found that solar electro-osmosis appears a better drainage performance over natural consolidation from the beginning to the end, and the drainages of solar electro-osmosis consolidation and natural consolidation are 605 ml and 48 ml in this case, respectively.

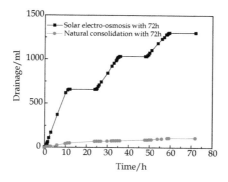

Fig. 4. Variation of drainage during Natural consolidation with 72 h and solar electro-osmosis with 72 h

The result illustrates the effectiveness of solar electro-osmosis technology. Furthermore, it can be easy to find that drainage of solar electro-osmosis is discontinuous while drainage of natural consolidation is continuous.

The average drainage rate of solar electro-osmosis can be calculated from Eq. (1):

$$v_s = \frac{Q}{T_s} \tag{1}$$

Where v_s is the average drainage rate, Q is the total of drainage, T_s is the actual electro-osmosis time. And the calculated result of average drainage rate of solar electro-osmosis with 72 h is 36.19 ml/h.

3.3   Power Consumption

As mentioned above, both electric current and applied voltage of solar electro-osmosis vary with the external environment including the small variation during a day and the large variation between different seasons. In this work, the electric current and voltage in September and November were showed in Fig. 5(A) and (B). The result reveals that the total of electric current and voltage in September is larger than that in November. However, the maximum of voltage is similar in September and November while the maximum of current in September is much larger than that in November and this is because the voltage is affected slightly by lightness while the electric current is influenced heavily. Besides, in actual construction of electro-osmosis, the stability of solar electro-osmotic voltage can be fully used when the voltage is dominant.

Furthermore, the drainage in September is larger on account of the larger energy consumption according to Fig. 5(C) and (D). Considering the drainage and the energy consumption, energy consumption coefficient is defined as the ratio of energy consumption to drainage which can be calculated from Eqs. (2) and (3).

$$W = \int_0^T u_t i_t dt \tag{2}$$

Where W is the total of energy consumption, T is the total of actual electro-osmotic time, u_t is the voltage during the time of t, i_t is the electric current during the time of t.

$$w = \frac{W}{Q} \tag{3}$$

Where w is the energy consumption coefficient, W is the total of energy consumption, Q is the total of drainage.

The calculated result showed a 0.20 w·h/ml energy consumption coefficient in September while 0.21 w·h/ml in November. Thus, the drainage efficiency of solar electro-osmosis is influenced slightly by different seasons, and the energy consumptions in summer and winter are the same when the same water volume is discharged. However, the longer electro-osmotic construction period is required in winter.

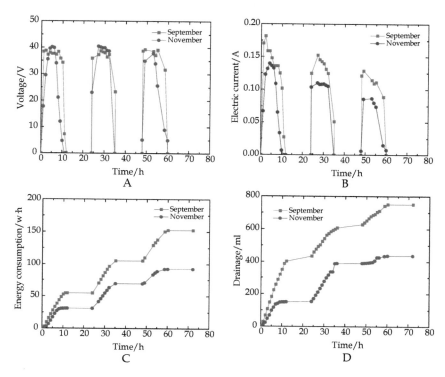

Fig. 5. (A) Evolution of voltage of solar electro-osmosis in September and November. (B) Evolution of current of solar electro-osmosis in September and November. (C) Evolution of energy consumption of solar electro-osmosis in September and November. (D) Evolution of drainage of solar electro-osmosis in September and November

3.4   Carbon Emission

The high energy consumption is one of the factors limiting the widely using of electro-osmosis technology in practice. At the same time, a large number of emissions of carbon dioxide will be produced with the high energy consumption. However, almost all the scholars only pay great attention to improving the effect of electro-osmosis [17, 18] and it is rare to make a comprehensive assessment of electro-osmosis technology from the carbon emission.

According to the assessment of life cycle, this work has made a calculation of carbon emission in solar electro-osmosis. The life cycle of solar electro-osmosis is consisted of 6 stages mainly, including production of industrial silicon, highly polysilicon refining, production of polysilicon wafer, production of polysilicon cell, production of photovoltaic module and transportation and erection of material and equipment.

Table 2. Emission of carbon dioxide in life cycle

Procedure	Production of industrial silicon	Highly extraction of pure polysilicon	Production of polycrystalline silicon slice	Production of polysilicon cells	Production of photovoltaic module	Total
Carbon emission	0.0043	0.0176	0.0014	0.0041	0.0006	0.0279

Based on the data of emission of carbon dioxide in life cycle in Table 2, the carbon emission unit drainage of solar electro-osmosis can be calculated from Eq. (4).

$$E_w = w \sum_{i=1}^{n} E_i \tag{4}$$

where E_w is the carbon emission unit drainage, w is the energy consumption coefficient, $\sum_{i=1}^{n} E_i$ is the total of emission of carbon dioxide in life cycle.

The result showed that the carbon emission unit drainage of solar electro-osmosis is 0.03×10^{-3} kg/ml.

4   Conclusions

(1) The electric current of solar electro-osmosis appears an intermittent variation with rising stage, steady stage and descent stage during a day. The applied voltage of solar electro-osmosis varies during the daytime which rapidly increases after the sunrise, peaks and keeps stable during the day, and decreases close to sunset. Solar electro-osmosis method presents an unsteady voltage as well as non-constant electric current.

(2) It is found that solar electro-osmosis appears a better drainage performance over natural consolidation, and the drainages of solar electro-osmosis consolidation and natural consolidation are 605 ml and 48 ml in this case, respectively. Thus, the solar electro-osmosis technology is effective to improve the consolidation of soft clay, and the significant drainage rate of solar electro-osmosis is 36.19 ml/h.

(3) The total of solar electro-osmotic electric current and voltage in September is larger than that in November. However, the maximum of voltage is similar in September and November while the maximum of current in September is much larger than that in November. Therefore, in actual electro-osmotic construction, the stability of solar electro-osmotic voltage can be fully used when the voltage is dominant.

(4) An almost similar energy consumption coefficient is found in September and November which reveals the slight influence on the drainage efficiency of solar electro-osmosis by different seasons. At the same time, the energy consumption is equal when the same water volume is discharged no whether in summer or in

winter and the only distinction is the longer electro-osmotic construction schedule relatively during winter.

(5) According to the assessment of life cycle, the carbon emission unit drainage of solar electro-osmosis is 0.03×10^{-3} kg/ml.

References

1. Bruell, C.J., Segall, B.A., Walsh, M.T.: Electroosmotic removal of gasoline hydrocarbons and TEC from clay. J. Environ. Eng. **118**(1), 68–83 (1992)
2. Casagrande, L.: Electro-osmosis stabilization of soils. J. Boston Soc. Civ. Eng. **39**(1), 51–83 (1952)
3. Bjerrum, L., Moum, J., Eide, O.: Application of electro-osmosis to a foundation problem in a Norwegian quick clay. Géotechnique **17**, 214–235 (1967)
4. Chappell, B.A., Burton, P.L.: Electro-osmosis applied to unstable embankment. J. Geotech. Eng. Div. **101**(8), 733–740 (1975). ASCE
5. Jones, C.J.F.P., Lamont-Black, J., Glendinning, S.: Electrokinetic geosynthetics in hydraulic applications. Geotext. Geomembr. **29**, 381–390 (2011)
6. Hamir, R.: Some aspects and applications of electrically conductive geosynthetic materials. Ph.D. thesis, University of Newcastle upon Tyne, UK (1997)
7. Yoshida, H., Shinkawa, T., Yukawa, H.: Comparison between electroosmotic dewatering efficiencies under conditions of constant electric current and constant voltage. J. Chem. Eng. Jpn **13**(5), 414–417 (1980)
8. Lockhart, N.C.: Electroosmotic dewatering of clays. I. Influence of voltage. Colloids Surf. **6**(3), 229–238 (1983)
9. Sprute, R.H., Kelsh, D.J.: Limited field tests in electrokinetic densification of mill tailings. Report of Investigations 8034. USBM (1975)
10. Micic, S., Shang, J.Q., Lo, K.Y., Lee, Y.N., Lee, S.W.: Electrokinetic strengthening of a marine sediment using intermittent current. Can. Geotech. J. **38**(2), 287–302 (2011)
11. Lo, K.Y., Inculet, I.I., Ho, K.S.: Electroosmotic strengthening of soft sensitive clays. Can. Geotech. J. **28**(1), 74–83 (1991)
12. Yoshida, H., Kitajyo, K., Nakayama, M.: Electroosmotic dewatering under A.C. electric field with periodic reversals of electrode polarity. Drying Technol. **17**(3), 539–554 (1999)
13. Estabragh, A.R., Naseha, M., Javadi, A.A.: Improvement of clay soil by electro-osmosis technique. Appl. Clay Sci. **95**, 32–36 (2014)
14. Wu, H., Hu, L.M., Wen, Q.B.: Electro-osmotic enhancement of bentonite with reactive and inert electrodes. Appl. Clay Sci. **111**, 76–82 (2015)
15. Xue, Z.J., Tang, X.W., Yang, Q.: Influence of voltage and temperature on electro-osmosis experiments applied on marine clay. Appl. Clay Sci. **141**, 13–22 (2017)
16. Tao Y.L., Zhou J., Gong X.N., Chen Z., Hu P.C.: Influence of polarity reversal and current intermittence on electro-osmosis. In: Ground Improvement and Geosynthetics, pp. 198–208. Geo-Shanghai (2014)
17. Rhodes, J.D., Raats, P.A.C., Prather, R.J.: Effects of liquid-phase electrical conductivity, water content, and surface conductivity on the bulk soil electrical conductivity. Soil Sci. Soc. Am. J. **40**(5), 651–656 (1976)
18. Rinaldi, V.A., Cuestas, G.A.: Ohmic conductivity of a compacted silty clay. J. Geotech. Geoenviron. Eng. **128**(10), 824–835 (2002)

Experimental Investigation on Compressive Deformation and Shear Strength Characteristics of Steel Slag in the Geotechnical Engineering

Li-yan Wang$^{(\boxtimes)}$, Qi Wang, Xiang Huang, and Jia-tao Yan

Jiangsu University of Science and Technology, Zhenjiang 212003, China
wly_yzu@163.com

Abstract. Steel slag has similar characteristics with sand, and pure steel slag used as geotechnical backfill material is an effective way to improve the comprehensive utilization rate of waste steel slag. In order to study characteristics of pure steel slag for application in geotechnical engineering, the compressive deformation characteristics and shear strength characteristics of steel slag under different water contents were explored by unidirectional consolidation compression tests and direct shear tests. Change rules of densities, compressive moduli, cohesion and internal friction angles of steel slag under different water contents were analyzed. It can be concluded that the compressive modulus of steel slag decreased with increasing the water content. The compressive modulus of dry steel slag was larger, reaching the level of medium sand. Compression moduli at different water contents were higher than those of silt soil and soft clay. Steel slag can be used as backfill material instead of mud soil and soft clay in the treatment of soft soil foundation. The cohesion of steel slag increased and the internal friction angle decreased with increasing the water content. The shear strength of dry steel slag was similar to that of medium sand, indicating that steel slag was more suitable as roadbed filling above the groundwater level.

Keywords: Steel slag · Moisture content · Compression deformation
Shear strength · Unidirectional consolidation compression test · Direct shear test

1 Introduction

Steel slag is waste residue produced in metallurgical industry, and its production rate is 8%–15% of crude steel output that was about 800 million tons in 2015 [1]. But the utilization rate of waste steel slag is only 10%, and 90% of steel slag has not been effectively utilized. Its stacking not only takes up lands, but also pollutes the environment seriously, so improving application value of waste steel slag is an important research content on its development and application.

The steel slag after aging can be used as roadbed filling and backfill material due to its high compressive strength and the basically stable performance after its aging. It is more advantageous to build roads with steel slag in marshlands because it has high activity and is easy to clump together forming a hard lump. At present, the reuse of waste steel slag

© Springer Nature Singapore Pte Ltd. 2018
L. Li et al. (Eds.): GSIC 2018, *Proceedings of GeoShanghai 2018 International Conference: Ground Improvement and Geosynthetics*, pp. 194–202, 2018.
https://doi.org/10.1007/978-981-13-0122-3_22

has become one of the research directions of many scholars. Wang et al. uses a mixture of discarded steel slag and sand as roadbed filling [2]. Huang et al. initially applied the steel-ash pile to the composite ground [3]. Li et al. applied the mixture of steel slag and sand in soft ground, which can effectively enhance the ground bearing capacity and reduce the foundation settlement [4]. Yadu et al. holds that the physical characteristics and strength characteristics of waste steel slag are similar with those of sand [5]. Compression tests were conducted with uniaxial compression apparatus to study the compression deformation characteristics of the new geotechnical material. In addition, tests were carried out with static tri-axial apparatus to study the strength characteristics of steel slag.

2   Test Materials

Steel slag used in tests was produced by the Yonggang Company in Zhangjiagang, China, with a density of 1.94 g/cm^3, as shown in Fig. 1. The waste steel slag is the blast furnace slag and can be applied in geotechnical backfill. Its material gradation is shown in Fig. 2 below. What can be obtained from the gradation curve of steel slag are as follows: d60 = 1.30 mm, d30 = 0.42 mm, d10 = 0.09 mm, the coefficient of uniformity Cu being 14.4 and the coefficient of curvature Cc being 1.51, thus the steel slag is classified as well-graded.

Fig. 1. Steel slag

The preparation of samples affected the reliability of final parameters, and correct preparation methods should be adopted for the preparation of good soil samples. The quality of every single sample was calculated by different relative densities of pure steel slag during sample preparation, and samples were compacted in layers. The maximum dry density was determined by using hammering method according to the standard for soil test method [6], and the minimum dry density was determined by using funnel and cylinder. Parallel experiments were conducted for each kind of test, and two average values were taken as the maximum dry density and the minimum dry density, respectively. The minimum dry density and the maximum dry density were tested to be 1.40 g/cm^3 and 1.94 g/cm^3, respectively.

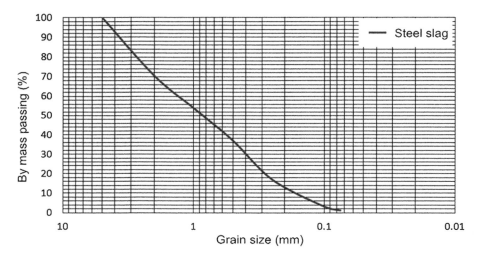

Fig. 2. Grading curve of steel slag

The relationship between density and water content of steel slag is shown in Fig. 3 below. It can be seen from the figure that the density increased with increasing water content in early stage, and the steel slag reached its maximum density when the water content reached 20%, which mainly because pores of samples with higher water contents were filled with more water when compaction conditions and porosities were both the same, thus densities of samples were larger. However, densities of samples decreased when the water content was as high as 25%, mainly because the overflow of pore-water appeared to be obvious and a part of water was lost in the process of sample preparation when water contents of samples were more than 20%. It was indicated that the steel slag reached saturation when the water content was 20%. Therefore, the highest water content was set at 20% in the subsequent study on compression deformation characteristics and strength characteristics of steel slag.

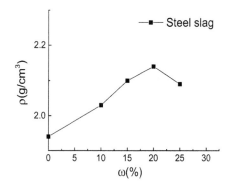

Fig. 3. Relationship between density and water content

3 Experimental Study on Deformation Characteristics of Pure Steel Slag

3.1 Effect of Water Content on Compression Deformation Characteristics of Steel Slag

It can be seen from Fig. 4 that water content is an important factor affecting the compression characteristics of steel slag. The strain of steel slag increased with increasing the water content, mainly because the water in steel slag overflowed with increasing vertical stress applied on the sample, thus the compression deformation increased.

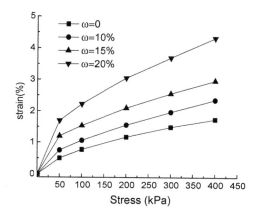

Fig. 4. Stress-strain curves of different water content

The compression moduli Es of steel slag at different stress segments p under different initial water contents ω are shown in Table 1 below. The Es-p curves of steel slag at different water contents are shown in Fig. 5. The Es-w curves at different p segments are shown in Fig. 6.

Table 1. Compression modulus of ω-p (MPa)

P (kPa) Water content ω (%)	0	10	15	20
50–100	19	16	15	10
100–200	25	20	18	12
200–300	31	24	22	15
300–400	39	26	24	16

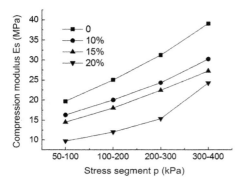
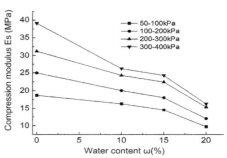

Fig. 5. Curves of Es-p with different water content

Fig. 6. Curves of Es-ω with different p

It can be seen from Fig. 5 that compression modulus Es of the steel slag increased with increasing the stress segment p when water contents ω were same. Es was the largest when ω = 0, indicating that steel slag has high internal friction strength and strong compression resistance under dry condition. The Es-p curve changed gently with increasing water content ω, indicating that the compression moduli of steel slag changed more slightly with increasing vertical stress p when water contents of steel slag were higher. It can be seen from Fig. 6 that the compression modulus Es of steel slag decreased with increasing water content ω in the same stress segment p. The compression modulus Es decreased by 32.9% when the water content increased from 0 to 10 and the p segment was 300 kPa–400 kPa, moreover, Es decreased by 22.1% when the p segment was 200 kPa–300 kPa, Es decreased by 20% when the p segment was 100 kPa–200 kPa, Es decreased by 12.8% when the p segment was 50 kPa–100 kPa, showing that the greater the p segment was, the greater the attenuation of Es was. The compression modulus decreased with increasing water content ω, mainly because the water formed the lubricant on the surface of steel slag particles and then reduced the frictional resistance, making the particles easy to slide and the steel slag easy to deform, thus the compression modulus decreased accordingly.

3.2   Comparison with Compression Moduli of Traditional Soils

The compression moduli of different soils were found by referring to the handbook of engineering geology, as shown in Table 2 below [7]. According to the requirements on compression modulus in the book of standard for soil test method, the compression modulus under the vertical stress segment of 100–200 kPa was taken as the compression modulus of the material. Therefore, the compression modulus of steel slag at the water content of ω = 0 was 25 MPa obtained by the previous experimental analysis, which was large and almost reached the level of medium sand. The compression modulus gradually decreased with increasing the water content, but the overall level exceeded that of silt and clay. In the treatment of soft soil foundation, steel slag can be used as backfill material instead of mud soil and soft clay, which not only increases the stability of foundation, but also saves money and protects the environment.

Table 2. Compression modulus of different soils

Soils	Mucky Soil	Clay	Fine sand	Medium sand	Pure steel slag ω (%)			
					0	10	15	20
Compression modulus (MPa)	1–5	4–15	19–31	31–42	25	20	18	12

4 Direct Shear Tests on Strength Characteristics of Pure Steel Slag

4.1 Test Schemes and Principles

It was difficult to conduct tri-axial tests on strength characteristics of dry steel slag because samples could not be molded for tests. Therefore, direct shear tests were conducted to study the strength characteristics of dry steel slag with water content ω of 0. The influence of water content on the shear strength of steel slag was studied by controlling the water content. The specific influencing factors are shown in Table 3 and the direct shear apparatus is shown in Fig. 7 below. The experiment adopted the method of fast shearing,

Table 3. Direct shear testing parameters table

Parameters	Numerical value
Water content ω (%)	0, 10, 15, 20
Vertical stress (kPa)	50, 100, 200, 300

Fig. 7. Direct shear apparatus

and the loading order was 50 kPa, 100 kPa, 200 kPa and 300 kPa. The strength envelopes were determined by shear strengths of samples under different levels of loads, and then the shear strength parameters (c, φ) were obtained.

4.2   Shear Strength Characteristics of Steel Slag

Figure 8 shows the relationship between shear strength and shear displacement under different normal stresses of 50, 100, 200 and 300 kPa etc. with water contents of steel slag are 0, 10, 15 and 20. As can be seen from the figure, the shear strength of steel slag increased with increasing the shear displacement, and there were no peak values. The steel slag exhibited hardening characteristics similar to loose sand.

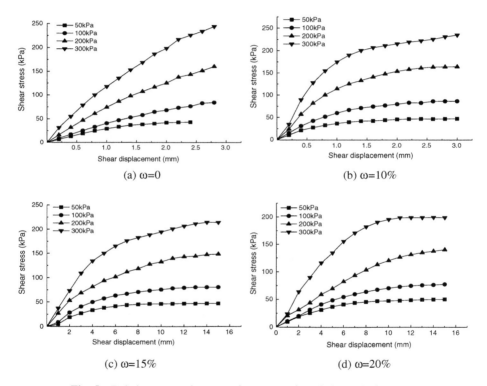

Fig. 8. Relation curves between shear strength and shear displacement

The relationship between the shear strength and the normal stress of the dry steel slag is shown in Fig. 9, and the cohesion c and the internal friction angle φ can be obtained from the figure. The fitting parameters of the curve are shown in Table 4.

The shear strength of dry steel slag was similar to that of medium sand by checking the handbook of engineering geology. The cohesion c and internal friction angle φ of steel slag are shown in Table 5 below. It can be seen from the table that the cohesive c of steel slag increased and the internal friction angle φ decreased with increasing the

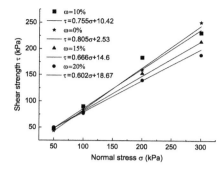

Fig. 9. Relation curves between normal stress and shear strength

Table 4. Fitting parameters of exponential curves

Water content ω (%)	a	b	R^2
0	0.805	2.53	0.92
10	0.755	10.42	0.96
15	0.666	14.6	0.98
20	0.602	18.67	0.94

water content, mainly because the water formed the lubricant on the surface of the steel slag particles when the water content increased, making particles easy to slide, thus the friction angle φ decreased. It can be seen from the gradation of steel slag that particle sizes less than 1 mm accounted for 60% of the total mass, and steel slag particles exhibited good cohesion when mixed with a certain percentage of water, thus the cohesion increased with increasing the water content.

Table 5. Cohesion and friction angles of steel slag with different water content

Water content ω (%)	c (kPa)	φ (°)
0	5	39
10	24.17	37
15	26.19	34
20	31.05	31

5   Conclusions

The compression deformation characteristics and strength characteristics of steel slag were studied by unidirectional compression tests and direct shear tests. The change rules of densities, compression moduli, cohesion and internal friction angles of steel slag under different water conditions were analyzed. Several conclusions could be drawn as follows:

(1) The density of steel slag increased with increasing the water content, and the density of steel slag reached its peak when water content was increased to 20%, indicating that the steel slag reached saturation. It provided the basis about selecting the highest water content for later study on the compression deformation characteristics and strength characteristics of steel slag.

(2) The strain of steel slag increased with increasing the water content, and the compression modulus of steel slag increased with increasing the stress segment at the same water content. The compression modulus decreased with increasing the water content. The compression modulus of steel slag was the largest when the water content was 0.

(3) The compression modulus of steel slag was relatively large when the water content of steel slag was 0, reaching the level of medium sand, and the compression modulus gradually decreased with increasing the water content, but the overall level exceeded that of silt and clay. In the treatment of soft soil foundation, steel slag can be used as backfill material instead of mud soil and soft clay.

(4) The shear strength of steel slag increased with increasing the shear displacement, and there was no peak values. The steel slag exhibited hardening characteristics similar to loose sand. The cohesive of steel slag increased and the internal friction angle decreased with increasing the water content. The shear strength of dry steel slag was similar to that of medium sand.

References

1. Zhang, Z., Liao, J., Ju, J., Dang, Y.: Treatment process and utilization technology of steel slag in China and abroad. J. Iron Steel Res. **07**, 1–4 (2013). (in Chinese)
2. Wang, L., Gao, P., Chen, G., Fu, R.: Experimental study on deformation behavior and shear strength of mixed soil blended with steel slag. Chin. J. Geotech. Eng. **S2**, 126–132 (2013). (in Chinese)
3. Huang, T., Wang, X., Wu, Y., Chen, C.: Experimental research and application of composite ground of steel-ash pile within a steel factory in North Henan Province. Chin. J. Geotech. Eng. **03**, 72–77 (1999). (in Chinese)
4. Li, W., Wang, H., Li, F., Zhang, X., Song, W.: Test study on the mechanical characteristics of soil mixed with steel slag and sand. J. Shenyang Jianzhu Univ. (Nat. Sci.) **05**, 794–799 (2008). (in Chinese)
5. Yadu, L., Tripathi, R.K.: Effect of the length of geogrid layers in the bearing capacity ratio of geogrid reinforced granular fill-soft subgrade soil system. Procedia Soc. Behav. Sci. **104**, 225–234 (2013)
6. GB/T 50123-1999 Standard for soil test method: The Ministry of Water Resources of the People's Republic of China. China Planning Press, Beijing (1999). (in Chinese)
7. Zhang, S., Chang, S.: Handbook of Engineering Geology, 4th edn. China Construction Industry Press, Beijing (2007). (in Chinese)

Predicting Compression Index Using Artificial Neural Networks: A Case Study from Dalian Artificial Island

Zhijia Xue[✉], Xiaowei Tang, and Qing Yang

State Key Laboratory of Coastal and Offshore Engineering,
Dalian University of Technology, Dalian 116024, China
xuegeneral@126.com

Abstract. Compression index is very important in the design of geotechnical engineering such as consolidation settlement prediction and construction design. However, measuring compression index is very complex and time-consuming. In addition, it is very difficult to collect unbroken core samples from underground. Artificial neural network has been adopted in some geotechnical applications and has achieved some success. In this paper, artificial neural network (ANN) models are developed for estimating compression index by basic soil parameters based on 2859 soil test data. All of the marine soil samples, which are divided into three subsets to train the optimum model, are collected from Dalian Artificial Island and compression index and other parameters are measured in soil mechanical laboratory as well. At last, the optimized ANN model structure and suitable inputs are determined followed by the comparison between empirical formulas predictions and ANN models output. It is revealed that ANN models perform better than empirical formulas with respect to the accuracy of compression index prediction.

Keywords: Compression index · Artificial neural network · Artificial island
Prediction

1 Introduction

Compression index can be used to calculate consolidation settlement issues in some geotechnical engineering designs, which can affect the security and economy of construction projects. It is obtained from slope of the curve of void ratio versus logarithm of effective pressure and is conventionally determined by oedometer test. Moreover, compression index reflects compressibility of soil. Void ratio change of specimen occurs as a consequence of dissipation of pore water pressure under various applied stresses of oedometer test condition, in which the lateral strain is limited. Though oedometer test is widely used, it is more time-consuming than tests to obtain basic parameters such as unit weight, plastic index and so on. Beyond that, it is difficult to obtain good quality samples from silted floor and deep soils in oedometer test.

It is generally known that the relationship between compression index and soil basic parameters (water content, density, specific gravity, void ratio, saturation, dry density, liquid limit, plastic limit, plastic index) is nonlinear and complex. Thus,

© Springer Nature Singapore Pte Ltd. 2018
L. Li et al. (Eds.): GSIC 2018, *Proceedings of GeoShanghai 2018 International Conference: Ground Improvement and Geosynthetics*, pp. 203–211, 2018.
https://doi.org/10.1007/978-981-13-0122-3_23

indirect prediction of compression index is useful, time saving, and easy to conduct. Through previous geotechnical engineers' and researchers' exploration and research from amounts of tests, some empirical formulas have been developed. As we know, if an issue can not be built a physical model, a regression model is an effective method. However, most of the regression models simplify the number of parameters, thus only adopting single or dual parameter models to predict compression index. As a result, formulas' prediction accuracy is limited.

More recently, artificial neural networks (ANNs) as a form of artificial intelligence have been used to some geotechnical engineering and structure engineering problems [1–3], the performance of which is more superior than the traditional method. Many parameters, such as preconsolidation pressure [1], unsaturated shear strength [2], pile settlement [4], California bearing ratio of fine grained soils [5], the unconfined compressive strength of compacted granular soils [6], subsurface soil layering and landslide risk [7], compaction parameters of fine-grained soils [8], and stress-strain relationship of geomaterials [9]. In addition, Ozer and Park predicted compression index based on other parameters using ANN model [3, 10]. However, these two studies faced the native construction, and few studies focus on Chinese marine soils to predict their compression index. Thus, this paper will adopt ANN method to predict compression index of Dalian marine soils. In addition, the ANN model in this paper is more steady because that there is 2859 data (much more than other studies) in this paper.

In this paper, ANN models are constructed to forecast the compression index of Dalian marine soils. We treat ANN as an indirect tool to measure the compression index of Dalian marine soils from other basic parameters. And this paper simplify and optimize parameters to predict compression index. Results indicate that the ANN model is an effective method to predict compression index issue and then to guide the geotechnical design.

2 Data and Preprocessing

2.1 Case Study and Sample Preparation

All specimens of marine soils in this paper are taken from Dalian Artificial Island. This project locate in Jinzhou Gulf and floor area is 24 km^2. The distance from the artificial island to coastline is 3000 m. To complete this reclamation project, 60 million m^3 of muck, 300 million m^3 of soils and stones were filled in this area. And an airport will be constructed in the future. Thus, it is necessary to evaluate the soil mechanical property to meet the construction requirement. By means of drilling to obtain specimens and then put them into a prefabricated metal cylinder as SL237-199 (China) recommends.

2.2 ANN Model

The advantage of ANN is that it can solve complex nonlinear issues. BP neural network is one of widely used ANN models, which information transfer direction is back propagation. What's more, the initial weight and bias are given random and treat MSE (mean square error) as cost function, weight's and bias's adjusting are based on MSE value and corresponding Levenberg-Marquardt. Classical ANN model has three kinds of layers: input layer, hidden layer and output layer, and every layer has its own neural nodes. The number of input layer and output layer base on concrete issue, in addition, the hidden layer's node number depends on computational training.

2.3   Preprocessing of Database

To predict the compression index more accurately, this paper explores the factors affecting the compression index by reviewing literature. This study takes water content, density, specific gravity, void ratio, saturation, dry density, liquid limit, plastic limit, and plastic index into account as model inputs. In this paper, 2859 data are divided into training, validation, and testing subsets, with proportions of 70% (2003 data), 15% (428 data), and 15% (428 data), respectively. ANN models get the best result when their training set has the most extensive data range in general. To achieve this demand, this paper employs the random method to divide sample data as proportions mentioned above. Training set is used to train network and obtain weights and bias between layers. Validation set is used to assess network's performance and avoid over fitting. Testing set is adopted to test the network's prediction availability.

3   Result and Discussion

3.1   Results of ANN Model with Nine Parameters as Inputs

According to previous training projects, a series of computational experiments have been completed to predict the compression index. Moreover, the result of each ANN model has been summarized in Table 1. As shown in Table 1, R^2 value of training subset is larger than validation's and testing's, while the RMSE value is opposite. Basically, training subset performs better than validation subset and testing subset. Specifically, model result of 8 nodes in the single hidden layer is better than others under logsigmoid-linear transfer function. And the model with 9 nodes performs best under tansigmoid-linear transfer function. In addition, the model with 13 nodes presents the best performance when using logsigmoid-tansigmoid function. Comparing the three models mentioned above, the one with 13 nodes under logsigmoid-tansigmoid function is more superior than the other two, with R^2 equaling 0.948023 and RMSE equaling 0.044533 in the testing subset.

Table 1. Results of the single hidden layer ANN model with nine parameters

Transfer-function	No.	R^2best			RMSE$_{best}$		
		Training	Validation	Testing	Training	Validation	Testing
log-lin	8	0.959491	0.954203	0.94757	0.040022	0.03875	0.044727
tan-lin	9	0.958464	0.954889	0.947852	0.040527	0.038459	0.044606
log-tan	13	0.960923	0.953384	0.948023	0.039309	0.039095	0.044533

Note: logsigmoid is abbreviated to log; linear is abbreviated to lin; tansigmoid is abbreviated to tan. No. is the node number of hidden layer's.

To explore the performance of ANN model with double hidden layers, Firstly, the nodes of single hidden layer's ANN model is optimized, which suppose number N. Secondly, the N is set as the nodes of first layer of two hidden layers' ANN model. Thus, this paper sets 13 nodes in the first hidden layer and 8–25 nodes in the second hidden layer respectively by using different transfer functions: logsigmoid-tansigmoid-(linear/tansigmoid). After ten times' computational experiments for each model, we summarize the results and list the best one under different transfer functions, which are shown in Table 2. It can be seen that the model with 8 nodes in its second hidden layer has the highest R^2 (0.948749) and lowest RMSE when the transfer function is logsigmoid-tansigmoid-tansigmoid. As expected, the predicting ability of double hidden layers is better than the single one.

Table 2. Results of the double hidden layer ANN model with nine parameters

Transfer function	SNo.	R^2best			RMSE$_{best}$		
		Training	Validation	Testing	Training	Validation	Testing
log-tan-lin	21	0.962038	0.953336	0.947373	0.038788	0.039463	0.045369
log-tan-tan	8	0.95998	0.953235	0.948749	0.03978	0.039158	0.044221

Note: SNo. means Second hidden layer's node No.

Observing the performance of each ANN model, we discover that the gap among R^2 and RMSE is small. Besides, all models have high R^2 and small RMSE. Both of observations above indicate that data are large enough to present a statistical regularity between the compression value and basic parameters.

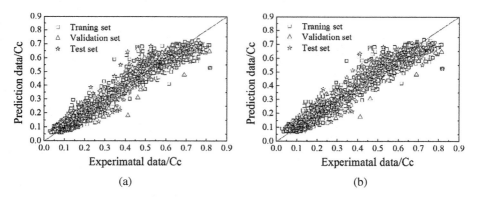

(a)   (b)

Fig. 1. Comparison of experimental and predicted compression index (a) (9-13-1 logsigmoid-tansigmoid ANN model) and (b) (9-13-8-1 logsigmoid-tansigmoid-tansigmoid ANN model)

Figure 1(a) and (b) depict the comparison results of experimental and predicted compression index. We find that most of data points are distributed around the y = x function line apart from a few data points, which reveals the fact that test values of Cc is very close to its prediction values. Therefore, we can conclude that ANN can build a relatively accurate model to predict Cc value through soil basic parameters. In addition, the performance of ANN model with double layers is better than the model with single layer when inputs contain nine parameters.

3.2   Results of ANN Model with Four or Five Parameters as Inputs

Although the ANN model with nine parameters as inputs has acquired a series satisfactory results, its structure is too complex to get the simple weight and bias. Consequently, it is difficult to extend these ANN models. In order to simplify the ANN model, this paper just adopts density, void ratio, water content, liquid limit, and plastic limit as model inputs. As we know, soil contains air, water, and soil grain. So parameters of saturation, dry density, and specific gravity can reflect the soil's physic properties. However, we do not have to spare time to measure them since they can be derived based on density, void ratio, and water content, which can be obtained easily from soil experiments and have been proven that they have strong correlation with the compression index. Thus, these three parameters (density, void ratio, and water content) are included as inputs in the model. Moreover, liquid limit and plastic limit reflect the soil's grain composition and thus affect the soil's compressibility. In conclusion, all of these five parameters have strong representative. To explore the performance of ANN model and compare the performance of different model inputs components, this paper conducts ANN training in a way of following computational scheme. Results of ANN models with best performance under different transfer functions and model inputs are described in Tables 3, 4 and 5. According to them, the ANN model with 15 nodes in its hidden layer achieves the best performance when model inputs include water content, density, void ratio, and liquid limit. Moreover, it is also better than the 9-13-8-1 (logsigmoid-tansigmoid-tansigmoid) ANN model, which testing subset's prediction was $R^2 = 0.949486$ and RMSE = 0.043902. And 9-13-8-1 means that there are 9 nodes in input layer, 13 nodes in the first hidden layer, 9 nodes in the second hidden layer, and 1 node in the output layer. Next we will emphasize the situation of the double hidden layers ANN model with these 4 parameters (w_n, e_o, ρ, w_l).

Table 3. Results of the single hidden layer ANN model with four parameters (w_n, e_o, ρ, w_l)

Transfer function	No.	R^2best			RMSE$_{best}$		
		Training	Validation	Testing	Training	Validation	Testing
log-lin	13	0.960816	0.950101	0.948171	0.039297	0.04049	0.044659
log-tan	14	0.95935	0.950943	0.948538	0.040092	0.040106	0.044312
tan-lin	15	0.960176	0.950736	0.949486	0.039683	0.040191	0.043902

Table 4. Results of the single hidden layer ANN model with four parameters (w_n, e_o, ρ, w_p)

Transfer function	No.	R^2best			RMSE$_{best}$		
		Training	Validation	Testing	Training	Validation	Testing
log-lin	4	0.955105	0.94625	0.947968	0.042133	0.04198	0.044556
log-tan	8	0.959699	0.9518	0.947674	0.039919	0.039754	0.044682
tan-lin	11	0.960695	0.954051	0.948251	0.039423	0.038815	0.044435

Table 5. Results of the single hidden layer ANN model with five parameters (w_n, e_o, ρ, w_l, w_p)

Transfer function	No.	R^2best			RMSE$_{best}$		
		Training	Validation	Testing	Training	Validation	Testing
log-lin	4	0.955105	0.94625	0.947968	0.042133	0.04198	0.044556
log-tan	8	0.959699	0.9518	0.947674	0.039919	0.039754	0.044682
tan-lin	11	0.960695	0.954051	0.948251	0.039423	0.038815	0.044435

For estimating performance of the double hidden layers ANN model with the 4 parameters (w_n, e_o, ρ, w_l), node number of the first hidden layer is determined as 15. In terms of node number of the second hidden layer, it ranges from 4 to 15, with one added once. Moreover, the transfer function between the second layer and output layer is tansigmoid or linear. Then computational experiments are conducted. The best results of double hidden layer ANN models under different transfer functions are listed in Table 6.

Table 6. Results of double hidden layer ANN model with four parameters (w_n, e_o, ρ, w_l)

Transfer function	No.	R^2best			RMSE$_{best}$		
		Training	Validation	Testing	Training	Validation	Testing
tan-lin-tan	8	0.960917	0.952057	0.947089	0.039312	0.039648	0.044931
tan-lin-lin	10	0.960344	0.952532	0.947951	0.039599	0.039451	0.044564

From Table 6, we can see that the model with tansigmoid as the transfer function performs best with R^2 equaling 0.947089 and RMSE equaling 0.044931 in the testing subset when the node number of the second hidden layer is 8. For the situation, in which the transfer function is linear, the model result is optimal with R^2 equaling 0.947951 and RMSE equaling 0.044564 in the testing subset when the node number of the second hidden layer is 10 and testing subset's $R^2 = 0.947089$ and RMSE = 0.044931. However, comparing with the ANN model whose structure is 4-15-1 and transfer function is tansigmoid-linear, the prediction ability of both of them is a little bit weaker. And 4-15-1 means that there are 4 nodes in input layer, 15 nodes in the first hidden layer, and 1 node in the output layer.

Figure 2 is the comparison outcome between predicted values of the compression index and experimental values. Comparing with Figs. 1(a) and 2(b), despite of similar prediction ability exhibited by these three ANN models, 4-15-1 transigmoid-linear

ANN model is more superior to the other two due to its simplicity. And its ANN architecture is given in Fig. 3.

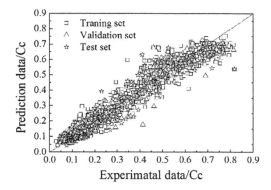

Fig. 2. Comparison of experimental and predicted compression index (4-15-1 tansigmoid-linear ANN model)

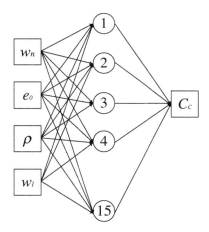

Fig. 3. Optimal ANN architecture (4-15-1 tansigmoid-linear)

4   Conclusion

1. In addition to those phenomenon, prediction ability of the ANN model with double hidden layers is better than that of single layer model when nine parameters as ANN inputs. And 9-13-8-1 structure with logsigmoid-tansigmoid-tansigmoid as transfer function obtains the best prediction ability, which R^2 equaling 0.948749 and RMSE equaling 0.044221.

2. For simplifying the ANN model inputs, thereby making it more accessible, we constrain parameters in ANN models just to be density, water content, void ratio, and liquid limit. Consequently, the model with $R^2 = 0.949486$ and RMSE = 0.043902 is proved to has the strongest prediction ability, better than the best outcome in 9 inputs ANN model structure And this ANN model's structure is 4-15-1 and transfer function is tansigmoid-linear.
3. As can be seen from the result of every ANN model, ANN is an useful tool to solve the nonlinear geotechnical issues. Overall, it is necessary to establish a nationwide database to analyze the relationship of parameters and furtherly guide the practical project applications.

Appendix (Weights and Bias of 4-15-1 Tansigmoid-Linear ANN Model)

W_1			
2.622903	4.083057	−1.22764	0.896845
1.350291	−2.33406	−2.70143	−3.50748
−2.33836	0.22522	−4.40407	1.415465
0.909055	−2.85348	−6.9055	−2.61151
3.461982	−4.91683	0.374592	0.850418
3.920666	0.637272	4.992674	−0.48619
−0.19546	−3.75934	1.610851	−5.08277
1.247785	−0.73106	1.083275	−1.26217
−1.05275	−1.7319	2.852696	3.852191
−5.53328	−1.61691	−4.20678	6.063494
−0.08983	−1.33851	−4.45048	3.358993
−2.61411	−0.58812	−2.32164	4.178343
−1.96377	0.219849	−1.37967	−4.30137
6.408356	−0.77561	2.902883	0.558393
−3.26617	−1.40982	3.581199	−0.82867

B_1	W_2	B_2
−1.57088	−6.10517	−0.02331
0.054728	−5.15034	
1.280595	4.743402	
−4.86217	−6.31247	
2.788758	−7.68178	
1.180403	−1.18919	
1.596082	−1.75728	
−2.98951	0.639066	
−1.87916	2.290802	

(continued)

(continued)

B_1	W_2	B_2
−1.14495	−2.59815	
2.56238	−1.5928	
−1.47991	−3.87658	
1.482598	−4.67425	
0.774307	2.109405	
−0.10101	−5.39566	

References

1. Çelik, S., Tan, Ö.: Determination of preconsolidation pressure with artificial neural network. Civil Eng. Environ. Syst. **22**(4), 217–231 (2005)
2. Lee, S.J., Lee, S.R., Kim, Y.S.: An approach to estimate unsaturated shear strength using artificial neural network and hyperbolic formulation. Comput. Geotech. **30**(6), 489–503 (2000)
3. Park, H.I., Lee, S.R.: Evaluation of the compression index of soils using an artificial neural network. Comput. Geotech. **38**(4), 472–481 (2011)
4. Pooya, N.F., Jaksa, M.B., Kakhi, M., et al.: Prediction of pile settlement using artificial neural networks based on standard penetration test data. Comput. Geotech. **36**(7), 1125–1133 (2009)
5. Taskiran, T.: Prediction of California bearing ratio (CBR) of fine grained soils by AI methods. Adv. Eng. Softw. **41**(6), 886–892 (2010)
6. Kalkan, E., Akbulut, S., Tortum, A., et al.: Prediction of the unconfined compressive strength of compacted granular soils by using inference systems. Environ. Geol. **58**(7), 1429–1440 (2008)
7. Farrokhzad, F., Barari, A., Ibsen, L., et al.: Predicting subsurface soil layering and landslide risk with Artificial Neural Networks: a case study from Iran. Geologica Carpathica **62**(5), 477–485 (2011)
8. Sivrikaya, O., Soycan, T.Y.: Estimation of compaction parameters of fine-grained soils in terms of compaction energy using artificial neural networks. Int. J. Numer. Anal. Meth. Geomech. **35**(17), 1830–1841 (2011)
9. Zhao, H., Huang, Z., Zou, Z.: Simulating the stress-strain relationship of geomaterials by support vector machine. Math. Prob. Eng. **2014**, 1–7 (2014)
10. Ozer, M., Isik, N.S., Orhan, M.: Statistical and neural network assessment of the compression index of clay-bearing soils. Bull. Eng. Geol. Env. **67**(4), 537–545 (2008)

Scale Influence of Treated Zone
Under Vacuum Preloading

Liwen Hu[(⊠)], Fuling Yang, and Zhan Wang

CCCC Fourth Harbor Engineering Institute Co., Ltd., Guangzhou 510230, China
hliwen@cccc4.com

Abstract. The distributions of deformation of soft soil subjected to vacuum preloading have been presented in many case studies; however, the explanations for that deformation behavior have not yet been investigated thoroughly. The influences by the scale are depicted in this paper based on simple numerical analyses and statistical presentation of case studies. Both the width and the depth of treated zone affect the deformation of soft soil in treated zone, which the reason is the development of the effective stresses and stress path. Due to boundary constraints, the increase in effective stresses may not be isotropic under vacuum preloading in the center of large area treated zone, and the maximum settlements and inward lateral displacements have been significant affected by the treated width and depth. K_0 consolidation condition under vacuum preloading may be estimated of 2.5 times treated depth away from the treated edge.

Keywords: Vacuum preloading · Deformation · Influence zone
Stress path

1 Introduction

Vacuum preloading has been commonly taken as an isotropic incremental loading process since it's proposed by Kjellman (1952). It is widely recognized that the total stress remains unchanged during vacuum preloading, thus the reduction of pore water pressure equals to the increase in effective stress at the same magnitude. The concept of isotropic or nearly isotropic incremental loading had been accepted (Cognon 1991; Qian et al. 1992; Leong et al. 2000; Chai et al. 2005; Tran and Mitachi, 2008). Consequently, inward lateral displacement was overestimated by Chai et al. (2005); the reason might be superimposition over a large area of treated zone without considering attenuation. While, Hu et al. (2017) discovered by centrifuge modeling that inward lateral displacement was developed mainly within limited distance from the treated edge, namely soil element without horizontal strain could be found far away from the edge in treated zone. From settlement point of view, Choa (1989), Tang and Shang (2000) and Chai et al. (2005) observed less settlement in vacuum preloading, whereas Woo et al. (1989), Shang et al. (1998), Mohamedelhassan and Shang (2002) presented a similar settlement by vacuum preloading and surcharge preloading at the same magnitude of pressure.

© Springer Nature Singapore Pte Ltd. 2018
L. Li et al. (Eds.): GSIC 2018, *Proceedings of GeoShanghai 2018 International Conference: Ground Improvement and Geosynthetics*, pp. 212–220, 2018.
https://doi.org/10.1007/978-981-13-0122-3_24

To explain the aforementioned contradiction and rethink the deformation behavior of soil subjected to vacuum preloading, numerical method is employed to depict the effective stress and the deformation of soil based on suitable initial conditions and boundary constraints. To explore the influence area for lateral displacement in treated zone, systematic numerical results are plotted based on different treated depth and width of soft soil, thus the behavior of soil can be discovered how sensitive of the scale effect, further verified by observations based on case studies.

2 Numerical Simulations on Vacuum Preloading

2.1 Numerical Model and Parameters

To investigate the deformation behaviour of soft soil subjected to vacuum preloading due to different treated breadths and depths, a series of numerical simulations are conducted with different dimensions of treated zone.

Numerical Cases. For simplicity and considering symmetry, plane strain problem will be taken into account to analyze the different deformations developed with changed treated dimensions. The soil is clay for the treated zone and untreated zone, and the half breadth of the treated zone B is set as 1 m, 5 m, 10 m, 20 m, 40 m, 80 m and 120 m, respectively. The thickness of the treated soil H is 10 m, 20 m, 30 m and 40 m, respectively. A piece of membrane is covered on the soil in treated zone and the penetration depth of membrane MD in the soil at the treated edge is varied according to the thickness of soil H. The typical values for MD are set as 0 m, 5 m, …, H. Also, for reference, a model case is set for surcharge at the same magnitude of vacuum with dimension of treated zone of $B = 120$ m and $H = 40$ m. The number of numerical modeling cases is 163 and a typical model dimension and mesh with $B \times H = 80$ m $\times$ 20 m and $MD = 10$ m (model code B80H20MD10) is shown in Fig. 1.

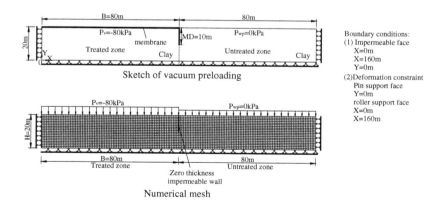

Fig. 1. Numerical modeling dimension and mesh for a typical case B80H20MD10

Soil Parameters and Numerical Idealization. Modified Cam-clay model is employed for the clay soil. The parameters are chosen as: $N = 2.212$, $\lambda = 0.25$, $\kappa = 0.05$, $v = 0.366$, $M = 0.98$, dry density = 1000 kg/m^3, $\gamma_w = 10$ kN/m^3. The initial void ratio is $e_0 = 1.6$ for the entire soil and the permeability is constant as $k = 0.000268$ m/d. The coefficient of earth pressure at rest is $K_0 = 0.577$, and the theoretical value is $K_{0MC} = 0.755$ deduced based on modified Cam-clay model. Since membrane is too thin to being simulated as a material compared to the soil, it is dealt with numerical idealization for its impermeable property.

Plane strain problem is considered in ABAQUS programming and the element type is CPE4P for the soil. The membrane on the top surface is modeled using the pore water pressure boundary during vacuum preloading process. The membrane inserted into the soil at the edge is modeled as a zero thickness impermeable wall by linking the adjacent two nodes of treated zone and untreated zone at the same position. This approach is carried out by equating the displacements of the adjacent elements nodes at the treated edge to fore deformation continuity, excluding pore water pressure.

The soil is fully saturated and the initial pore water pressure is $P_{wp} = 0$ kPa on the top surface. The side faces and base boundary are impermeable. Pin support is set to the base boundary, so the movements are constrained. On the left and right face, roller supports are applied. The top surface is allowed to move during preloading process.

Numerical Procedure. The soil is balanced under gravity firstly, thus the initial stress state of soil could be met, and the initial pore water pressure is imposed in the soil to meet the fully saturated state. After that, a vacuum pressure is applied on the top surface of treated zone by changing the surface pore water pressure as $P_v = -80$ kPa, while $P_{wp} = 0$ kPa is still acted on the top surface of the untreated zone, the vacuum consolidation is therefore conducted. For the other cases, the dimensions of the treated zone as well as the membrane penetration depths are changed in other numerical models. While taking the surcharge as a reference, a pressure of 80 kPa is applied on the top surfaces of the treated zone and untreated zone. At these above processes, crack is not taken into account, since the purposes of this paper are to investigate the influence of scale of treated zone based on the soil behavior in treated zone.

2.2   Numerical Results

The final horizontal effective stress 0.5 m below the soil surface is plotted with horizontal distance as shown in Fig. 2(a) at $MD = 0$ m. The half breadth of treated zone B varies from 10 m to 120 m. It is observed that all the horizontal effective stresses are nearly equal in untreated zone. In treated zone, when $B = 10$ m, the horizontal effective stress varies from 80 kPa at the corner of treated zone to 95 kPa at the symmetry boundary, namely at the centre of treated zone. A passive compression zone is presented. When B increases to 80 m, the horizontal effective stress decreases at horizontal distance about 10 m. At horizontal distance of 100 m, the horizontal effective stress nearly equal to the indication by theoretical K_0 condition. This means from effective stress point of view that the inward lateral displacement in treated zone can be zero when B is larger than 100 m.

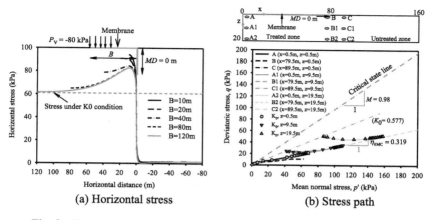

Fig. 2. Variation of horizontal effective stresses with distances (H = 20 m)

Figure 2(b) illustrates the effective stress path followed at different horizontal locations and vertical depths during vacuum preloading. At shallow depth, z = 0.5 m, the stress follows theoretical K_0 line in the centre at position A, while it departs the theoretical K_0 line near the edge of treated zone at position B and out of the treated zone at position C. At middle depth at z = 9.5 m, the stress in the centre (A1) and near the edge (B1) follows the theoretical K_0 line, whereas the stress out of treated zone at C1 is above the theoretical K_0 line. At deep depth at z = 19.5 m, all the stresses at different horizontal locations (A2, B2, C2) follow a theoretical K_0 line.

The surface settlement and horizontal strain along horizontal distance is shown in Fig. 3. Figure 3(a) shows the settlement at the centre line increases with the increase in treated width. The maximum settlements are 1.25 m, 1.58 m, 1.81 m, 1.89 m and 1.90 m for the case with B equals to 10 m, 20 m, 40 m, 80 m, and 120 m, respectively. Compare to surcharge case under K_0 condition, the maximum settlement ratios are 0.66, 0.83, 0.95,

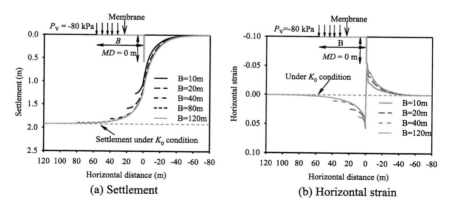

Fig. 3. Variation of surface settlements and horizontal strains with distances (H = 20 m)

0.99 and 1. From settlement point of view, the K_0 condition could be acceptable if the distance is larger than 60 m from the edge of treated zone.

The horizontal strain for soil element 0.5 m below surface is plotted in Fig. 3(b). The horizontal strain in treated zone is positive, i.e., the soil element is compressed, and at a distance larger than 60 m, the strain tends to be zero, thus the K_0 consolidation condition appears. While in untreated zone, the horizontal strain is negative, namely, the soil element is stretched. Also the tension strain appears to be zero when the distance is large enough from the treated edge.

The relationship between the dimensionless inward lateral displacement and half treated width is demonstrated in Fig. 4. Membrane penetration depths *MD* are allowed to vary from 0 m to 40 m. Take the surface lateral displacement at treated edge in case B120H40MD40 as a reference, the ratio of lateral displacement is defined as the ratio between the maximum lateral displacement and the reference lateral displacement. In Fig. 4(a), the ratio increases with the increase in treated breadth. When *B* is larger than 60 m, the ratio is very close to 1. When the treated thickness *H* increases, a similar trend is observed with that shown by treated thickness of *H* = 10 m. All the ratios are close to 1 when the treated breadth is larger than 60 m with little differences. At given treated breadth and treated thickness, the ratio increases with increase in membrane penetration depth. However, the magnitude of increase is limited.

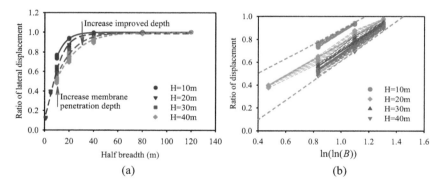

Fig. 4. Relationship between ratio of inward lateral displacement and treated width

Figure 4(b) depicts the ratio between maximum lateral displacement at the edge and the largest lateral displacement varies with treated width. At given treated thickness, the ratio increases with increase in breadth. At given breadth, the ratio decreases slightly with reduction of membrane penetration depth. All the ratios could reach 1 potentially. This means an influence zone can be determined and no lateral displacement can be estimated when the ratio equals to 1. The dash line intersecting the abscissa at ratio equal to 1 indicates the least *B* for each case to obtain the largest lateral displacement at given thickness. For *H* = 10 m, 20 m, 30 m, and 40 m, the least widths are 2.5*H*, 2.25*H*, 1.9*H* and 1.5*H*, respectively. This means the influence zone is less than 60 m for all treated thickness. On the other hand, an isotropic incremental loading condition could be induced when the treated width as well as the treated thickness is too small.

In general, from deformation point of view, the no lateral strain zone can be determined. The zone is about $2.5H$ from the edge of treated zone for commonly soil strata.

3   Field Observation of Soil Under Vacuum Preloading

It is investigated from above numerical results that the calculated settlement based on K_0 condition is applicable for large area vacuum preloading. However, for large area vacuum preloading, the calculated lateral displacement was overestimated compared to the measured based on isotropic incremental loading assumption (Chai et al. 2005). It is questionable that under what condition the deformation is developed mainly vertically during vacuum preloading for in site soil. Assuming no interaction between soil strata, an average lateral strain is gross simplified at given depth as

$$\varepsilon_{ha} = \delta_h / B \tag{1}$$

Where, δ_h = the maximum lateral displacement measured near the edge.

Similarly, within treated depth (vertical drain penetration depth) H, an average vertical strain is estimated roughly as

$$\varepsilon_{va} = (S_1 - S_2)/(h_2 - h_1) \tag{2}$$

Here, the settlement S_1 and S_2 are subsurface settlement at depth h_1 and h_2 developed in the centre of treated zone.

The relationship between strain ratios ($\varepsilon_{ha}/\varepsilon_{va}$) and depths is plotted in Fig. 5. The symbols with black fill represent a site with B/H less than 2.

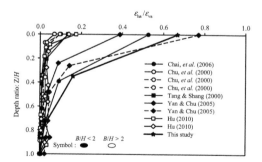

Fig. 5. Relationship between strain ratios and depths

It shows that the strain decreases with an increase with depth, except for that induced in Nansha site (GZPG 2004). The strain decreases with an increase with depth within 10 m. Then the strain ratio increases to a value greater than 0.1 at depth about

15 m for all the 4 lines. It is induced that a soil layer with low compressibility existed at that depth. The lateral displacement was developed due to interaction of soil strata. Generally speaking, the lateral strain is larger at shallow depth with respected to vertical strain at the same depth. It is also found that when $B/H > 2$, the strain ratio is normally less than 0.1.

Figure 6 illustrates various strain ratios induced from literatures. For ease of comparison, the ground surface settlement in the centre and the lateral displacement near the edge at ground surface are adopted for estimation of overall vertical strain and overall lateral strain. It is observed that all the strain ratios are less than 0.8. At B/H equals to 1, the strain ratio is in the range of 0.1 to 0.7. It is suggested that an overall plane strain condition or near isotropic incremental loading condition is followed. At $B/H > 2.5$, the overall strain ratio is less than 0.1. Actually, the element strain ratio is larger at the corner close to the boundary and attenuates to a lower value in the centre. This means the strain ratio at the centre of treated zone may be less than 0.05 at $B/H > 2.5$. It is believable that, from settlement point of view, a nearly K_0 condition remains, in other word, the influence zone for inward lateral displacement in the treated zone is about $2.5H$.

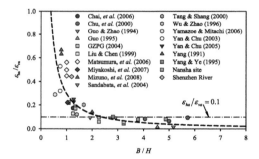

Fig. 6. Various strain ratios against width-depth ratio deduced from literatures

Actually the strain ratio is governed by soil properties and effective stress. Despite of width-depth ratio, the lateral displacement is also attributable to membrane penetration depth as discussed previously, which results different increase in effective stress around the boundary.

The above observations are confirmed by preceding numerical parametric studies. However, from effective stress point of view, the influence zone is larger. In the case of prediction of settlement, it is confident that the conventional C_c–C_r method is applicable for large area treated zone. For those with a treated width ($2B$) less than $4H$, a discount factor may be required for the estimation of settlement.

4 Conclusions

The behavior of soft soil subjected to vacuum preloading is strongly influenced by the scale of treated zone, especially with treated width. The larger the treated area is, the less the horizontal stress is produced by vacuum loading, thus the stress path follows nearly on K_0 consolidation line for soil element at the centre of large area treated zone or at deep depth.

The larger the vacuum preloading area is, the larger the settlement in the centre is predicted by numerical simulation, so does the inward lateral displacement at the edge of treated zone, and the maximum settlement in the centre is close to settlement under K_0 consolidation condition when the distance from the edge of treated zone is larger than $2.5H$, while lateral displacement is negligible at the centre.

For compressible soil subjected to vacuum preloading, the ratio of average horizontal strain over vertical strain usually decreases with depth if no existence of intercalated incompressible soil. At $B/H < 2.5$, the strain ratio of average horizontal strain over vertical strain is in the range of 0.1 to 0.7. At $B/H > 2.5$, the strain ratio is less than 0.1. This indicates that the soil behavior follows nearly K_0 consolidation condition $2.5H$ far away from the edge of treated zone.

References

Chai, J.C., Carter, J.P., Hayashi, S.: Ground deformation induced by vacuum consolidation. J. Geotech. Geoenvironmental Eng. (ASCE) **131**(12), 1552–1561 (2005)

Choa, V.: Drains and vacuum preloading pilot test. In: Proceedings of the 12th ICSMFE, Rio De Janeiro, pp. 1347–1350 (1989)

Cognon, J.-M.: Vacuum consolidation (La consolidation atmosphérique). Révue Française Géotechnique **57**, 37–47 (1991). (in French)

GZPG: Report on improvement of reclaimed soft soil in Nansha Harbour, Guangzhou, Guangzhou Port Group (2004)

Hu, L.W., Ng, C.W.W., Wang, Y.P.: Centrifuge modelling on underwater soft soil subjected to vacuum preloading. In: Proceedings of the 3rd International Conference on Ground Improvement and Ground Control, 27–29 Oct 2017, Hangzhou, China, pp. 123–136 (2017)

Kjellman, W.: Consolidation of clay by means of atmospheric pressure. In: Proceedings of the Conference on Soil Stabilization, Boston, 18–20 June, Massachusetts Institute of Technology, pp. 258–263 (1952)

Leong, E.C., Soemitro, R.A.A., Rahardjo, H.: Soil improvement by surcharge and vacuum preloadings. Geotechnique **50**(5), 601–605 (2000)

Mohamedelhassan, E., Shang, J.Q.: Vacuum and surcharge combined one-dimensional consolidation of clay soils. Can. Geotech. J. **39**(5), 1126–1138 (2002)

Qian, J.H., Zhao, W.B., Cheung, Y.K., Lee, P.K.K.: The theory and practice of vacuum preloading. Comput. Geotech. **13**(2), 103–118 (1992)

Shang, J.Q., Tang, M., Miao, Z.: Vacuum preloading consolidation of reclaimed land: a case study. Can. Geotech. J. **35**(5), 740–749 (1998)

Tang, M., Shang, J.Q.: Vacuum preloading consolidation of Yaoqiang Airport runway. Geotechnique **50**(6), 613–623 (2000)

Tran, T.A., Mitachi, T.: Equivalent plane strain modeling of vertical drains in soft ground under embankment combined with vacuum preloading. Comput. Geotech. **35**(5), 655–672 (2008)

Woo, S.M., Van Weele, A.F., Chotivittayathanin, R., Trangkarahart, T.: Preconsolidation of soft Bangkok clay by vacuum loading combined with non-displacement sand drains. In: Proceedings of 12th ICSMFE, Rio De Janeiro, pp. 1431–1434 (1989)

Research on Laboratory Mixing Trial of Marine Deposit and Cement in Hongkong

Yingxi He[1(✉)], Heping Yang[2], Jiayong Yang[3], Haipeng Liu[2], and Pengcheng Zhao[4]

[1] Second Engineering Company of CCCC Fourth Harbour Engineering Co., Ltd., Guangzhou 510000, China
190172945@qq.com
[2] Changsha University of Science and Technology, Changsha, China
[3] China Forest Exploration and Design Institute in Kunming, Kunming, China
[4] China Design Group Co., Ltd., Shanghai, China

Abstract. For improving soft foundation of marine deposit in a seawall project in Hongkong, ground improvement of Deep Cement Mixing ("wet method") was firstly researched and applied for the first time. A series of testing including physical & mechanical characteristics and chemical components for marine deposit, and researches on laboratory mixing trial of marine deposit and cement was carried out. By analyzing testing results from laboratory trial, several main factors of soil properties (e.g. natural moisture content, sand fraction, organic content of soil) and cement content, curing age influencing the strength of cement-soil mixture were researched. The correlation of cement-soil mixture strength and different curing age are emphatically analyzed. And it's found that strength of cement-soil mixture increased with the increase of cement content, and the strength increase with curing age is more dominant for the stabilized soil with a larger amount of cement (max. content 280 kg/m3). Testing results demonstrated that strength of cement-soil from 3 kinds of cement content in lab mix meet design requirements. Based on above research results, the feasibility of using DCM to improve the marine deposit in Hongkong is preliminarily verified, and important technical parameters (e.g. economical cement content) were suggested to be applied in following field DCM field trial.

Keywords: Laboratory mixing trial · Deep cement mixing · Marine deposit
Ground improvement

1 Introduction

The deep mixing method is a deep in-situ admixture stabilization technique using cement or cement-based special binders. Compared to the other ground improvement techniques deep cement mixing has advantages such as the considerable strength increase within short period, few adverse impacts on environment and high applicability to any kind of soil if binder type and amount are properly selected. The application covers on-land and in-water constructions ranging from strengthening the foundation ground of buildings, embankment supports, earth retaining structures, and urban infrastructures, liquefaction hazards mitigation, artificial island constructions, and etc.

© Springer Nature Singapore Pte Ltd. 2018
L. Li et al. (Eds.): GSIC 2018, *Proceedings of GeoShanghai 2018 International Conference: Ground Improvement and Geosynthetics*, pp. 221–229, 2018.
https://doi.org/10.1007/978-981-13-0122-3_25

Due to the various advantages of DCM, the total volume of stabilized soil by the mechanical deep mixing method from 1975 to 2010 reached 72.3 million m^3 for the wet method of deep mixing in the Japan. In the 1990s deep mixing gained popularity also in Southeast Asia, the United States of America and central Europe.

In Hongkong, Deep Cement Mixing by wet method was researched and applied to improve soft ground of marine deposit in a marine project for first time from 2016. Sole binder of cement was adopted to improve soft marine sediments with high moisture content and high void ratio. For verification of the feasibility of using cement to improve the strength characteristics of marine deposit for the foundation in a seawall project in Hongkong, the laboratory mixing trials for cement and soft marine silt/clay were carried out.

2 Insitu Soil Conditions

Based on information from geotechnical investigation conducted in the project, characteristic of marine deposits is described as following: (1) Marine Deposits of the Hang Hau Formation form the seabed; (2) these deposits comprise mostly very soft to soft, grey clayey silt and are relatively homogeneous throughout the area; (3) Shear strengths in the undrained state vary from less than 3 to 20 kPa (CEDD 2002); minor silt and sand lenses locally present throughout the sequence; (4) disarticulated and articulated bivalves are common, as shell debris ranging in size from less than 0.1 to 20 mm (CEDD 2002). The physical and chemical properties of the soils are tabulated in Stabilization Mechanism of cement and soil

Table 1. Physical and chemical properties of the soils

	Particle Size Distribution				Thickness (m)	Liquid limit, w_L (%)	Plastic limit, w_p (%)	Plasticity index, I_p	Natural water content, w (%)	Void ratio (%)	Dry density ρ_d (g/cm^3)	Chemical properties	
	Gravel (%)	Sand (%)	Silt (%)	Clay (%)								Organic content	PH
Min.	0.0	1.0	21.0	23.0	6.0	52	31	16	51	1.35	0.79	0.7	6.5
Max.	14.0	23.0	66.0	51.0	19.0	61	40	24	105	2.86	1.12	3.5	8.1
Average	1.8	17.1	45.1	35.9	13.0	56	34	19	83	2.21	0.89	2.2	7.3

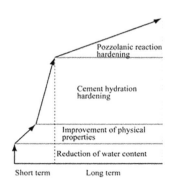

Fig. 1. Mechanism of cement stabilization

The mechanism of stabilizations by cement to improve soft soil is illustrated in Fig. 1, which consists of four steps: the hydration of binder, ion exchange reaction, formation of cement hydration products, and formation of pozzolanic reaction products (Kitazume and Terashi 2013).

3 Stabilization Mechanism of Cement and Soil

The mechanism of stabilizations by cement to improve soft soil is illustrated in Fig. 1, which consists of four steps: the hydration of binder, ion exchange reaction, formation of cement hydration products, and formation of pozzolanic reaction products (Kitazume and Terashi 2013).

4 Lab Mixing Trial for Cement and Natural Soil

4.1 Method of Laboratory Mixing

The laboratory mix trial was conducted according to the Japanese Geotechnical Society guideline 0821-2009: Practice for Making and Curing Stabilized soil Specimens without Compaction (Japanese Geotechnical Society 2009).

The detailed mixing procedure and Sample testing procedure adopted by laboratory is shown in Fig. 2.

In the laboratory mix trial, a test specimen is produced, cured and tested by the following steps:

(1) Sampling natural soil from site,
(2) disaggregation and homogenization of original soil in lab,
(3) preparation of binder-water slurry at prescribed water/binder ratio,
(4) mixing of soil and binder-slurry to prepare uniform soil-binder mixture (about 12 min),

Fig. 2. Procedure of cement-soil laboratory mixing

(5) rest time before the molding,
(6) filling the soil-binder mixture into the molds,
(7) test specimen curing for 28, 56, 90 day.
(8) **Unconfined Compressive Strength (UCS)** testing for curing age of 28, 56, 90 day.

4.2 Adopted Parameters of Laboratory Mixing Trial

The lab mixing trial were conducted on marine deposit (marine silt/clay) from a marine project in Hongkong, and summary of soil physical and chemical properties is shown in Stabilization Mechanism of cement and soil.

The mechanism of stabilizations by cement to improve soft soil is illustrated in Fig. 1, which consists of four steps: the hydration of binder, ion exchange reaction, formation of cement hydration products, and formation of pozzolanic reaction products (Kitazume and Terashi 2013).

Table 1. For laboratory mixing trial, 5 different content of cement, D of 160, 180, 200, and 220 kg/m^3 with a special highest content of 280 kg/m^3, and 3 different curing age of 28, 56 and 90 days were adopted. The cement content, D is defined as a dry weight of cement added to 1 m^3 of natural soil.

In the laboratory mix trial, the marine silt/clay with an initial water content of 40–90% was stabilized with ordinary Portland cement (type: EN 197-1- CEM I, class 32.5) with water to cement ratio, W/C of 80%.

4.3 Analysis of Testing Results

The strength of the stabilized soil is influenced by many factors including original soil properties, type and content of cement, curing age and mixing process, etc.

Correlation Between Cement Content and UCS

Figure 3 shows the influence of the cement content of, D on the unconfined compressive strength, q_u, in which the marine silt/clay was stabilized with ordinary Portland cement, and tested at 3 curing periods. The unconfined compressive strength increases with the amount of cement. Testing results show that q_{ul28} (UCS of age 28 day in lab mix) ranges within 1.15 to 1.83 MPa and with mean value of around 1.51 for content of 160 kg/m^3, and q_{ul28} ranges within 1.59 to 4.83 MPa and with mean value of around 2.66 MPa for content of 220 kg/m^3, and q_{ul90} (UCS of age 90 day in lab mix) ranges within 1.85 to 3.24 MPa with mean value of around 2.43 MPa for content of 160 kg/m^3, and q_{ul90} ranges within 1.21 to 5.25 MPa with mean value of around 3.05 MPa for content of 220 kg/m^3, and for highest content 280 kg/m^3 the average q_{ul90} of 3.37 MPa in curing age 90 days. In Fig. 4, percentage of UCS larger than 2 MPa is 38.5%, 74.4%, 82.8% and 88.5% respectively for content 160, 180, 200 and 220 kg/m^3.

As the minimum design UCS for field DCM columns is 1.5 MPa in curing age of 90 days, based on the testing results from lab mixing trial, 3 nos. cement content (160, 180, and 200 kg/m^3) are suggested to be applied in following field DCM trial carried out by DCM specialist barges.

Correlation Between UCS and Curing Age

Figure 5 shows the influence of curing age on the mixture strength of cement and marine silt/clay at the content of 160–280 kg/m^3. The testing results illustrate that UCS of marine silt/clay stabilized by sole cement increases with increasing curing age.

The relationships between the strength of stabilized soil at two different curing periods have been studied. In Fig. 6, strength ratio q_{ul56}/q_{ul28} (q_{ul56}: UCS of age 56 day in lab mix) ranges within 1.04 to 1.8 with mean value of 1.31 for four kinds of cement content (from 160–220 kg/m^3), while q_{ul90}/q_{ul28} ranges within 1.08 to 1.84 with mean value of 1.45. Although strength ratio, q_{ul90}/q_{ul28}, depends on the soil properties and amount of binder, the q_{uf90}/q_{uf28} (q_{uf90}: UCS of age 90 day in field mixing) of 1.45 was tentatively adopted for prediction of q_{uf90} using q_{uf28} in following field DCM trial.

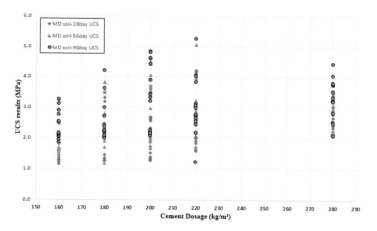

Fig. 3. UCS vs cement content in curing age of 28, 56 and 90 days

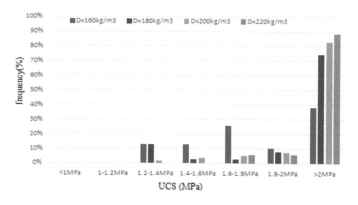

Fig. 4. Frequency vs UCS range in curing age of 28, 56 and 90 days

From Fig. 5, for content 280 kg/m^3, the results show that the average q_{ul90} of 3.37 MPa in curing age 90 days increase by 22% compared with the average q_{ul28} of 2.76 MPa in curing age 28 days, while average q_{ul28} of 2.76 MPa for content 280 kg/m^3 increase slightly by about 0.1 MPa compared with average q_{ul28} of 2.66 MPa for content 220 kg/m^3. It can be concluded that the strength increase with curing age is more dominant for the cement-soil mixture with a larger amount of cement (max. content 280 kg/m3).

Correlation Between UCS and Density of Mixture

In Fig. 8, the correlation of unconfined compressive strength q_u and density of cement-soil mixture shows good regulation, and q_u increase with mixture density increase, with q_u range from 1.15 MPa to 4.1 MPa for mixture density from 1540 to 1870 kg/m^3 for cement content of 160–220 kg/m^3 (Fig. 7).

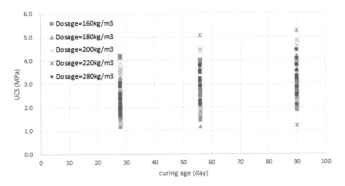

Fig. 5. UCS vs curing age for content 160–220 kg/m^3

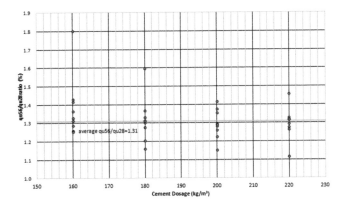

Fig. 6. Cement content vs strength ratio q_{u56}/q_{u28}

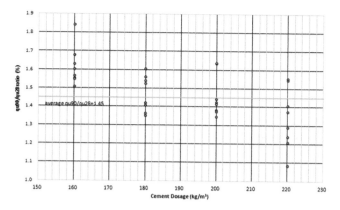

Fig. 7. Cement content vs q_{u90}/q_{u28}

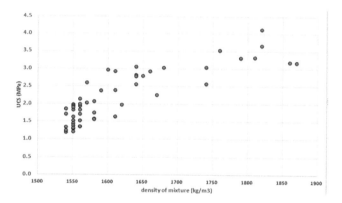

Fig. 8. Density of cement-soil mixture vs UCS

Correlation Between UCS and Properties of Natural Soil

The large variation of strength found from Figs. 9, 10 and 11 clearly shows that the strength gain by cement stabilization heavily depends upon the type and properties of natural soil. The organic content and pH of original soil are the most dominant factors influencing the strength (Kitazume and Terashi 2013).

The influence of the initial water content of natural soil on the unconfined compressive strength, q_u is shown in Fig. 9. The unconfined compressive strength decreases with increasing initial water content irrespective of the soil properties and cement content.

Figure 9 shows that the strength of stabilized soils decreases rapidly with the total water content. For marine silt/clay with water content higher than 78% increase of cement content does not lead to evident strength increase.

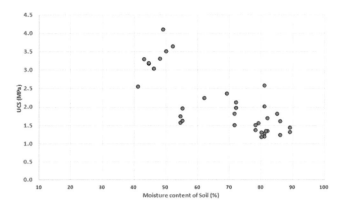

Fig. 9. 28d UCS vs natural moisture content of soil

In Fig. 10, the unconfined compressive strength for mixture of marine deposit and cement, q_u is dependent upon the sand content in natural soil and increase with sand content increase, with q_u range from 1.1–2.0 MPa and 2.3–3.2 MPa respectively for sand fraction 7% and 15%.

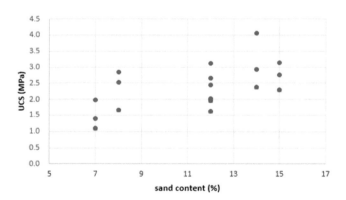

Fig. 10. Sand content of soil vs UCS

In Fig. 11, the correlation of unconfined compressive strength q_u and organic content in natural soil is not in good regulation, but also qu decrease with organic content increase, with q_u range from 2.3–4.1 MPa and 1.5–2.6 MPa respectively for organic content with 1.8% and 3.2%.

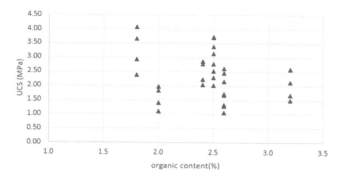

Fig. 11. Organic content of soil vs UCS

5   Conclusion

(1) Find the valuable and important correlation between the UCS of cement-soil mixture and cement content.

(2) The UCS of cement-soil mixture increases with the curing period increasing, and the strength increase with curing age is more dominant for the stabilized soil with a larger amount of cement. The strength ratio, q_{uf90}/q_{uf28}, of 1.45 was tentatively adopted for prediction of q_{uf90} using q_{uf28} in following field DCM trial.

(3) The properties of marine silt/clay, mainly including natural moisture content, sand fraction, organic content of soil, influence significantly the strength of cement-soil mixture and also results in the big variation of cement content.

(4) 3 nos. economical cement content (160, 180, and 200 kg/m³) are adopted to be applied in following field DCM trial carried out by DCM specialist barges.

(5) Above research achievements in laboratory mixing trial verify preliminarily the feasibility using DCM ("wet method") to improve the marine deposit in Hongkong.

References

Civil Engineering and Development Department (CEDD). Geology of Tung Chung and northshore Lantau Island. Geotechnical Engineering Office, Civil Engineering and Development Department, Hong Kong (2002)

Kitazume, M., Terashi, M.: The Deep Mixing Method. Taylor & Francis Group, London (2013)

Practice for Making and Curing Stabilized Soil Specimens without Compaction (JGS 0821-2009). Japanese Geotechnical Society (2009)

Experimental Investigation of Black Cotton Soil Stabilized with Lime and Coconut Coir

A. U. Ravi Shankar[✉], B. J. Panditharadhya, Satish Karishekki,
and S. Amulya

National Institute of Technology Karnataka, Surathkal, India
aurshankar@gmail.com

Abstract. Expansive soil occurring above the water table undergo volume changes with change in moisture content. In expansive soils, increase in water table causes swelling–shrink behaviour which leads to cracks and differential settlement resulting in several damages to the pavements, canal beds and linings, foundations, buildings, etc. An attempt is made in this paper to study the effect of adding lime-coir fiber on geotechnical properties of black cotton soil.

In the present study an effort is made to obtain the optimum dosage of lime for stabilization of black cotton soil abundantly available in Karnataka state of India. The study incorporates investigation of basic geotechnical properties like grain size distribution, specific gravity, consistency limits and engineering properties like Maximum Dry Density (MDD), Optimum Moisture Content (OMC), Unconfined Compressive Strength (UCS) and California Bearing Ratio (CBR). Swelling properties have been determined by conducting Free Swell Index (FSI) test. Durability of the soil is studied by conducting wet-dry cycle and freeze-thaw cycles (WD and FT tests). Fatigue test has been conducted to determine the fatigue life of treated and original soil. Further chemical analysis was conducted to determine the chemical composition of untreated and treated soil. The optimum dosage of lime obtained was 4%. The investigations were carried out to study the effect of addition of coir fibers which are obtained from local market to evaluate the extent of modification on MDD, OMC, UCS and CBR of the soil. Maximum improvement in UCS and CBR values are observed when 1% of coir are mixed with the soil. Soil stabilized with Lime-Coir fiber has shown better results when compared to soil stabilized with lime alone. It is concluded that the proportion of 1% coir fiber in a soil is the optimum percentage of materials having maximum soaked CBR value. Hence, this proportion may be economically used in road pavement and embankments.

Keywords: Black cotton soil · Free swell index (FSI) · Lime-coir fiber
Unconfined Compressive Strength (UCS) · California Bearing Ratio (CBR)
WD and FT tests

1 Introduction

Expansive soil is one among the problematic soil that has a high potential for shrinking or swelling due to change of moisture content. Destructive results caused by this type of soil have been reported in many countries. Rural roads, national and state highways

© Springer Nature Singapore Pte Ltd. 2018
L. Li et al. (Eds.): GSIC 2018, *Proceedings of GeoShanghai 2018 International Conference: Ground Improvement and Geosynthetics*, pp. 230–243, 2018.
https://doi.org/10.1007/978-981-13-0122-3_26

in the North Karnataka are facing problem due to early distress in the pavement due to the poor strength of soil under wet condition. Sub-grade is the major component of a pavement. The performance of the pavement is dependent on the type and properties of the sub-grade soil. Various techniques are used for stabilization of sub-grade soil. Hence, it is necessary to have a proper diagnostic study of the soil to be used as sub grade.

Soil stabilization is the process of creating or improving certain desired properties in a soil material so as to render it stable and useful for a specific purpose. Since the inception of this process of stabilization, most soil materials which have been thought not useful have found application in many areas of engineering. Thagesen (1996) defines stabilization as any process by which a soil material is improved and made more stable. Garber and Hoel (Garber and Hoel 2000) described soil stabilization as the treatment of natural soil to improve its engineering properties. McNally et al. (1998) states that the improvements in engineering properties caused by stabilization can include the following: increases in soil strength (shearing resistance), stiff-ness (re-sistance to deformation) and durability (wear resistance), reductions in swelling potential or dispersivity (tendency to deflocculate) of wet clay soils and other desirable characteristics, such as dust proofing and water proofing unsealed roads. Stabilization of soil is employed when it is more economical to overcome a deficiency in a readily available material than to bring in one that fully complies with the requirements of specification for the soil (Ola 1975). It has been regarded as a last resort for upgrading sub-standard materials where no economic alternative is available. A continual refer-ence to economy here denotes a careful consideration of all costs that would be incurred by importation (not readily available) of a compliant soil and comparing this to the cost of improving the properties of an unstable but readily available soil. Interest in the art of soil stabilization grew with a better appreciation of the cyclic loading effects of heavy traffic which creates a need for stronger pavements that often cannot be provided by realistic thickness of unbound granular materials, and the availability of purpose built in–situ stabilization equipment that improves homogeneity of mix. Although road construction has been the major area of application of soil stabilization techniques, they have also been applied in soil foundation strengthening, although to a limited extent. The principal additives employed for soil stabilization are:

- gravel, crushed aggregate, grit and loan
- Portland cement and cement - slag blends
- Lime (quick - lime, hydrated lime) and gypsum
- Lime - pozzolan (lime plus fly ash or ground slag) mixtures
- Asphlt.

All agents for stabilization have particular soil material to which when they are applied would produce the required properties. McNally (1998) explained that the selection of an appropriate stabilizing agent and construction procedures involves a number of considerations.

1.1   Need for Soil Stabilization

Expansive soils are known to behave differently than conventional soils. As a result of seasonal moisture variations, these soils shrink in summer and swell in rainy season causing downward and upward movement of the structures. Such soil exhibits very low shear strength in saturated conditions, whereas very high shear strength in case of dry condition. These soils due to such behaviour are considered as problematic for usage in subgrade. There are many works in past that deals with stabilization of Black cotton soil with lime alone and also there are many works using Coconut Coir. In this investigation, the comparison is done to know the effect of lime addition to Black cotton soil and lime-coconut coir (two different proportions) addition to Black cotton soil. Thus construction cost can be considerably reduced by stabilizing the soil.

2   Literature Studies

In India, large tracts are covered by expansive soil referred as "Black Cotton Soil" (BC soils). They are so named because of their suitability for growing cotton. Black cotton soils have varying colors ranging from light grey to dark grey and black. BC soils are inorganic clays of medium to high compressibility and form a major soil group in India. They are characterized by high shrinkage and swelling properties. Because of its high swelling and shrinkage characteristics, the BC soils have been a challenge to the highway engineers. The BC soils are very hard when dry, but lose its strength completely in wet condition. It was observed that on drying, the BC soils develop cracks of varying depth. The mineralogy of the soil is dominated by the presence of montmorillonite which is characterized by large volume change from wet to dry season and vice versa. Deposits of BC soils in the field show a general pattern of cracks during the dry season of the year. Cracks measuring 70 mm wide and over 1 m deep have been observed and may extend up to 3 m or more in case of high (Adeniji 1991). As a result of wetting and drying process, vertical movement takes place in the soil mass. All these movements lead to failure of pavements, in the form of settlement, heavy depression, cracking and unevenness.

2.1   Stabilization with Lime

Lime stabilization is one of the oldest process of improving the engineering properties of soils and can be used for stabilizing both base and sub base materials (Garber and Hoel 2000). The need to replace clay soil in the embankment/sub grade layer is eliminated/reduced with Lime treatment. When lime reacts with soil, there is exchange of cations in the adsorbed water layer and a decrease in plasticity of the soil occurs. The resulting material is more friable than the original clay, and is, therefore, more suitable as sub-grade. Lime can be used to treat soils in order to improve their workability and load-bearing characteristics in a number of situations. Lime is frequently used to dry wet soils at construction sites and elsewhere, reducing downtime and providing an improved working surface. An even more significant use of lime is in the modification and stabilization of soil beneath road and similar construction projects. Lime can substantially increase the stability, impermeability, and load-bearing capacity of the subgrade. Both quicklime and hydrated lime may be used for this purpose. Application of lime to sub-grades can provide significantly improved engineering properties. The black cotton soil treated with 4% to 6% of lime can be safely used as subgrade and subbase material as the soaked CBR increases considerably, thereby cutting down the thickness of the upper crust of the road (Gaulkar 1991).

2.2   Stabilization with Coconut Fibers

The outer covering of fibrous material of a matured coconut, termed coconut husk, is the reject of coconut fruit. The fibers are normally 50–350 mm long and consist mainly of lignin, tannin, cellulose, pectin and other water soluble substances. Coir retains much of its tensile strength when wet. It has low tenacity but the elongation is much higher (Babu and Vasudevan 2008). Mainly, coir fiber shows better resilient response against - synthetic fibers by higher coefficient of friction. For instance, findings show that coir fiber exhibits greater enhancements (47.50%) in resilient modulus or strength of the soil than the synthetic one (40.0%) (Ayyar et al. 1988). Viswanadham (1989) reported that the efficacy of randomly distributed coir fibers in reducing the swelling tendency of the soil. for coir-stabilized lateritic soils, the maximum dry density (MDD) of the soil decreases with addition of coir and the value of optimum moisture content (OMC) of the soil increases with an increase in percentage of coir. The compressive strength of the composite soil increases up to 1% of coir content and further increase in coir quantity results in the reduction of the values. The percentage of water absorption increases with an increase in the percentage of coir. Tensile strength of coir-reinforced soil (oven dry samples) increases with an increase in the percentage of coir. The degradation of coir depends on the medium of embedment, the climatic conditions and is found to retain 80% of its tensile strength after 6 months of embedment in clay. Unlike synthetic reinforcing materials, degradation of coir takes place much more slowly due to its high lignin content (about 40–60%) (Vinod et al. 2009). So, the fiber is also very long lasting, with infield service life of 4– 10 years. The water absorption of coir fiber is about 130–180% and diameter is about 0.1–0.6 mm soil (Rowell et al. 2000). Coir is a cheap and abundant waste material in India, Indonesia, Brazil, Sri Lanka and in some other Asian countries where coconuts are grown and subsequently processed. Coir geo-textiles are presently available with wide ranges of properties which can be economically utilized for temporary reinforcement purposes (Subaida et al. 2009).

Lime treatment in BC soil improves strength but it imparts brittleness in soil specimen. Black cotton soil treated with 4% lime and reinforced with coir fiber shows ductility behaviour before and after failure. An optimum fiber content of 1% (by weight) with aspect ratio of 20 for fiber was recommended for strengthening the BC soil (Ramesh et al. 2010a, 2010b).

2.3   Durability Studies

Variations in climatic conditions have been recognized by pavement engineers as a major factor affecting pavement performance. These variations resulting from freeze–thaw and wet–dry actions, or a combination of these actions, have been presented in a number of previous studies. Importance of climatic conditions has been emphasized by Little (2005). The influence of such actions on a pavement structure indicates possible changes in the engineering properties of associated pavement materials. In this regard, several studies have been undertaken to evaluate the performance of pavement materials under these actions. Specifically, during the last few decades increased emphasis has been placed by transportation agencies and researchers to better understand the behaviour of stabilized aggregate bases and subgrade soil under freeze–thaw and wet-dry cycles. This research area, however, is still not fully explored and additional studies are needed. Little (2005) investigated the effect of cyclic wetting and drying on the expansive characteristics of clays. In this study, six expansive soils were obtained from various locations in Irbid (a city in northern Jordan). After each cycle the swell potential and swell pressure were measured. The experimental data indicated that upon repeated wetting and drying the soil showed sign of fatigue after every cycle resulting in decreased swelling ability. Furthermore it was noted that the first cycle causes the most reduction in swelling potential. As the number of cycles increases additional reduction was observed until an equilibrium state is reached.

Scanning electron micrographs clearly showed a continuous re-arrangement of particles during cyclic wetting and drying. This led to lower structural element orientation due to the integration of structure along the bedding resulting in correspondingly lower water absorption thus reducing swelling ability. Cyclic wetting and drying results in particle aggregation as demonstrated by the reduction in clay content and plasticity (Liquid Limit and Plasticity Index) between the initial and final cycles. This inevitably caused a reduction in swelling characteristics (Homoud et al. 1995).

3 Experimental Investigations

3.1 Materials

In the present study, commercially available Lime is used to stabilize the soil. Lime is a whitish grey solid having a crystalline structure. Lime is highly reactive with water, generating considerable heat in the hydration process. This material will react with the moisture in the air, and as such, it can be used as a desiccant. In the presence of moisture, lime reacts slowly with carbon dioxide in the air, reforming calcium carbonate. As a chemically active material it is desirable to reduce atmospheric exposure during handling and storage to a minimum. Hydrated lime, though only slightly soluble in water, forms suspensions easily; the resulting solution and suspension is strongly alkaline, possessing a pH of 12.4. In this study, 2, 4, 6 and 8% lime were used for the preparation of lime-stabilized samples.

After fixing the Lime dosage, the locally available coir is collected and fibers were taken for stabilizing the mix in the percentages of 0.25, 0.50, 0.75 and 1.00%. Aspect ratio of 20 and maximum of 1.00% addition of coir fibers to the soil is considered based on the literature studies.

3.2   Experimental Investigations

Standard and Modified Proctor Tests

There is optimum moisture content for a soil, at which maximum dry density is attained for a particular type and amount of compaction. To assess the amount of compaction and water content required in the field, compaction tests are conducted. In the present study Standard Proctor Compaction test as per IS: 2720 (Part VII) 1980 was conducted for untreated and treated BC Soil. The Results obtained for Standard Proctor test for both untreated and treated soil are given in Table 1. In the present study Modified Proctor Compaction test as per IS: 2720 (Part VIII) 1983 was conducted on untreated soil and as per IS: 4332 (Part 3) 1967 for soil treated with varying percentages (2%, 4%, 6% and 8%) of Lime and treated soil (with optimum lime content 4%) with varying percentages (0.25%, 0.50%, 0.75%, 1.0% Coir by weight). The compaction test is done immediately after treating it with the stabiliser. The Results obtained for Modified Proctor test for both untreated and treated soil are given in Table 2.

Table 1. Standard Proctor Test Results for untreated and treated BC soil

Sample	OMC (%)	MDD (g/cc)
BC Soil	18.42	1.62
BC Soil + 2% Lime	19.25	1.616
BC Soil + 4% Lime	20.64	1.602
BC Soil + 6% Lime	22.06	1.59
BC Soil + 8% Lime	23.87	1.582
BC Soil + 4% Lime + 0.25% coir	20.74	1.598
BC Soil + 4% Lime + 0.5% coir	22.52	1.58
BC Soil + 4% Lime + 0.75% coir	23.69	1.567
BC Soil + 4% Lime + 1.0% coir	24.35	1.554

Table 2. Modified Proctor Test Results for untreated and treated BC soil

Sample	OMC (%)	MDD (g/cc)
BC Soil	17	1.703
BC Soil + 2% Lime	17.83	1.692
BC Soil + 4% Lime	18.76	1.68
BC Soil + 6% Lime	20.15	1.674
BC Soil + 8% Lime	21.16	1.662
BC Soil + 4% Lime + 0.25% coir	19.5	1.68
BC Soil + 4% Lime + 0.5% coir	21.02	1.664
BC Soil + 4% Lime + 0.75% coir	22.21	1.648
BC Soil + 4% Lime + 1.0% coir	23.14	1.635

The results show that the addition of lime tend to increase the OMC and reduce the MDD.

Table 3. UCS results of untreated and treated soil for different curing period

Lime dosage (%)	UCS in kPa				
	Curing period (Days)				
	0 days	7 days	14 days	28 days	60 days
0	266	298	310	324	358
2	307	484	550	615	686
4	434	788	968	1056	1142
6	508	933	1071	1188	1313
8	611	1017	1162	1307	1492

Table 4. UCS values for addition of various percentages of Coir + Lime 4% for different curing period

Coir (%) by weight	BC Soil + Lime 4% + Coir %			
	UCS in kPa			
	Curing period (Days)			
	0 days	7 days	14 days	28 days
0	266	298	324	358
0.25	318	485	536	586
0.5	457	641	725	785
0.75	534	856	965	1038
1	680	1222	1382	1534

UCS Test

After treating the soils with stabilizer, the UCS test was carried as per pre-scribed standards (IS: 2720 (Part 5) – 1973) for different curing periods - 0 days, 7 days, 14 days, 28 days and 60 days. The results of UCS test are given below in Tables 3 and 4.

The relationship between UCS values can be observed from Tables 3 and 4. As the curing period increases, the UCS value of all the soil mixes with lime increases. The soil with 8% lime as shows higher UCS values at all curing periods. For normal soil there is no change in UCS as curing period increases. For 4% lime the UCS value after 60 days curing was 1142 kPa whereas for normal soil it was 358 kPa. Also, IRC:37 – 2012 says that stabilized soil having minimum UCS value of 1.5 MPa (1500 kPa) can be used as sub-base and minimum UCS value of 4.5 MPa (4500 kPa) can be used as base material as per strength criteria. Beyond 1% dosage of coir fibers, required mix is not achieved and chances of segregation of fibers is more. Though the UCS values are increasing as lime content increases, in the present investigation keeping economy in mind and the strength requirement, 4% lime is considered as lime fixation point and 1% in case of coir fibers.

The increase in UCS after lime application is as a result of flocculation, cation exchange and the formation of various cementitious compounds i.e., calcium silicate hydrates and calcium aluminate hydrates due to pozzolanic reaction between silica present in soil and lime. The reaction is initiated immediately after mixing and thereby shows strength gain even at 0 days of curing or initial testing condition without curing.

CBR Test

CBR tests were conducted for soaked and unsoaked condition for heavy compaction results. CBR tests for untreated soils, soil with varying percentage of Lime and soil with varying percentage of Coir + Lime 4% were conducted for different moist curing period for soaked conditions. For untreated BC soil, CBR values obtained were 24% and 2% for unsoaked and soaked conditions. The results for treated soil samples are given in Tables 5 and 6.

Table 5. CBR values of Lime stabilised BC soil for different curing period

Lime dosage (%)	Soaked CBR value (%)			
	Curing period (Days)			
	0 days	7 days	14 days	28 days
0	2	3	3	3
2	3	4	4	5
4	3	7	8	8
6	4	7	9	10
8	4	9	10	11

Table 6. CBR values of Coir + Lime stabilised BC soil for different curing period

Coir content (%)	BC Soil + Lime 4% + Coir fiber			
	Soaked CBR Value (%)			
	Curing period (Days)			
	0 days	7 days	14 days	28 days
0	3	3	3	3
0.25	5	6	7	8
0.5	5	8	9	9
0.75	6	9	10	12
1	6	12	13	15

The soaked CBR increased to a value of 8.27% at 28 days of curing at 4% addition of lime, which is 2.7 times that of the natural soil. In a similar manner when soil is treated with the various percentage of lime, strength is increased with the curing period.

Free Swell Index

The free swell index test is conducted to determine the amount of swelling in treated and untreated soil. The procedure followed is as per IS: 2720 (part-10) – 1977. The procedure involves in taking two oven dried soil samples (passing through 425μ IS sieve), 10 g each which are placed separately in two 100 ml graduated soil sample. Water is filled in one cylinder and kerosene (non-polar liquid) in the other cylinder up to 100 ml mark. The final volume of soil is read after 24 h to calculate free swell index. The free swell index of the soil shall be calculated as follows:

$$\text{Free swell index, percent} = ((Vd - Vk)/Vk) * 100$$

Where, Vd = the volume of soil specimen read from the graduated cylinder containing distilled water, and Vk = the volume of soil specimen read from the graduated cylinder containing kerosene (Table 7).

Table 7. Free swell index test results of untreated and treated soil

Property	BC soil	Lime content %			
		2	4	6	8
FSI (%)	40	34	26	19	12

Free Swell Index values were determined immediately after adding the lime without allowing the mix to become stiffer if left for curing.

Durability Tests

Soil samples have been prepared by varying lime percentage as 2, 4, 6 and 8%. Two samples were prepared for each percentage of lime. The durability tests, i.e., Wet-Dry (WD) and Freeze-Thaw (FT) procedures were adopted as per ASTM D559 and 560. Soil specimens with 76 mm height and 38 mm diameter were prepared and then they were subjected to 7 days moist curing. The test contains 12 cycles of each WD and FT. In wet cycle, specimens were submerged in water at room temperature for 5 h, then its dimensions and weight were taken. In dry cycle, the specimens were dried at a temperature of 71 °C for 42 h and specimens were thoroughly brushed parallel and again dimensions and weight were taken. This procedure is repeated for 12 cycles. In Freeze cycle, samples were placed in water-saturated felt pads and stood on carriers in a freezer at a temperature not higher than −10 °C for 22 h. Thawing was done by keeping them in a moisture room for 22 h and dimensions and weight were taken after brushing. The weight loss of specimen for WD and FT should not be more than 14% after 12 cycles.

When the untreated soil samples were immersed in water for the first cycle of wetting, the samples couldn't withstand and samples got collapsed within a fraction of second. The data collected had permitted calculations of volume and moisture changes of specimens after the 12 cycles of test. The percentage weight loss after each cycle are tabulated in below Tables 8 and 9.

Table 8. Percentage weight loss during alternate cycles of wetting and drying for different percentages of Lime

No. of cycles	Percentage weight loss							
	2% Lime		4% Lime		6% Lime		8% Lime	
	Wetting	Drying	Wetting	Drying	Wetting	Drying	Wetting	Drying
1	2.21	−13.47	1.34	−9.01	1.46	−7.77	0.86	−13.16
2	Collapsed		1.42	−10.27	1.69	−14.82	0.68	−19.11
3			Collapsed		1.08	−16.06	−1.98	−21.83
4					0.64	−17.57	−2.75	−15.58
5					Collapsed		−4.36	−18.38
6							−5.94	−20.28
7							−9.04	−21.57
8							Collapsed	
9								
10								
11								
12								

Table 9. Percentage weight loss during alternate cycles of freezing and thawing for different percentages of Lime

No. of cycles	Percentage weight loss							
	2% Lime		4% Lime		6% Lime		8% Lime	
	Freeze	Thaw	Freeze	Thaw	Freeze	Thaw	Freeze	Thaw
1	7.34	6.9	2.39	2.04	4.46	3.93	0.7	0.17
2	9.66	8.88	4.91	4.12	6.43	5.61	2.95	2.23
3	11.1	10.21	6.48	5.58	7.61	6.37	4.61	3.74
4	11.45	10.59	7.02	6.16	7.63	6.47	5.1	4.21
5	12.07	10.71	7.76	6.53	7.86	6.67	5.94	4.69
6	12.26	11.3	7.85	7.14	8.15	7.01	6.07	5.31
7	12.55	11.4	8.57	7.49	8.21	7.04	6.72	5.58
8	12.77	11.59	8.79	8.08	8.64	7.23	6.85	5.8
9	13.05	11.8	9.09	8.33	8.79	7.4	7.53	6.11
10	13.24	12.06	9.3	8.71	8.91	7.55	7.93	6.38
11	13.39	12.24	9.53	8.96	9.14	7.75	8.35	6.8
12	13.65	12.53	9.8	9.19	9.33	8.06	8.63	7.04

When the untreated samples were immersed in water for the first cycle of wetting, the samples couldn't withstand and samples got collapsed within a fraction of a second, while soil treated with 2% of lime survived one complete cycle but collapsed during the second wetting cycle. The soil treated with 4% of lime survived 2 complete cycles before collapsing. The soil treated with 6% of lime survived 4 complete cycles before collapsing. The soil treated with 8% of lime lasted 10 complete cycles before collapsing. From the Table 8, it is evident that the durability of the stabilised soil increases with the increase in the percentage of lime.

After the 12 cycles of freezing and thawing UCS test was conducted on the samples. It was found that UCS values of these samples were more than UCS values of samples cured for 60 days for all percentages of lime except 2%. This makes us to falsely assume that the strength gain is due to the freeze-thaw effect. But, freeze-thaw cycle is a weathering action which reduces the strength practically. Probably the strength gain is due to the hydration of lime with number of increasing days. Hence it can be satisfactorily used in cold countries too.

3.3   Flexural Fatigue Life

Fatigue life is the number of load cycles corresponding to the failure of the specimen under repeated loading or number of loading. To investigate fatigue behavior of lime stabilized soils, specimens are exposed to the repeated loading in the laboratory. For this purpose the laboratory experiments are conducted in a fatigue testing apparatus and the specimens are subjected to number of repeated loads.

The type of specimen tested for fatigue capacity of the stabilized specimen is similar to the one tested for their unconfined compression test. A cylindrical specimen of length to diameter ratio of 2 is used and the treated soil samples are tested for 7 days curing. Fatigue test was conducted on UCS samples which were cured for 7 days. Load given in fatigue test was $1/3^{rd}$, $1/2^{nd}$ and $2/3^{rd}$ of 7 days UCS Strength. The results are tabulated in Table 10.

All the fatigue loading tests are conducted on cylindrical specimens using fatigue testing equipment. For this propose the following testing procedure is adopted:

- The cylindrical specimen is mounted on the loading frame and the deflection sensing transducers are set to read the deformation of the specimen. The load cell is brought in contact with the specimen surface.
- In the control unit through the dedicated software, the selected loading stress level, frequency of loading and the type of wave form are fed in to the loading device
- The loading system and the data acquisition system is switched on simultaneously and the process of fatigue load application on the test specimen is initiated.
- The repeated loading, at the designated excitation level (i.e., at the selected stress level and frequency) is continued till the failure of the test specimen.
- The failure pattern of the test specimen is noted down manually.

From the Table 10 we can observe that number of cycles has increased greatly for treated soil and number of cycles has increased with increase in curing period. It can also be observed that number of cycles of loading also depend on amount of stress applied. Number of cycles has reduced with increase in applied stress.

Table 10. Fatigue test results of untreated and treated BC Soil

Sample	7 Days curing			Fatigue life (No of cycles)
	UCS (N)			
	Total UCS strength (kPa)	Load	Applied load (N)	
BC Soil	298	340	110	20480
			170	16759
			230	13372
BC Soil + Lime 2%	484	550	180	122650
			270	96204
			370	74658
BC Soil + Lime 4%	788	890	300	150000
			450	150000
			600	150000
BC Soil + Lime 6%	933	1060	350	150000
			530	150000
			710	150000
BC Soil + Lime 8%	1017	1150	380	150000
			580	150000
			770	150000
BC Soil + Lime 4% + Coir 0.25	485	550	180	125425
			270	102624
			370	76873
BC Soil + Lime 4% + Coir 0.5%	641	730	240	150000
			370	150000
			490	150000
BC Soil + Lime 4% + Coir 0.5%	856	970	320	150000
			490	150000
			650	150000
BC Soil + Lime 4% + Coir 1.0%	1222	1390	460	150000
			700	150000
			930	150000

4   Conclusions

The following conclusions were made on the basis of the laboratory tests and analysis of results:

(1) As the lime content and coir content increases, OMC increases and MDD decreases. The increase in OMC is due to hydration process of pozzolanic material.

(2) The UCS of lime treated samples increase with increase in curing period. The increase in UCS value of treated soil is more than 3 times than the original soil.

(3) As the curing period increases the CBR values increases. At 4% lime the CBR value after 4 weeks curing was 8% for modified proctor density, whereas for original soil was 3%.

(4) From freeze and thaw test it has been observed that the percentage loss in weight after the 12th cycle is within 14%. As per ASTM D 559 and 560, the soil is durable.

(5) From Free Swell Index test it was observed that swelling property of BC soil has reduced greatly after stabilizing the soil with lime. This shows that the lime is very effective in reducing the swelling properties of BC soil. As per IS: 1498 – 1970, the Free Swell Index has reduced from medium to low.

(6) The fatigue life of stabilized soil is 7 times more than that of the original soil (20480 to 150000).

(7) Maximum improvement in UCS and CBR values are observed when optimum value of 1% of coir was considered as per the literatures and mixed with the soil.

(8) The UCS of stabilized BC soil with 1% coir modified proctor compaction after 4 weeks curing was 1535 kPa, whereas for original soil it is 358 kPa.

(9) As the coir percentage increases the CBR value increases after 28 days curing with 1% coir the CBR value increased to 15% from the original soil CBR of 3%.

(10) From the present study, it can be concluded that the proportion of 1% coir fiber and 4% lime in a BC soil is optimum percentage of materials having maximum soaked CBR value. Hence, this proportion can be economically used in road pavements as subgrade and in embankments.

References

Adeniji, F.A.: Recharge function of vertisolic vadose Zone in sub-sahelian Chad Basin. In: First International Conference on Arid Zone Ideology, Hydrology and Water Resources, Maduguri (1991)

ASTM D 559: Standard test method for Wetting and Drying of compacted Soil-Cement mixtures. American Society for Testing and Materials (2015)

ASTM D 560: Standard test method for Freezing and Thawing of compacted Soil-Cement mixtures. American Society for Testing and Materials (2016)

Garber, N.J., Hoel, L.A.: Traffic and Highway Engineering, 2nd edn. Brooks/Cole Publishing Company, London (2000)

Gaulkar, M.P.: Construction of Roads in Black Cotton Soil. Indian Highways, New Delhi (1991)

Al-Homoud, A.S., Basma, A.A., Husein Malkawi, A.I., Al Bashabsheh, M.A.: Cyclic swelling behaviour of clays. J. Geotech. Eng. **121**(7), 562–565 (1995)

IRC: 37: Guidelines for design of Flexible Pavements. Bureau of Indian Standards, New Delhi (2012)

IS: 1498: Indian Standard for Classification and Identification of Soils for General Engineering Purposes. Bureau of Indian Standards, New Delhi (1970)

IS: 2720 Part VII: Determination of Moisture Content and Dry Density using Light Compaction. Bureau of Indian Standards, New Delhi (1980)

IS: 2720 Part VIII: Determination of Moisture Content and Dry Density using Heavy Compaction. Bureau of Indian Standards, New Delhi (1983)

IS: 2720 Part V: Determination of Unconfined Compressive Strength. Bureau of Indian Standards, New Delhi (1973)

IS: 2720 Part X: Determination of Free Swell index of soils. Bureau of Indian Standards, New Delhi (1977)

Little, D.N.: Handbook for Stabilization of Pavement Subgrades and Base Courses with Lime. National Lime Association Kendall Publishing Company (2005)

McNally, G. H.: Soil and Rock Construction Materials, pp. 276–282, 330-341. Routledge, London (1998)

Ola, S.A.: Stabilization of Nigerian lateritic soils with cement, bitumen, and lime. In: Proceedings of the 6th Regional Conference for Africa on Soil Mechanics and Foundation Engineering (1975)

Ayyar, T.S.R., Joseph, J., Beena, K.S.: Bearing capacity of sand reinforced with coir rope. In: Proceedings of the First Indian Geotextiles Conference on Reinforced Soil. pp. 11–16, Bombay, India (1988)

Ramesh, H.N., Manoj, K.K.V., Mamatha, H.V.: Compaction and strength behaviour of lime-coir fiber treated Black cotton soil. Geomech. Eng. 2(1), 19–28 (2010a)

Ramesh, H.N., Manoj, K.K.V., Mamatha, H.V.: Effect of lime-coir fiber on geotechnicalproperties of black cotton soil. In: Indian Geotechnical Conference, IGS Mumbai Chapter &IIT Bombay (2010b)

Rowell, M., Han, S., Rowell, S.: Characterization and factors effecting fiber properties. Nat. Polymer Agr. Compos., pp. 115–134 (2000)

Babu, G.L.S., Vasudevan, A.K.: Strength and stiffness response of coir-fiber reinforced tropical soil. J. Mater. Civil Eng. 20(9), 571–577 (2008)

Subaida, E.A., Chandrakaran, S., Sankar, N.: Laboratory performance of unpaved roads reinforced with woven coir geotextiles. Geotext. Geomembr. 27(3), 204–210 (2009)

Thagesen, B.: Tropical rocks and soils. In: Highway and Traffic Engineering in Developing Countries. Chapman and Hall, London (1996)

Vinod, P., Bhaskar, A.B., Sreehari, S.: Behaviour of a square model footing on loose sand reinforced with braided coir rope. J. Geotext. Geomembr. 27, 464–474 (2009)

Viswanadham, B.V.S.: Bearing capacity of geosynthetic reinforced foundation on a swelling clay. Master of Technology Dissertation, Indian Institute of Technology Madras, India (1989)

Influence of Damping Forms on the Behavior of Sand Under Dynamic Compaction

Yuqi Li$^{(\boxtimes)}$, Jun Chen, and Jiejun Zhu

Shanghai University, Shanghai 200444, China
liyuqi2000@shu.edu.cn

Abstract. Based on Mohr-Coulomb theory, a centrifuge test of dry sand was simulated using FLAC3D. The local and Rayleigh damping were applied respectively to numerical model. The result shows that the ground settlement around tamping pit under local damping is smaller than that under Rayleigh damping. There is little difference between the maximum vertical stresses of the first impact under local damping and Rayleigh damping. The maximum vertical stresses both appear at 1–2 m depth under the tamping pit. However, the calculation efficiency under local damping is higher than that under Rayleigh damping. The dissipation of energy grows apparently as the depth of foundation is augmented. The study on the simulation of different damping forms in FLAC3D is helpful to further research about dynamic compaction.

Keywords: Dynamic compaction · Mohr–Coulomb theory · FLAC3D
Local damping · Rayleigh damping

1 Introduction

Dynamic compaction (DC) is introduced as a routine method of site improvement by Menard in 1969. Due to the advantages such as a wide application of foundation soil, simple equipment and low cost, DC has been studied by scholars for years. The possibility of improving the mechanical characteristics of fine saturated soils to a great depth using DC has been researched by Menard and Broise [1]. The low-energy dynamic compaction process and the centrifuge test to monitor the process of compaction were studied by Merrifield and Davies [2]. To optimize the DC design of loose backfill with heterogeneity and saturated silt, field test was conducted by Feng et al. [3].

It is too complicated to use the analytical method explaining the mechanics of DC. With the development of computer technology, numerical methods such as finite element method and finite difference method have been programmed to solve engineering problems. Ground response to dynamic compaction has been simulated using two-dimensional finite element method [4]. Borg and Volger [5] discussed the differences between two- and three-dimensional simulations, as well as investigated the effect of stiction and sliding grain-on-grain contact laws on the dynamic compaction of loose dry granular materials. Dynamic interaction of two closely spaced embedded square or rectangular foundations under the action of machine vibration was analyzed by Ghosh [6] using FLAC3D. Using comparative analyses with different damping ratio values the behavior of structure for soils and reinforcement was studied by Mendonca and Lopes [7].

© Springer Nature Singapore Pte Ltd. 2018
L. Li et al. (Eds.): GSIC 2018, *Proceedings of GeoShanghai 2018 International Conference: Ground Improvement and Geosynthetics*, pp. 244–251, 2018.
https://doi.org/10.1007/978-981-13-0122-3_27

However, there is few discussions about the influence of damping forms on simulating DC. In this paper, Mohr-Coulomb mathematical model was chosen in FLAC3D to simulate the Takada and Oshima's centrifuge test [8]. And two damping forms which are local and Rayleigh dampings were applied to the numerical model to analyze their impacts to the settlement and dynamic stress.

2 Numerical Simulation

2.1 Constitutive Model

Mohr–Coulomb elastoplastic constitutive model was applied to the simulation of soil mass in this paper considering that dry sand is compacted by impact load. Maximum tensile-stress and Mohr-Coulomb criterions are adopted in the constitutive model. The yield surface equations are shown in the followings:

$$f' = \sigma_1 - \sigma_3[(1 + \sin\varphi)/(1 - \sin\varphi)] + 2c\sqrt{(1 + \sin\varphi)/(1 - \sin\varphi)} \quad (1)$$

$$f' = \sigma_3 - \sigma' \quad (2)$$

where c is the cohesion force and φ is the internal friction angle. The principal stress σ_1 is no less than σ_3, and σ' is the tensile-stress limit value.

2.2 Soil Parameters and Mesh Generation

Centrifuge model test is a validated method to simulate in-situ soils under centrifugal force. According to the Oshima and Takada's centrifuge model test [8], the model ground was a semicylindrical column whose diameter was 30 cm, and the base area, mass and drop height of tamper were 8 cm^2, 80 g and 43.8 cm, respectively. The maximum and minimum dry densities of a 13 cm thick model ground were 1.82 t/m^3 and 1.37 t/m^3 and the relative density was 30%. A 5 cm thick base ground was compacted to relative density of 95%. Under a centrifugal acceleration field of 50 g, the centrifuge model simulated a prototype ground of 15 m in diameter and 6.5 m thick. The mechanical parameters were determined after calculation [9]. The cohesion forces are zero in both soil layers, and other soil parameters, as listed in Table 1 [9], were input into FLAC3D to simulate the process of centrifuge test. In order to obtain precise results, the length of mesh was 0.55 m according to wave frequency and velocity. In FLAC3D model, the base area, height, mass and drop height of tamper are 4 m^2, 0.64 m, 20 t and 21.9 m respectively according to the similarity principle. Figure 1 shows the cylinder foundation model whose radius and height are 11 m and 15 m, respectively.

Since the elasticity modulus of soil will improve with the increase of tamper time during the dynamic compaction process, Qian and Shuai [10] obtained the following formula:

$$E = E_0 N^{0.516} \quad (3)$$

where E_0 is the initial elasticity modulus of soil; N is tamping time and is the elasticity modulus of soil after tamping for N times.

According to the above formula, the elasticity modulus, shear modulus and bulk modulus of soil stratum 1 after tamping for seven times were obtained and the relationship between the mechanical parameters of soil stratum 1 and tamping time is listed in Table 2.

Table 1. Mechanical and physical parameters used in FLAC3D.

Soil stratum	1	2	Soil stratum	1	2
Thickness: m	6.5	8.5	Elasticity modulus: MPa	3.75	20.83
Density: kg/m³	1484	2560	Shear modulus: MPa	1.44	8.33
Modulus of compression: MPa	5.05	25.0	Internal friction angle: °	30	42
Bulk modulus: MPa	3.125	13.89	Poisson's ratio	0.30	0.25

Fig. 1. Geometric model of soil mass in FLAC3D.

Table 2. Relationship between mechanical parameters of soil stratum 1 and tamping time.

Tamping time	Elasticity modulus: MPa	Shear modulus: MPa	Bulk modulus: MPa
1	3.75	1.44	3.13
2	5.36	2.06	4.47
3	6.61	2.54	5.51
4	7.67	2.95	6.39
5	8.61	3.31	7.17
6	9.45	3.64	7.88
7	10.24	3.94	8.53

2.3   Boundary Condition

The top boundary of the model is free, the bottom and lateral boundaries are assumed to be quiet boundaries.

2.4 Relationship Between Contact Stress and Time

The shape of the stress-time plot is similar to a triangle [11], as is shown in Fig. 2, in which P_{max} is peak stress, t_R is loading time, t_N is the total contact time of tamper and $t_R - t_N$ is unloading time. According to the dynamic stress equilibrium equation of tamper movement [9], the peak contact stress, loading time, unloading time and total contact time of tamper under the different tamping time can be obtained, as listed in Table 3.

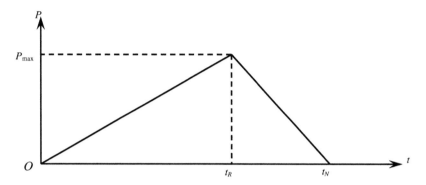

Fig. 2. Relationship between contact stress and time.

Table 3. Relationship between peak stress, loading/unloading time and tamping time.

Tamping time	Peak stress: MPa	Loading time: s	Unloading time: s	Total time: s
1	1.96	0.07278	0.02573	0.09851
2	2.35	0.06086	0.02152	0.08238
3	2.61	0.05482	0.01938	0.07420
4	2.82	0.05090	0.01799	0.06889
5	2.99	0.04805	0.01699	0.06504
6	3.14	0.04584	0.01621	0.06205
7	3.27	0.04405	0.01558	0.05963

2.5 Numerical Modeling Procedure

The modeling procedure is outlined as follows:

(1) A foundation model of 11 m in radius and 15 m in depth was generated and the initial stress field under gravity was applied.
(2) The dynamic and large strain modes were on, and quiet boundaries and local damping were set. The contact stress of tamper was applied on the foundation model according to Fig. 2 and Table 3 and the loading time and unloading time were set according to Table 3.

(3) After tamping for once, the shear modulus and bulk modulus of soil stratum 1 were given new values, as listed in Table 2. The contact stress, loading time and unloading time were set according to Table 3.

(4) The tamping was finished by successively giving the new mechanical parameters and setting contact stress, loading time and unloading time for seven times.

(5) Rayleigh's damping can be considered by applying Rayleigh's damping and repeating steps (2), (3) and (4).

3   Results and Analysis

3.1   Crater Depth

The improvement of soil mass can be assessed by crater depth. As shown in Fig. 3, the result of simulation at the first tamping is approximate to the field test, where the crater depth of centrifuge test is a little smaller than the others. The subsequent blows show a good consistency in crater depth. Meanwhile, the cumulative influence depth of simulation, shown in Fig. 4, is pretty much the same as the field test. Therefore, the numerical results verify the feasibility of simulation in FLAC3D and the numerical model can be used in dynamic analysis of DC.

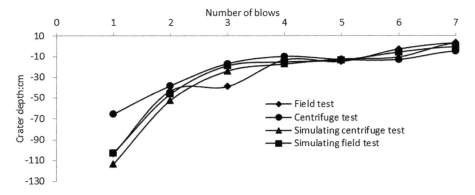

Fig. 3. Crater depth of single tamping versus blow count.

3.2   Damping Forms

Different damping forms have a certain effect on the improvement of soil mass. The range of the critical damping ratio is normally 2%–5%, and 5% is taken in this paper. As can be seen from Figs. 5 and 6, the crater depth under local damping is slightly less than that under Rayleigh's damping. The results under Rayleigh's damping are closer to the field test data (see Figs. 3 and 4) than those under local damping. However, the dynamic computing time of Rayleigh's damping in FLAC3D is very long for numerical results in terms of plenty of meshes. Therefore, the local damping should be selected for a more efficient calculation. In engineering application reasonable damping form is very important for simulating projects.

The first blow is taken as an example to research the stress. From Figs. 7 and 8, the stress contours both show up a "pear" shape. The maximum vertical stress under Rayleigh's damping is 20 kPa bigger than that under local damping. The maximum vertical stress appears in 1–2 m depth range under the tamping hammer's bottom.

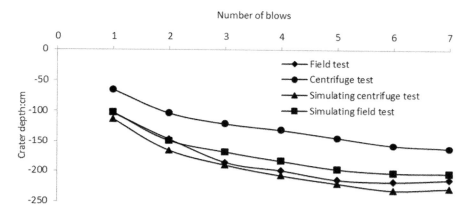

Fig. 4. Cumulative increase in crater depth with blow count.

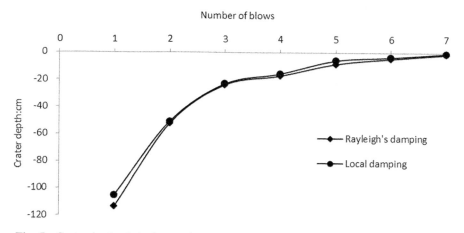

Fig. 5. Crater depth of single tamping versus blow count with different damping forms.

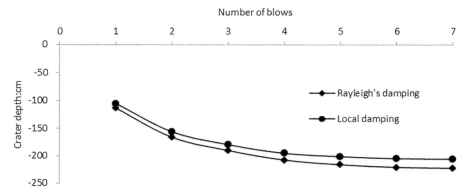

Fig. 6. Cumulative crater depth versus blow count with different damping forms.

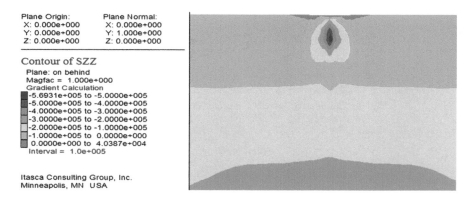

Fig. 7. Contour of vertical stress under Rayleigh's damping after the first blow.

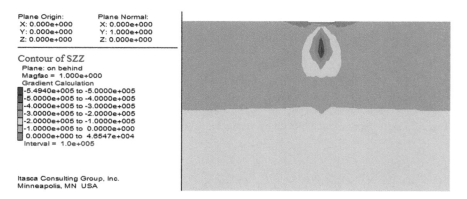

Fig. 8. Contour of vertical stress under local damping after the first blow.

4 Conclusions

Through the numerical simulations of Takada and Oshima's DC centrifuge test in FLAC3D under the different damping forms the following conclusions were drawn:

(1) The simulating crater depths of dry sand under dynamic compaction are in accord with those of centrifuge test and field test, which proved the validity of numerical calculation.
(2) After changing damping form, the crater depth and vertical stress become different. Meanwhile, contour of vertical stress shows a "pear" shape after the first blow. The calculation efficiency under local damping is higher, however, the results under Rayleigh's damping are closer to field test data. Therefore, in engineering practice reasonable damping form should be considered when simulating in FLAC3D.

References

1. Menard, L., Broise, Y.: Theoretical and practical aspects of dynamic consolidation. Geotechnique 25(1), 3–18 (1975)
2. Davies, M.C.R., Merrifield, C.M.: A study of low-energy dynamic compaction: field trials and centrifuge modeling. Geotechnique 50(6), 675–681 (2000)
3. Feng, S.J., Shui, W.H., Gao, L.Y., et al.: Field evaluation of dynamic compaction on granular deposits. J. Perform. Constructed Facil. 25(3), 241–249 (2011)
4. Gu, Q., Lee, F.H.: Ground response to dynamic compaction of dry sand. Geotechnique 52(7), 481–493 (2002)
5. Borg, J.P., Vogler, T.J.: Rapid compaction of granular material: characterizing two- and three-dimensional mesoscale simulations. Shock Waves 23(2), 153–176 (2013)
6. Ghosh, P., Kumar, R.: Seismic interaction of two closely spaced horizontal square and rectangular ground anchors in layered soil. Int. J. Geotech. Eng. 11(1), 1–10 (2017)
7. Mendonca, A., Lopes, M.L.: Role of the damping ratio of reinforcement on the behaviour of geogrids-reinforced systems. Geotech. Geol. Eng. 29(3), 375–388 (2011)
8. Takada, N., Oshima, A.: Comparison between field and centrifuge model tests of heavy test. In: Proceedings of the International Conference Centrifuge, vol. 94, pp. 337–342. CRC Press, Singapore (1994)
9. Zhu, J.J.: Numerical Simulation Study on Dynamic Compaction of Dry Sand. Shanghai University, Shanghai (2017). (in Chinese)
10. Qian, J.H., Shuai, F.S.: Application of boundary element method in dynamic compaction. China Sci. 3, 329–336 (1987). (in Chinese)
11. Wu, M.B., Wang, Z.Q.: Numerical analysis of the mechanism of dynamic compaction. Geotech. Invest. Surveying 3, 1–5 (1989). (in Chinese)

Numerical Investigation on the Effect of Saturated Silty Soils Under Multi-location Tamping

Wei Wang[1](✉), Jianhua Wang[2], and Qingsong Feng[1]

[1] Engineering Research Center of Railway Environment Vibration and Noise, Ministry of Education, East China Jiao Tong University, Nanchang 330013, China
rebwg05@163.com
[2] State Key Laboratory of Ocean Engineering, Department of Civil Engineering, Shanghai Jiao Tong University, 800 Dong Chuan Road, Shanghai, China

Abstract. Dynamic compaction (DC) is a well-known method of ground improvement used for its efficiency and operability. Currently, the majority of studies have focused on the analysis of dynamic compaction in dry soils, and only a few investigations have been conducted for saturated soil responses caused by DC, despite the fact that most application of DC in saturated deposits. Therefore, it is necessary to understand the behavior of saturated soils during DC, particularly in the saturated soils with poor permeability under multi-location tamping, in order to enrich the detailed design of DC. In this study, a numerical investigation on the responses of saturated silty soils under multi-location tamping is carried out using a coupled hydro-mechanical model. Verification of the proposed model is performed against the previous test data. Then, a series of parametric studies have been performed to examine the effects of tamping distance, tamping order and tamping patterns on the improvement degree of saturated granular soils under DC. Finally, some suggestions are presented to improve dynamic compaction effect in saturated deposits.

Keywords: Dynamic compaction · Soil improvement
Multiple tamping locations · Coupled hydro-mechanical analysis

1 Introduction

Dynamic compaction (DC) is a widely used ground improvement method. Originally, the predominant soil types considered for DC treatment are dry sandy and granular materials. As a result of its simplicity, cost-effectiveness and speed of construction, in recent years, DC method has been extended to many major engineering projects at coastal area, such as offshore land reclamation (Perucho et al. (2006), port construction (Feng et al. 2011), and highway construction (Miao et al. 2006). However, as the ground water table is near the ground surface in coastal areas, a significant increase of pore water pressure is noticed after each impact, which results in liquefaction and a reduction of bearing capacity during DC in saturated soils, especially in the saturated soils with poor permeability.

© Springer Nature Singapore Pte Ltd. 2018
L. Li et al. (Eds.): GSIC 2018, *Proceedings of GeoShanghai 2018 International Conference: Ground Improvement and Geosynthetics*, pp. 252–259, 2018.
https://doi.org/10.1007/978-981-13-0122-3_28

For the saturated soils with poor permeability, e.g. silty soil, clayed soil, due to the failure of the pore water pressure to be dissipated in time, the strength recovery of soil is slower, which greatly influences the soil reinforcement effect and construction progress in DC. Actually, dynamic compaction site is consisted of multiple tamping locations. The tamping distances and pattern present a variety of forms, which directly affects the quality of dynamic compaction treatment of soil, the construction efficiency, the uniformity of strengthening foundation and economic benefits. So far, its focus most is the research of single location tamping. For the multi-location tamping, it mostly discusses the effect of tamping distance on sandy soils. However, the research of the soil reinforcement effect with poor permeability under multi-location tamping is rare. Different with soil reinforcement effect with well permeability, the soil reinforcement effect with poor permeability is involved with dissipation of pore water pressure. Proper tamping pattern and tamping distance not only affect the quality of reinforcement in DC, but also the construction efficiency. Therefore, in practical engineering applications, mastering rule of soil reinforcement under multi-location tamping with poor permeability is important.

In this study, a numerical investigation of multi-location tamping in saturated silty soils was carried out. The behaviors of silty soil are described by means of the viscoplastic cap model. This soil constitutive relationship is implemented into the LS-DYNA and is then used to integrate with a 3D finite element model for numerical investigation. Then utilizing the field data from the previous studies, the paper investigated the effect of the construction parameters on soil compaction in order to provide technological supports for detailed design and guiding construction in site.

2 Coupled Hydro-Mechanical Model for Dynamic Compaction

2.1 Governing Equations of Coupled Hydro-Mechanical Analysis

The mechanical behavior of saturated soils subjected to dynamic loads, is governed by the interaction of the solid skeleton with the pore fluid. In this paper, the extended Biot's dynamic equations (Zienkiewicz et al. 1980), are adopted for modeling of DC. The governing equations, which represent equilibrium of a porous medium, equilibrium of fluid phrase and the mass conservation of fluid phase, are as follows:

$$\sigma_{ij,j} - \rho \ddot{u}_i - \rho_f \ddot{w}_i + \rho g_i = 0 \tag{1}$$

$$-p_{,i} - \frac{\rho_f g_i}{k} \dot{w}_i - \rho_f \ddot{u}_i - \frac{\rho_f}{n} \ddot{w}_i + \rho_f g_i = 0 \tag{2}$$

$$\dot{u}_{i,i} + \dot{w}_{i,i} + \frac{n}{K_f} \dot{p} = 0 \tag{3}$$

where σ_{ij} is total stress which is further separated into the effective stress σ'_{ij} acting on the solid skeleton by means of soil constitutive model, and pore pressure p for the fluid in the pores. g_i *is* body force acceleration, u_i is displacement of the solid skeleton, w_i is displacement of pore fluid, k is permeability, n is porosity of the mixture. The density ρ_f is that of the pore fluid and ρ is the density of the mixture written as $\rho = (1 - n)\rho_s + n\rho_f$, where ρ_s is the density of the solid skeleton. K_f is bulk modulus of the fluid.

2.2 Soil Constitutive Law

The choice of constitutive model is based on the consideration that the soil improvement effect is caused by transient effective stress induced by impact load. In this paper, the applied viscoplastic cap model was developed on the basis of inviscid cap model by introducing an associative viscous flow rule from Perzyna's theory (Tong et al. (2007)) and has been applied by Wang et al. (2017b).

2.3 Modeling of Impact Load

Usually, there are two accepted methods to calculate dynamic load (Pan et al. (2002)). The first is to apply vertical velocity to the hammer nodes as an initial condition of the problem, which can be determined from the free fall equation. Another method is to simplify the dynamic load as a normal distribution curve or a damped half-sine wave. In this study, the first method is used and a symmetric penalty function algorithm is adopted to obtain the contact force in the numerical model.

3   Verifications of the Present Model

Numerical calculation results are compared with field test data in order to verify the validity of the numerical model. In this section, a field case of dynamic compaction treatment in a highway construction reported by Miao et al. (2006) is used as verification of the present model.

3.1 Numerical Model and Material Parameters

According to geotechnical investigation, this site is mainly composed of non-plastic silt with initial relative density of 35%. Figure 1 shows the 3-D finite element mesh used in this analysis. The overall model size is 15 m in radius and 14 m in high, and the model consists of three layers of soil. The first two layers are viscoplastic soil, and most inviscid cap parameters derived from Wang et al. (2017b), the rest of viscous soil parameters have been chosen based on the similar simulations of impact problems reported by Tong et al. (2007) and dynamic cone test studies from Kim et al. (2015). And model parameters are listed in Table 1. The third soil layer is modeled as an elastic material as limited plastic deformation may occur in this soil layer. The Young's modulus E of the third layer is assumed to be 7×10^3 kPa. The Poisson ratio v is assumed to be 0.3. The bottom boundary is fixed along both horizontal and vertical directions. The boundary on the cylindrical side is fixed along a horizontal direction, and only the vertical direction is free. Moreover, non-reflecting boundaries are used on the exterior boundaries around the soil model to simulate an infinite domain.

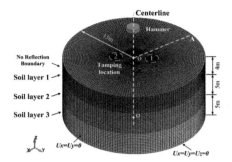

Fig. 1. FEM mesh and boundary conditions

Table 1. Material parameters for dynamic compaction analysis

Soil layer	K (kPa)	G (kPa)	α (kPa)	γ	β	θ	W	R	D (MPa^{-1})	T (kPa)	η (s^{-1})	f_0 (MPa)	N
1	10600	2610	17.8	0	0	0.25	0.5	4.33	0.3	0	0.1	0.5	1.0
2	27300	2960	21.2	0	0	0.21	0.5	4.33	0.3	0	0.1	0.5	1.0

3.2 Comparison with Field Data

In order to verify the numerical model, comparison was made with field data of tamping energy 2500 kN•m per blow. Figure 2 presents a comparison between the measured and calculated results of both pore water pressure (PWP) and ground lateral displacements. Figure 2(a) shows the simulated pore water pressure is close to the measured results. From Fig. 2(b), it can be observed that, the simulated lateral displacements show similar tendency to the measured results, although its value is slightly lower than the measured results in measured results, although its value is slightly lower

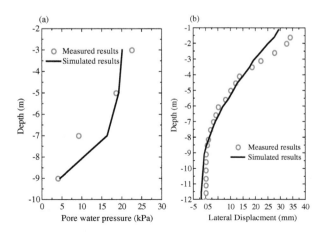

Fig. 2. Comparison of field measured and simulated (a) pore water pressure; (b) lateral displacement

than the measured results in shallow soils. Consequently, the established FE model is sufficient accuracy and applied in further simulation.

4   Results and Discussion

4.1   Effect of Tamping Distance

Figure 3 shows distribution of relative density and the normalized PWP for different tamping distances. Figure 3 shows the pore water pressure at the center of hammer bottom is the highest, which results in liquefaction and weakness of reinforcement effect. The soil on both sides of hammer bottom is strengthened because the foundation soil of hammer bottom is not easy to compress and then turn to the two sides. Further, if the tamping distance is 3.0·r, soil compaction degree under the tamping location 2 is better than location 1. Soil compaction degree under the tamping location 2 near the tamping location 1 is better than another side. But with the increase of tamping distance, the condition will be weakened. If the tamping distance is 6.0·r, the soil

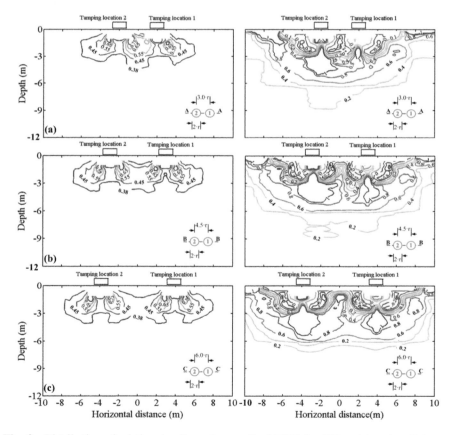

Fig. 3. Distribution of relative density and normalized PWP for different tamping distances

compaction degree under the tamping location 1 and 2 have no significant difference. On the contrary, with the increase of distance, the lateral restriction is gradually smaller and independent reinforcement. Therefore, with the smaller distance, the reinforcement effect between tamping locations is better, while it is weakened under hammer bottom. In addition, as shown in Fig. 3 that the smaller the distance is, the water head gradient of the foundation soil is evenly distributed, and it's not easy to make the pore water flow and dissipate.

4.2 Effect of Tamping Order

Figure 4 shows distribution of relative density and the normalized PWP for different tamping orders. As shown in Fig. 4(a), for tamping order 1122, because it reinforces the tamping location 2 after tamping location 1, it makes liquefaction area under tamping location 2 bigger than tamping location 1 and center soil compaction degree is lower than tamping location 1. Meanwhile, due to the lateral restriction provided by the

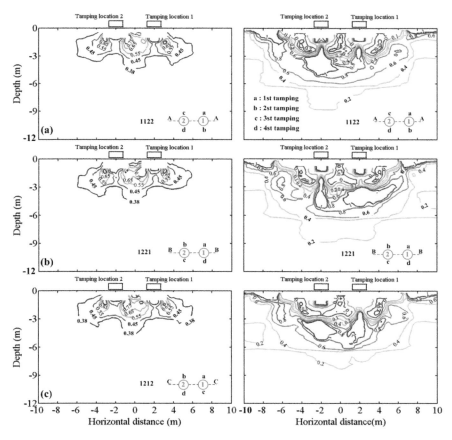

Fig. 4. Distribution of relative density and normalized PWP for different tamping orders

soil under the tamping locations, soil compaction degree under tamping location 2 near tamping location 1 is better than another side. Similarly, as shown in Fig. 4(b) and (c), the change of the tamping order results in different distribution of normalized PWP. For tamping order 1221, the liquefaction area under tamping location 1 presents lateral distribution which results in the soil compaction degree near tamping location 1 is higher than another side. Meanwhile, the tamping location 2 in vertical distribution which results in the soil compaction degree near tamping location 2 is higher than another side. For tamping order 1212, it is totally opposite to tamping order 1221. Comparing Fig. 4(b) and (c), the soil compaction degree near the middle of tamping locations is higher than another side.

4.3   Effect of the Relative Position Between Tamping Locations

Figure 5 shows distribution of relative density and the normalized PWP for different tamping patterns. In general, because it reinforces the tamping location 2 after tamping location 1, it makes pore water difficult to dissipate under tamping location 2 and the liquefaction area under tamping location 2 bigger than tamping location 1. At this time,

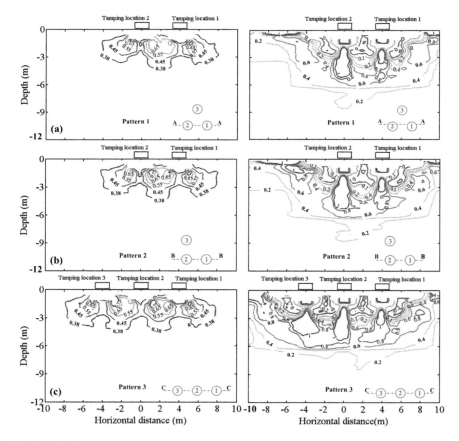

Fig. 5. Distribution of relative density and normalized PWP for different tamping patterns

the center soil of tamping location is more difficult to compress, and then close to the sides. Therefore, the center soil compaction degree of tamping location 2 is lower than tamping location 1, but the side degree is higher. Meanwhile, due to the lateral restriction provided by the soil under the tamping locations, soil compaction degree under tamping location 2 near tamping location 1 is better than another side. Further, comparing Pattern 1 and Pattern 2, it can be seen that in Pattern 2, tamping location 3 is close to tamping location 2 which results in the center soil compaction degree under tamping location 2 is lower but the side degree is higher. Similar to the above conclusions, the position of tamping location 3 makes the liquefaction area of tamping location 2 in Pattern 3 is smaller than Pattern 2 and soil compaction degree under tamping location 2 in Pattern 3 is higher.

5 Conclusion

This paper investigated the responses of saturated silty soils under multi-location tamping. Based on the numerical results, the following conclusions can be made:

(1) The drainage condition of foundation soil is the main factor which influences the strengthening effect of soil caused by DC.

(2) Increasing the tamping energy level will increase the influence of reinforcement, but also destroy the soil structure, causing the pore water pressure of the soil to dissipate slowly.

(3) As the distance of tamping location is shorten, the reinforcement effect of the soil is better, while the reinforcement of soil in the bottom of tamping will be weakened. Moreover, the difference of tamping order and tamping pattern can affect the redistribution of the pore pressure in the soil, thus affecting the uniformity of the foundation soil and the strengthening effect of deep soil.

References

Feng, S.J., Shui, W.H., Tan, K.: Field evaluation of dynamic compaction on granular deposits. J. Perform. Constructed Facil. Am. Soc. Civ. Eng. **25**, 241–249 (2011)

Kim, Y.H., Hossain, M.S., Wang, D.: Effect of strain rate and strain softening on embedment depth of a torpedo anchor in clay. Ocean Eng. **108**, 704–715 (2015)

Miao, L.C., Chen, G., Hong, Z.: Application of dynamic compaction in highway: a case study. Geotech. Geol. Eng. **24**, 91–99 (2006)

Pan, J.L., Selby, A.R.: Simulation of dynamic compaction of loose granular soils. Adv. Eng. Softw. **33**, 631–640 (2002)

Perucho, A., Olalla, C.: Dynamic consolidation of a saturated plastic clayey fill. Ground Improv. **10**(2), 55–68 (2006)

Tong, X.L., Tuan, C.Y.: Viscoplastic cap model for soils under high strain rate loading. J. Geotech. Geoenviron. Eng. **133**(2), 206–214 (2007)

Wang, W., Dou, J.Z,, Chen, J.J., Wang, J.H.: Numerical analysis of the soil compaction degree under multi-location tamping. J. Shanghai Jiaotong Univ. (Sci.) **22**(4), 417–433 (2017b)

Zienkiewicz, O.C., Chang, O.C., Bettess, P.: Drained, undrained, consolidating and dynamic behaviour assumptions in soils. Géotechnique **30**, 385–395 (1980)

A Numerical Back Analysis of a Ground Improvement Project on Underconsolidated Clay Under Combined Vacuum and Surcharge Preloading in Macau

Yue Chen[1(✉)], Thomas Man Hoi Lok[1], and Hei Yip Lee[2]

[1] University of Macau, Macau S.A.R, China
chengyue1999@gmail.com
[2] Civil Engineering Consultants Co. Limited, Macau S.A.R, China

Abstract. This paper describes the ground improvement monitoring and back analysis results of a well-documented project in the Cotai area in Macau. The improvement project covers a total area of 220000 m^2, with vacuum preloading in selected areas, and conventional surcharge in other areas of the site. Settlement achieved by soil improvement was between 0.8 m to 1.5 m, and undrained shear strength increased from 10 kPa to more than 25 kPa. A back analysis, using PLAXIS and calibrated with ground settlement records, provided the undrained shear strength values that are in general agreement with measurements.

Keywords: Ground improvement · Vacuum preloading · Settlement
Undrained shear strength

1 Introduction

In this case study, the soil improvement results and a back analysis using PLAXIS (2015) [3] to compute the shear strengths of a well-documented ground improvement projects, completed in 2013 in Cotai, Macau S.A.R., is described. The project covers a total area of 220000 m^2 on a reclaimed land on marine clay undergoing consolidation. Vacuum preloading in conjunction with prefabricated vertical drains (PVD) was applied on selected areas of about 70000 m^2, while conventional surcharge was applied on other areas.

The vacuum preloading method for soil improvement was first proposed by Kjellman [1]. Vacuum pressure is applied to a soil mass sealed with geomembrane at the top using suction pumps. Because of suction pump capacity limitations and possible leakage of geomembrane, the target vacuum pressure was set as 75 kPa in this project. This vacuum pressure is equivalent to about 4 m of soil as a conventional surcharge, which is applied on other areas. Comparison of these two methods was described by Rujikiatkamjorn et al. [4].

© Springer Nature Singapore Pte Ltd. 2018
L. Li et al. (Eds.): GSIC 2018, *Proceedings of GeoShanghai 2018 International Conference: Ground Improvement and Geosynthetics*, pp. 260–267, 2018.
https://doi.org/10.1007/978-981-13-0122-3_29

2 Soil Profile Description

As shows in Fig. 1, the construction covers an area of 220000 m^2 and is divided into different zones according to different preloading methods. The upper triangular shaded area of 70000 m^2 was improved by vacuum preloading method, and the rest by conventional surcharge method, different color represents different amount of surcharge applied. The subsurface conditions along the section shown in the insert figure were selected as the analysis profile of this study, and five settlement points are located along this section.

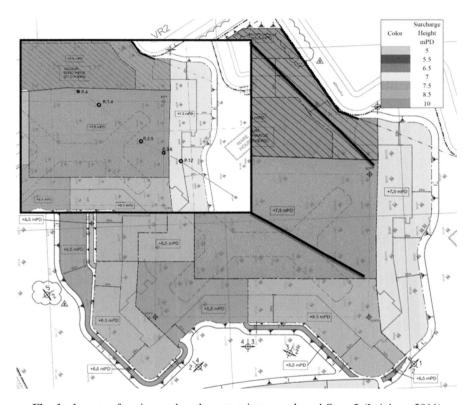

Fig. 1. Layout of project and settlement points on selected Sect. 2 (Leighton 2011)

Six boreholes located along the selected section are used to estimate the geological profile, which can be divided into the following layers as indicated in Fig. 2:

- Unsorted Fill: soft to lose construction waste, garbage, and other reclaimed material.
- Sand Fill: loose sands mainly for constructing a working platform.
- Marine Deposit Fill: soft marine clay fills mainly from the reclaimed project.
- Marine Deposit: soft soil sediments with S_u equals to about 10 kPa.
- Alluvium Clay layer: firm to stiff clay layer.
- Alluvium Sand layer: dense, fine to coarse sand layer.

The soil index and engineering properties obtained from laboratory and field testing are summarized in Table 1.

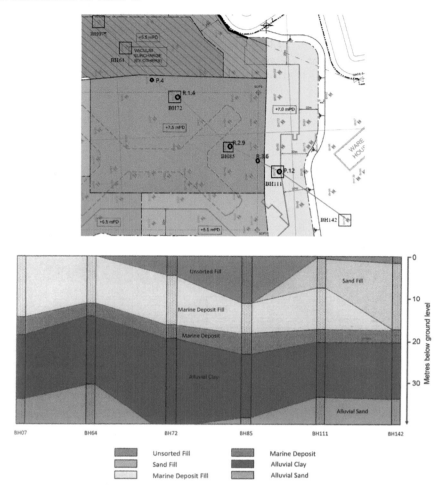

Fig. 2. Geological profile of selected Sect. 2 (Leighton 2011)

Table 1. Soil index and engineering properties

Layer	Dry unit weight kN/m³	Saturated unit weight kN/m³	Void ratio e	c' kPa	ϕ' deg.	k m/day	C_c
Sand fill	14.7	17.66	0.8	0	41.1	2.196	0.2
Unsorted fill	12.26	17.28	0.92	15	32	2.67	0.25
Marine deposit fill	12	17	1.31	0	25	4.32E-3	0.48
Marine deposit	10.79	16.7	1.67	0	25	3.88E-3	0.48
Alluvial clay	14.15	17.66	1.045	17.39	25	6.91E-2	0.47
Alluvial sand	14.42	15.7	1.072	11.1	41.25	2.59E-1	0.29

[*] Data provided by Leighton Ltd. [2].

3 Construction Sequence of the Soil Improvement Project

According to the construction report by Leighton Ltd. [2], soil improvements using vacuum and surcharge methods were performed according to the following sequence.

1. Soil improvement project was commenced on the 1/8/2012.
2. Wick Drains were inserted from the 9/8/2012 to the 20/9/2012.
3. The dewatering started with the Horizontal Drain installation on the 29/8/2012.
4. Trench work began on the 11/9/2012 and ended on the 19/10/2012.
5. The horizontal membrane was installed from the 10-26/10/2012 in parallel with the monitoring installations.
6. The vacuum pumping started on the 27/10/2012.
7. The vacuum preloading was completed on the 15/2/2013.

The above sequence is simulated in the numerical model, but the computation process was established with certain simplifications.

4 Numerical Modeling

Based on the geological profile in Fig. 2, a numerical model of the analysis section was established in PLAXIS and is shown in Fig. 3. Ground settlement points are also shown at corresponding locations along the section.

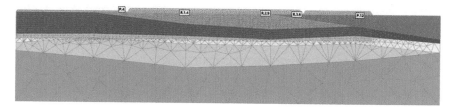

Fig. 3. Numerical model established in PLAXIS

The built-in Soft Soil model was adopted in this analysis. The modeling procedure and applied ramp loading was followed by the real construction sequence. Based on the PLAXIS model, settlements are computed and compared with site measurements.

Computation of settlements based on parameters in Table 1 was first performed and the settlements (unmodified parameters) are shown in Fig. 4. Significant differences are found between the calculated and measured settlements values. These results suggest that using the initial parameters by directly taking the average from laboratory testing would not provide reasonable predictions compared with the measurements.

Based on the assumption that the differences are mainly due to the inaccurate consolidation parameters of C_c and k obtained from laboratory tests, the consolidation parameters were then calibrated based on the observed settlements. The computations based on the calibrated parameters are also shown in Fig. 4 as the square block curve which was shifted closer to the measured curve. The modified parameters are summarized in Table 2:

Table 2. Consolidation parameters (Modified data/Raw data)

Layer	C_c	$k_x = k_y$ m/day
Sand fill	0.28/0.2	2.196
Unsorted fill	0.1/0.25	1.37/2.67
Marine deposit fill	0.11/0.48	2.14E-3/4.32E-3
Marine deposit	0.13/0.48	1.54E-3/3.88E-3
Alluvial clay	0.47	1.59E-1/6.91E-2
Alluvial sand	0.29	1.99E-1/2.59E-1

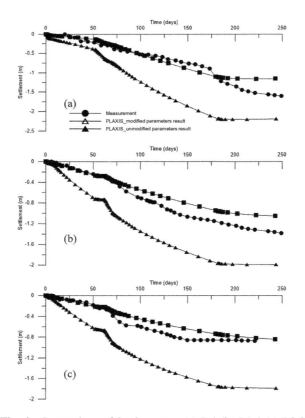

Fig. 4. Comparison of Settlement at (a) P.4 (b) R1.4 (c) R2.9

In Fig. 4, dots represent the site measurement data, the triangle represents the PLAXIS result based on Table 1 and square block represent the result based on Table 2.

5 Undrained Shear Strength Development

After the consolidation parameters were calibrated with the measured settlement, it was then used to calculate the changes of effective stress and undrained shear strength under surcharge loading conditions. Two sections shown in Fig. 5 are selected to examine the results of soil improvement by vacuum preloading and conventional surcharge. The comparison will be made between computations and measurements of undrained shear strength at the beginning and at the end of soil improvement periods.

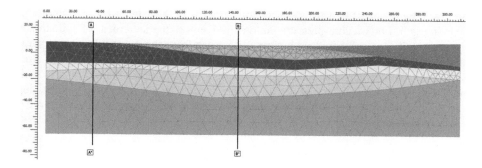

Fig. 5. Sections selected for vacuum preloading (A-A) and conventional surcharge (B-B)

5.1 At the Beginning of Improvement

Before surcharge loading was applied, vane shear tests were performed to obtain the undrained shear strength of the soft Marine clay layers. Figure 6 shows good agreement between the calculated and measured pre-surcharge undrained shear strength, although the measured values are generally more scattered than the predicted ones. The scatteredness may be due to the heterogeneity of the marine deposits and the measurement accuracy of the field vane shear tests. Nevertheless, the trends of both the calculated and measured values are generally in good agreement, indicating the numerical model has reasonably simulated the initial shear strength conditions prior to loading.

5.2 At the End of Improvement

In this part, the comparison is divided into vacuum preloading area and conventional surcharge area.

Vacuum Preloading Area

Figure 6(b) shows the undrained shear strength with depth after vacuum preloading. The calculated undrained shear strength shows the expected trend of increasing linearly with depth, but the measured values are more scattered and do not show the expected strength increase over depth. This may be due to the difficulties in applying uniform vacuum pressure to the soil and the degree of consolidation at different distances from the PVDs being different. However, the calculated values appear to fall within the average values of the measured undrained shear strength, which implies the computational model prediction of undrained shear strength due to vacuum preloading can be considered generally acceptable.

Conventional Preloading Area

The undrained shear strength profile with depth from conventional preloading area is shown in Fig. 6(c). Comparing Fig. 6(b) and (c), the undrained shear strength in the vacuum preloading (Fig. 6(b)) area is higher than that in conventional surcharge area (Fig. 6(c)) by about 10 kPa. This conclusion agrees with the actual measured undrained shear strength. For vacuum pressure preloading method, the increment of soil strength is more significant compared with similar pressure from the conventional surcharge. This is probably due mainly to the different load application procedure and the stress distribution from where the loads are applied. Finally, the calculated values of the undrained shear strength are generally in the average of the measured values, suggesting the prediction of undrained shear strength from conventional surcharge is generally reasonable.

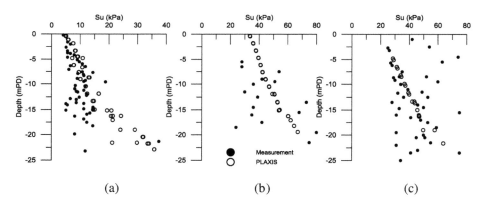

Fig. 6. Comparison of measured and calculated S_u. (a) initial condition. (b) after vacuum preloading. (c) after conventional preloading (Measured value in dots and simulation in open circles)

6   Conclusion

In this study, predictions of the initial and after improvement of undrained shear strength of the Macau Marine clays, using a computer software PLAXIS, under both vacuum preloading and conventional surcharge are presented. It is found that the numerical model calibrated by the observed settlement can be used to predict the development of effective stress and undrained shear strength under loadings. The predictions have been found to be in general agreements with the site measurements. Further research may focus on the relationship between settlement and undrained shear strength development. It may be possible that the estimation of undrained shear strength can be estimated from ground settlement measurement which is generally simpler to obtain than shear strength measurements.

It is also observed in the project that both vacuum preloading and conventional surcharge provide similar undrained shear strength development after soil improvement, but the vacuum preloading seems to have slightly higher strength increase from this study, due mainly to the loading applications and stress distribution from the loading points.

Acknowledgement. The authors would like to express their gratitude to Mr. Graeme Coles, Engineering Manager of Leighton Asia, for providing the valuable geotechnical data and comments during the course of this study.

References

1. Kjellman, W.: Consolidation of clayey soils by atmospheric pressure. In: Proceedings of a Conference on Soil Stabilization, pp. 258–263. Massachusetts Institute of Technology, Boston (1952)
2. Leighton: Wynn Resorts Macau – COTAI – Site Investigation, Falcon Foundation Engineering Limited (2011)
3. PLAXIS: Plaxis Material Models and Plaxis Reference Manuals, The Netherlands (2015)
4. Rujikiatkamjorn, C., Indraratna, B., Chu, J.: Numerical modeling of soft soil stabilized by vertical drains, combining surcharge and vacuum preloading for a storage yard. Can. Geotech. J. **44**, 326–342 (2007)

Geosynthetics

Application of Large-Size Sandbag Cofferdam in Land Reclamation Engineering

Weiping Peng[1], Lingwei Chen[1,2(✉)], and Xiaowen Zhou[2]

[1] Guangzhou Urban Planning & Design Survey Research Institute,
Guangzhou 510060, China
lingweichan@163.com
[2] South China University of Technology, Guanghzou 510641, China

Abstract. With the rapid economic and social development, land reclamation has been an increasingly urgent need to many countries. Land reclamation is indeed a preferential way to increase the land, but it may also result in some problems. Construction of cofferdam on coastal soft soil is one of the typical geotechnical problems in which many technical obstacles remain unsolved. A new technology of forming a cofferdam on soft soil by stacking super large sandbags has been developed in recent years. With its good integrity and strong adaption to ground deformation, the large-size sandbag cofferdam has been demonstrated an advanced technology. Although some reclamation engineering projects have adopted this new technology successively, we still saw accident cases where the cofferdam slope lost stability. It shows that the bearing capacity of soil bed and the slope stability of cofferdam cannot be estimated correctly in some situation. Under the action of super large bags, the theories about the critical stability and deformation limitation in design should be studied and developed. In this paper, two land reclamation projects completed in China are introduced. Among these two cases, one is totally successful but another has several slope failures during construction. Deformation and slope safety are analysed and compared to the actual behaviour of the cofferdams. Insights are summarized for better understanding of the large-size sandbag cofferdam technology.

Keywords: Land reclamation · Soft foundation
Large-size sandbag cofferdam

1 Introduction

In land reclamation engineering, people have to build an ambient cofferdam near coast firstly to form a closed region, and then fill the region with soil to obtain a land. The cofferdam is usually located in bad working conditions of weak seabed and strong water flow and wave actions. How to build a quick and safe cofferdam in sea is always a difficult challenge. In recent 10 years, a new method named "large-size sandbag cofferdam" is developed and applied to many practical projects in the world especially in China. The sandbag is made of geotextile with large sizes in length and width.

© Springer Nature Singapore Pte Ltd. 2018
L. Li et al. (Eds.): GSIC 2018, *Proceedings of GeoShanghai 2018 International Conference: Ground Improvement and Geosynthetics*, pp. 271–278, 2018.
https://doi.org/10.1007/978-981-13-0122-3_30

Sandbags are filled with pumped sea sand and stacked together layer by layer. In such way, a cofferdam embankment can be made very quickly. Although the small-size sandbag embankment has been widely used in projects such as coastal protection, flood fighting, erosion control, and land reclamation [1–3], but in offshore sea, small-size sandbag embankment is not applicable with its low stability in sea environment. To increase stability, the large-size sandbag has reached tens of meters in width and over hundred meters in length. With its good integrity and strong adaption to foundation deformation, the large-size sandbag cofferdam has rapidly become an competitive construction method in land reclamation engineering since its appearing [4, 5].

This paper introduces two actual land reclamation projects in China, one is located in Tianjin and another in Guangdong. These two projects were constructed and completed in 2013. The large-size sandbag cofferdam in the Guangdong behaved very well without any safety problem, but in the Tianjin project, several failure accidents appeared during the cofferdam construction. To repair the collapsed cofferdam embankment, huge cost and much time were paid. These examples indicate that the design and construction codes need to be improved to level up the engineering reliability. In this paper, some stability analyses based on limit equilibrium are carried out to discuss the causes of embankment failures. Through the comparison between the calculated safety and the in-situ situation, some insights are obtained which should be helpful for the better understanding of the large-size sandbag cofferdam behaviour.

2 Guangdong Cofferdam Project

2.1 Cofferdam Scale and Layout

Guangdong cofferdam project is located in the north of Donghai Island. The cofferdam needed to form a reclamation area of 9.27 km².

2.2 Geological Conditions

According to the geological investigation report, the seabed deposit can be divided into fourteen layers: ① medium coarse sand; ② mud and muddy-silty clay, ③ medium coarse sand mixed with mud, ④ mud and muddy clay, ⑤ clay and silty clay, ⑥ clay and muddy-silty clay, ⑦ medium coarse sand, ⑧ mud and muddy clay, ⑨ clay and silty clay, ⑩ medium sand layer, ⑪ clay and silt clay, ⑫ mud and muddy clay, ⑬ silt clay, ⑭ coarse sand. The characteristics of these fourteen soil layers are listed in Table 1.

In Tables 1 and 2, the cohesive force C_q and friction angle φ_q are the result of the quick direct shear test; the cohesive force C_{cu} and friction angle φ_{cu} are the result of consolidated quick shear test.

Table 1. Characteristic of soil layers (Guangdong project)

Soil layer number	Thickness (m)	Unit weight (kN/m³)	Soil strength			
			C_q (kPa)	φ_q (°)	C_{cu} (kPa)	φ_{cu}(°)
①	0.1–10.8	17.51	0.0	26	0.0	28
②	0.1–5.5	16.93	9.76	5.42	12.1	8.6
③	0.1–9.5	17.52	6.50	24	7.30	26
④	0.1–11.8	16.84	10.27	5.01	13.69	7.55
⑤	0.2–9.4	18.61	20.03	10.83	30.79	12.88
⑥	1.7–12.5	17.25	22.12	11.08	29.70	14.26
⑦	0.7–16.5	17.82	0.0	29	0.0	31.5
⑧	1.60–23.0	17.31	22.41	12.66	32.29	14.83
⑨	1.1–20.5	17.33	25.11	13.57	37.23	15.65
⑩	0.5–12.0	17.91	0.0	30.3	0.0	32.0
⑪	0.9–19.2	17.42	26.89	14.67	42.86	15.96
⑫	0.5–13.5	18.01	0.0	32.3	0.0	34.5
⑬	0.6–20.0	17.52	29.71	15.82	42.85	16.5
⑭	0.6–9.1	18.14	0.0	34.5	0.0	36.5

2.3 Cofferdam Design and Construction

The large-size sandbag is employed to build the cofferdam. The sandbags are made of geotextile, silt sand drawn from the seabed is used to fill the sandbags. The big sizes of sandbags are lifted horizontally layer by layer to construct the cofferdam embankment. The thickness of each layer is controlled at 500–800 mm. The typical cross-section of the cofferdam embankment is illustrated in Fig. 1.

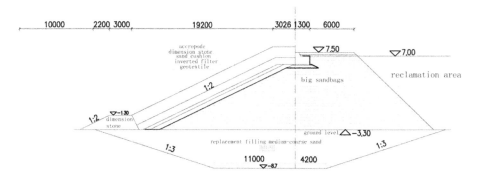

Fig. 1. Typical cross-section design of the cofferdam (Guangdong project).

The Guangdong cofferdam project began in February, 2011 and was completed in March, 2013, lasting about 25 months. In the construction process, beside artificial work of paving sandbag, ships were used to pump sand for filling the bags and fixing

| a. big sandbags before sand filling | b. big sandbags after sand filling |

Fig. 2. Construction of the big sandbags in site.

the positions of sandbags. Figure 2 gives the site pictures of sandbag cofferdam. In the whole construction process, the big sandbag cofferdam behaved very well with no any safety problem.

2.4   Stability Analysis

The simplified Bishop method is applied in stability analysis, the formulas are as follows:

$$F_s = \frac{M_{Rk}}{M_{sd}} \tag{1}$$

$$M_{Rk} = R \sum \frac{c_{ki}b_i + (q_{ki} + W_{ki} - u_{ki}b_i)tg\varphi_{ki}}{cos\alpha_i + sin\alpha_i th\varphi_{ki}\frac{1}{\gamma_R}} \tag{2}$$

$$M_{sd} = \gamma_s \left[\sum R(q_{ki}b_i + W_{ki})sin\alpha_i + M_p \right] \tag{3}$$

Where F_s is the safety factor; M_{Rk} is the standard value of anti-sliding moment along most dangerous slip surface; M_{sd} is the designed value of sliding moment acts along most dangerous slip surface; R is the arc radius of most dangerous slip surface; γ_s is the comprehensive partial coefficient, which equal to 1; γ_R is the resistance partial coefficient; W_{ki} is the gravity standard value of the ith soil strip; u_{ki} is the standard value of pore pressure; M_p is the sliding moment caused by other reasons; q_{ki} is the characteristic value of variable action; b_i is the width of the ith soil strip; α_i is the slope angle at the middle point of the slip surface for ith soil strip; φ_{ki} is the internal friction angle of soil; c_{ki} is the cohesive force of soil.

Uniformly distributed load of 10 kPa on the cofferdam crest and designed lowest water level of −0.43 m are considered in calculation. The strength parameters for the sandbags are chosen as c = 0, φ = 28°, the unit weight of sand is 18 kN/m³. For the foundation soils, quick direct shear strengths (C_q, φ_q) and consolidated quick shear strengths (C_{cu}, φ_{cu}) are used respectively for service stage and construction period of the cofferdam. The calculated safety factor of the cofferdam is 1.0585 and 1.2513 respectively for the two periods. The calculated safety factors are a little bigger than the required safety values in related Chinese design and construction specifications about embankment. But, the margins of safety seem to be very small, which means the cofferdam was at a state close to critical state. In other word, although this project was successful with no accident, the risk of accident always existed in it.

3 Tianjin Cofferdam Project

3.1 Cofferdam Scale and Layout

The Tianjin project is located 7 km away from the estuary of Duliujian River. The cofferdam needed to form a reclamation area of 4.324 km^2.

3.2 Geological Conditions

According to the geological investigation report with 64 drilling holes, the seabed deposit can be divided into seven layers: ①$_1$ mud, ①$_2$ muddy clay, ①$_3$ clay mixed with shell; ②$_1$ silty clay, ②$_2$ silt; ③$_1$ silty clay, ③$_2$ silt. The characteristic of main soil layers are listed in Table 2.

Table 2. Characteristic of soil layers (Tianjin project)

Soil layer number	Thickness (m)	Moisture content (W%)	Unit weight (kN/m^3)	Plasticity index I_P	Soil strength			
					C_q (kPa)	φ_q (°)	C_{cu} (kPa)	φ_{cu} (°)
①$_1$	3.0–8.0	60.7	16.21	23.5	6	0.2	13.8	16
①$_2$	3.0–8.0	45.5	17.42	21.2	13	2.5	12.7	16.1
①$_3$	0.5–2.0	27.4	18.61	12.8	16	12.6	27.5	27.0
②$_1$	0.5–3.0	26.6	19.13	13.3				
②$_2$	0.2–2.0	25.1	19.12	7.4			33.0	29.6
③$_1$	3.5–17.5	23.9	19.93	13.0	19	17.0	38.5	23.2
③$_2$	2.0–6.5	23.5		7.4				

3.3 Cofferdam Design and Construction

The large-size sandbag is employed to build the cofferdam. The sandbags are made of geotextile; silt sand drawn from the seabed is used to fill the sandbags. The big size sandbags are lifted horizontally layer by layer to construct the cofferdam embankment. The thickness of each layer is controlled at 500 mm. The typical cross-section of the cofferdam embankment is illustrated in Fig. 3.

The Tianjin cofferdam project begun in November, 2011 and was completed in November, 2013, lasting about 24 months. The cofferdam construction method is similar to the one applied in Guangdong project, But several slope failures appeared. On June 21th, 2013, the big sandbags cofferdam collapsed, where the stake mark is

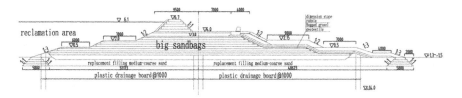

Fig. 3. Typical cross-section design of the cofferdam (Tianjin project).

Fig. 4. Pictures of collapse status in construction site (Tianjin project).

K2+765 - K2+835, the elevation is +2.5 m, and the tide level is 1.39 m. On July 27th, 2013, the big sandbags collapsed again, where the stake mark is K1+200 - K1+400, the elevation is +3.0 m, and the tide level is 1.19 m. Figure 4 is the pictures of the collapse status.

3.4   Stability Analysis

The Canadian software Geo-studio is adopted for stability analysis. Because the section is symmetrical along the axis, and the stress on both side is the same. Half of the section along the axis is selected for modelling. The limit equilibrium method is adopted in the analysis.

The big sandbags are regarded as the combination of two different materials, which include sandy soil and geotextile. The bags are treated as the reinforced material of filling sand, like the soil-nail braced structure. Through the action of friction between sandbag and sandy soil, the tensile properties are reflected and the slope stability is strengthened [6].

In order to find the failure causes in Tianjin cofferdam project, the cross-section of the collapse area is adopted to model and analyse. The elevation of the big sandbag is +3.0 m, and the strength parameters for the sandbags are chosen as c = 0.2 kPa, $\varphi = 30°$, the unit weight of sand is 16 kN/m^3. The tensile strength of 50 kN/m, acting on the contact surface of the sandbags, is considered in calculation. Three different conditions are calculated, the result is shown in Table 3. Figure 5 is the result of stability analysis.

Table 3. Influence of different foundation soil strength to cofferdam stability

Serial number	Thickness of foundation (m)	Strength of foundation soil		Safety factor K
		C (kPa)	$\varphi(°)$	
1	20	6	0.2	0.987
2		9	3.2	1.198
3		12	6.2	1.352

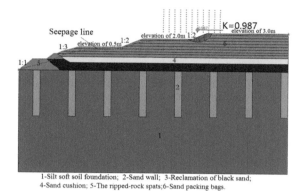

1-Silt soft soil foundation; 2-Sand wall; 3-Reclamation of black sand;
4-Sand cushion; 5-The ripped-rock spats;6-Sand packing bags.

Fig. 5. Result of stability analysis (the 1st condition in Table 3)

As shown in Table 3, we can find that the strength increment of the soil during embankment construction is very important to increase the stability of the big sandbag cofferdam. The inadequacy of foundation bearing capacity will result in serious collapse accident of cofferdam. Combine with the construction of Tianjin project, the reason for the collapse is the high stacking speed of sandbags. There is no enough time to let the soft soil consolidate to a necessary extent, which lead to the inadequacy of foundation bearing capacity.

4 Conclusion

The large-size sandbags cofferdam is widely adopted in land reclamation in China. The engineering practice develops faster than the theory. There is no specialized standard to guide the design and construction, and many key problems remain unsolved. Although the large-size sandbag cofferdam has a good integrity, the foundation bearing capacity and slope stability under the conditions of weak seabed and strong water flow and waves should be further studied.

According to the engineering failure accident happened and the stability analysis of the cofferdams in Guangdong and Tianjin reclamation projects, as the foundation is composed of thick soft soil, the inadequacy of foundation bearing capacity may result in serious collapse accident of cofferdam.

The strength increment of the soil with its consolidation during embankment construction is very important to increase the stability of the large-size sandbag cofferdam. In the construction, the stacking speed of sandbags should be controlled to let the soil consolidate to a necessary extent.

References

1. Chen, L., Zhou, X., et al.: Centrifugal modeling of big sandbag cofferdam on coastal soft soil. Chin. J. Rock Mech. Eng. **35**(s2), 4235–4240 (2016)
2. Chen, L.: Deformation and failure mode of the Large-scale geotextile sand container cofferdam on soft foundation. Ph.D. thesis, South China University of Technology (2016)
3. Zhou, X., Wang, X., et al.: Characteristics of failure zones of soft soil foundation under flexible load sandbag cofferdam. Chin. J. Geotech. Eng. **39**(Suppl. 2), 45–48 (2017)
4. Yang, M., Zang, D.: Application of sand-bag cofferdam by hydraulic filling for NANJMAT project in UAE. China Harbour Eng. **3**, 45–48 (2009)
5. Wang, X., Xu, J.: High efficiency construction technology of large-size sand-filled bags in closing levee project. China Harbour Eng. **2**, 58–59 (2010)
6. Zhou, X., Wang, L., et al.: Analysis of stability of dykes with fabriform sand on soft foundation. J. Hohai Univ. **42**(3), 243–245 (2014)

Centrifuge Model Tests of Basal Reinforcement Effects on Geosynthetic-Reinforced Pile-Supported Embankment

Chao Xu[1], Di Wu[2(✉)], Shitong Song[1], and Baochen Liu[2]

[1] Key Laboratory of Geotechnical and Underground Engineering of Ministry of Education, Tongji University, Shanghai 200092, China
[2] College of Architecture and Transportation Engineering, Guilin University of Electronic Technology, Guilin 541004, China
wudi@guet.edu.cn

Abstract. As an effective and economical structure, geosynthetic-reinforced pile-supported (GRPS) embankment has been used in the construction of roadbed of railway and highway in soft soil area. Its key load transfer mechanism includes the soil arching effect of the embankment fill and the tensioned membrane effect of reinforcement in the mattress. Via these two effects, more embankment load is carried by piles, and the total settlement and differential settlement of the roadbed can be under control. Four centrifuge model tests were conducted to study the reinforcement effect of geosynthetic on embankment stability and pile efficacy. The following results are achieved: the surface settlement and differential settlement of embankment could be effectively reduced through appropriate setting of reinforced mattress; Basal reinforcement increased the pile efficacy and was affected by the number of reinforcement layers; The tensile force of geogrid below the embankment shoulder was larger than that near the road center because of the lateral restraint of basal reinforcement.

Keywords: Geosynthetic-reinforced and pile-supported (GRPS) embankment
Geogrid · Basal reinforcement · Membrane effect · Arching effect
Centrifuge model test

1 Introduction

As compared with some other ground improvement methods, geosynthetic-reinforced pile-supported (GRPS) embankments can be constructed quickly, do not require soft soil replacement and staged construction, and meet strict settlement requirements. Therefore, they become one of the favorable technologies for the construction of embankments on soft soil foundations. GRPS embankment, consisting of embankment fill, geosynthetic reinforcement, piles, and foundation soils, is a complex soil-structure system. The interaction among these components determines the load distribution and the functions exerted by the reinforcement. Its key load transfer mechanisms include soil arching and tensioned membrane effects and subsoil resistance.

Soil arching effect has been one of the intensive research topics in the study of GRPS embankment. Hewlett and Randolph [1] observed arching through a glass case

© Springer Nature Singapore Pte Ltd. 2018
L. Li et al. (Eds.): GSIC 2018, *Proceedings of GeoShanghai 2018 International Conference: Ground Improvement and Geosynthetics*, pp. 279–287, 2018.
https://doi.org/10.1007/978-981-13-0122-3_31

and proved that the arch was close to hemisphere form between square lay-out piles. Other model tests (Low et al. [2]; Chew and Phoon [3]; Cao et al. [4]) were carried out to investigate the influence of embankment height, pile cap size and reinforcement on soil arching effect.

Centrifuge model test takes both advantages of the scaled model test and prototype experiment. Tests (Barchard [5]; Zhang et al. [6]; Wang et al. [7]) showed that in GRPS embankment load transfer largely depended on reinforcement function; in piled embankment without basal reinforcement, load transfer mainly depended on soil arching effect.

Existing researches concerned more about pile efficacy and pile-soil stress ratio (Han and Gabr [8]; Briancon et al. [9]; Van Eekelen et al. [10]), however considered little about the function of geosynthetic-reinforced mattress as a whole and its influencing factors. For these reasons, a series of 4 centrifuge model tests were applied to investigate the basal reinforcement effects on GRPS embankment. The results of load distribution and deformation were presented and analyzed. The influencing factors including reinforcement settings, position and the properties of reinforcing material were explored. The conclusions are used to provide guidance on reasonable establishment of GRPS embankment design.

2   Centrifuge Model Test

The prototype embankment was 6 m high, 16 m wide on the surface, and the slope ratio was 1:1.5. The filling process of the embankment was divided into three steps, and each step had a 2 m instantaneous load followed by 60-day repose as a transitional period. Reinforced concrete piles used in the GRPS embankment were 16 m in length, 0.5 m in diameter and 2.8 m in pile spacing with square disposing mode. Pile caps (1.6 m × 1.6 m) were fixed onto the piles. The basal reinforced cushion was 0.6 m thick and the tensile strength of bi-direction geogrid was 80 kN/m at a tensile strain of 5%. The thickness of soft soil foundation was 16 m, and the bearing stratum was sandy soil layer. The selected model ratio N equaled 80.

The number of geogrid layers was systematically analyzed in this experimental research plan. Four comparative tests were arranged including one non-reinforced piled embankment. Table 1 lists the tests that were conducted.

The model foundation of the centrifuge tests was composed of two layers, with soft soil layer over sand layer. The soft soil was silty clay which is a mixture of silty clay

Table 1. Test program of GRPS embankment.

Tests	Number of layers in the mattress	Layout of geogrid
G1	0	None
G2	1	Middle of the mattress
G3	2	Uniform distribution
G4	4	Uniform distribution

and mucky clay in Shanghai. The preparation of the soil included the process of drying, smashing, sieving and mixing with water into a paste. Then the soil sample was placed into the centrifugal field (80 g) to undergo the self-weight consolidation. The undrained shear strength (C_u) of the soil was 12.6 kPa and the plasticity index of the soil was 15, and the modulus of compression ($E_{s0.1-0.2}$) equaled 2.71. The permeability coefficient was 7.56×10^{-9} m/s. Sand with good grading distribution was used both for the mattress and the embankment fill. The particle size, less than 1 mm, was required to be as fine as possible because of the magnified effect in centrifuge model test. The model embankment was prepared by layers, and each layer was 1 cm thick. The volume of every layer was calculated through the embankment section area. The uniformity and compactness of the embankment were guaranteed through controlling the dry density of 1.62 g/cm^3 under compaction. The lower layer of soil foundation simulated the bearing stratum of piles.

Figure 1 shows the size of the model and the layout of the instrumentations. The measuring components included three displacement sensors (s1–3) to monitor settlement at the road center, road shoulder and slope toe, respectively. An eddy displacement sensor, combined with a pre-placed board (s4), was applied to measure the horizontal displacement at 25–50 mm below the foot of the slope. Five earth pressure cells (epc1–5) were embedded to measure the load on pile and soil pressure over and below the reinforced mattress, aiming to study the load distribution through soil arching and tensioned membrane effect. The strains of geogrid were detected by four grid strain gauges (gs1–4) fixed with geogrid using epoxy as medium. It should be noted that the strain gauges stuck on the geogrids were calibrated before the tests.

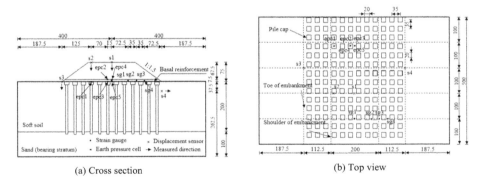

(a) Cross section   (b) Top view

Fig. 1. Dimensions of models and arrangement of instruments (unit: mm).

After the foundation soil was in place, the following steps were carried out in each test: (1) consolidating soft soil foundation in the 80 g-centrifugal field for 5 h; (2) inserting model piles into predetermined positions vertically and precisely; (3) burying the soil pressure cells flatly in the mattress which should be carefully paved; (4) preparing the embankment according to the scheduled size by layers; (5) positioning displacement sensors to measure the displacement and settlement of the specified locations. The variable acceleration loading method was taken to simulate the process of step-heaped loading exactly according to the prototype. Detailed time history of acceleration is shown in Fig. 2.

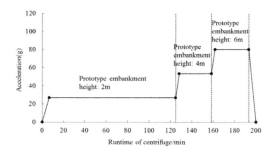

Fig. 2. Loading curve of models.

3 Analysis of Test Results

This section elaborates the analysis of the test results in prototype dimensions except the runtime of centrifuge for unified description.

3.1 Deformation and Stability of Embankment

Figure 3 presents the settlement at the center of the embankment surface. It proves that the center settlement of non-reinforced piled embankment was far greater than those of reinforced and piled embankments. The former reached 1.46 m, while the latter ranged between 0.20–0.31 m when runtime of centrifuge was 180 min, which means the construction of the embankment was completed. It shows that the adoption of geogrid helped to decrease the vertical displacement. Moreover, the embankment settlement decreased with adding reinforcement layers, but the reinforcement effect of the mattress with 4 layers of geogrids was not so obvious compared to that with 2 layers. For 1 layer

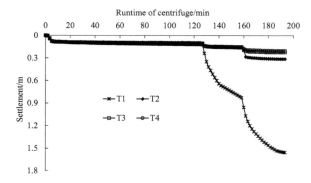

Fig. 3. Settlement curve of the center of the surface of the embankment.

of reinforcement, geogrid would have a stronger restriction effect on the settlement when paved in the middle of the mattress, where the soil-geogrid interaction could be better performed.

Figure 4 shows the differential settlement calculated by the embankment surface settlement of center (s1) and edge (s2), which depicted that the non-uniform settlement of embankment could be reduced effectively by the increasing of reinforcing layer number. Thus, the embankment integrity was also improved.

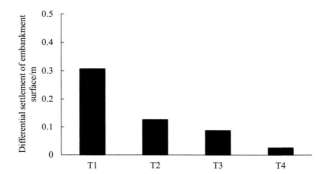

Fig. 4. Differential settlement of embankment surface between center and shoulder of the road (runtime of centrifuge: 180 min).

The stability of embankment on soft ground can be reflected by the ratio (η) of maximum lateral displacement at the foot of embankment slope to the maximum settlement of subsoil surface (Chai [11]; Fei and Liu [12]). A rapid increase of this ratio means that the embankment is close to failure. Hence, quantitative evaluation of embankment stability can be done through the experimental result analysis. The maximum vertical displacement and lateral displacement of foundation soil in the test were measured by s3 and s4 separately. Figure 5 demonstrates the relationship between the ratio (η) and the embankment filling height of prototype (H).

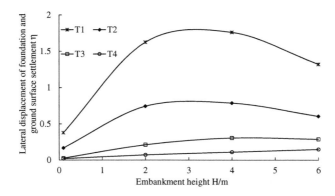

Fig. 5. Relationship between the ratio of lateral displacement of foundation and ground surface settlement to embankment height.

Figure 5 shows that with embankment height increasing, the trend of η–H curves of models with 2 layers (T3) and 4 layers (T4) of geogrids were similar, that is, the ratio η slightly increased with embankment height and then stayed stable. A sharp increase of η in the height of 0–3 m appeared in the non-reinforced piled embankment (T1), which means that instability and large deformation occurred. Although the ratio decreased afterwards, it still maintained a large value. The embankment reinforced with 1 layer geogrid in the mattress (T2) underwent a quick rise in the initial stage of heap loading, while the magnitude was not as large as that in non-reinforcement embankment.

The above results show that the reinforcement in the mattress could reduce the embankment surface settlement and differential settlement, improve the overall stability, reduce the deformation of the foundation, and play the role of lateral restriction. Different reinforcement layers had great impacts on the effect of basal reinforcement in GRPS embankment and consequently had significant influence on the deformation of piled embankment.

3.2 Soil Arching Effect and Tensioned Membrane Effect

Pile efficacy (Hewlett and Randolph [1]), the ratio of embankment load on one pile to the total load in the processing range of a single pile, is utilized frequently in recent years to evaluate the soil arching effect. But there was no geosynthetic in their paper. In this paper, geosynthetic was used, so pile efficacy is illustrated by two calculations. The equations are as follows,

Above the reinforced mattress:

$$E_{pa} = \frac{ep_2 \times a^2}{[\gamma \times (H - d) + q] \times s^2} \tag{1}$$

Below the reinforced mattress:

$$E_{pb} = \frac{ep_3 \times a^2}{[\gamma \times (H - d) + \gamma' \times d + q] \times s^2} \tag{2}$$

The explanation of each symbol is shown in Fig. 1. In fact, the pile efficacy E_{pa} calculated using the pressure measurements above the mattress simply demonstrates the role of soil arching, while E_{pb} calculated using the measurements below the mattress demonstrates both the soil arching effect and the tensioned membrane effect.

A relationship can be found between E_{pa} and the filling height of the embankment, and so as to E_{pb}, which are both illustrated in Fig. 6. The result of T1 is not presented, because there was no reinforcement in the embankment mattress. In the upper half of Fig. 6, with the filling height increasing, E_{pa} became greater. The results of models T2–T4 were relatively close, indicating that the different settings of geogrid had little influence on soil arching effect.

In the lower half of Fig. 6, E_{pb} increased with the filling height, and the values of models T2–T4 reached 56.9%–64.2% under full load. On one hand, the results reveal

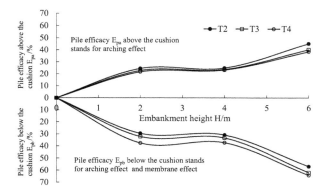

Fig. 6. Relationship between pile efficacy above the reinforcement and the increment in the center and embankment height.

that the role of reinforced mattress was to increase the load shared by the pile, and further exerted the bearing capacity of the pile. On the other hand, it indicates that the tensioned membrane effect of reinforcement was affected by the number of reinforcement layers.

3.3   Tensile Force of the Geogrid

The tensions of the geogrid were measured by four strain gauges (sg1–4) stuck on geogrids between the piles in the experiments. In mattress with more than one layer reinforcement, the strain gages were pasted on the bottom layer of the geogrids. The sg4 measurement of T4 was abnormal and was not adopted for analysis.

Figure 7 shows the distribution of reinforcement tension after full load of the embankment. It reveals that the tensile force measured by sg1 at the center of embankment and sg4 close to the foot of the slope were relatively small, while the tensile forces measured close to road shoulder by sg2 and sg3 were larger, which corresponded with the results reported by Zhang et al. [6]. The tensile force of the geogrid was influenced by two factors: the self-weight of the embankment fill between piles and the lateral slip of the embankment slope. The lateral slip of the embankment slope generated less tension of geogrid at the embankment center. The effect of self-weight of the embankment fill between piles was less obvious for geogrid close to the slope foot. Both of the two factors caused the large tensile force of geogrid under the road shoulder. The test results shown in Fig. 7 illustrate that, for the tensile force measured in the same location and with the same tensile modulus of the geogrid, its value got largest in the case with one reinforced layer at the middle of mattress (T2), and the case with multi-layers (T3, T4) came last. It is believed that in multi-layer reinforced embankment, the load was undertaken by all the layers and the tensile force of each single layer was reduced.

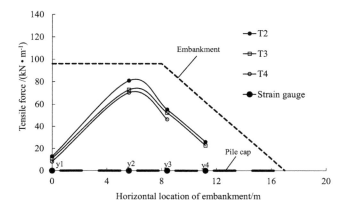

Fig. 7. Distribution of tensile force in belt (runtime of centrifuge: 180 min).

4 Conclusions

Four centrifuge model tests were performed on the role of basal reinforcement in GRPS embankment. The model test results were presented and analyzed, and the following conclusions can be drawn:

(1) The surface settlement and differential settlement of embankment could be effectively reduced through appropriate setting of reinforced mattress.
(2) The role of reinforced mattress was to increase the load shared by the pile and further exerted the bearing capacity of the pile. The tensioned membrane effect of reinforcement was affected by the number of reinforcement layers.
(3) Geosynthetic basal reinforcement contributes to a GRPS embankment system in two ways, i.e., the lateral restraint and tensioned membrane effect. The former restricts the embankment lateral deformation and improved its stability, while the latter acts as an uplifting force to the overlying filling between piles, and transfers the load it takes onto the piles through tensioned membrane effect. Due to the lateral restraint of basal reinforcement, the tensile force of geogrid below the embankment shoulder was larger than that in the road center.

Acknowledgment. Financial support for this work is gratefully acknowledged from the Natural Science Foundation of China Grant 41772284 and Natural Science Foundation of Guangxi Grant 2015GXNSFBA139236. All the support is greatly appreciated.

References

1. Hewlett, W.J., Randolph, M.F.: Analysis of piled embankments. Ground Eng. **22**(3), 12–18 (1988)
2. Low, B.K., Tang, S.K., Chao, V.: Arching in piled embankments. J. Geotech. Geoenvironmental Engineering, ASCE **120**(11), 1917–1938 (1994)

3. Chew, S.H., Phoon, H.L.: Geotextile reinforced piled embankment-full scale model tests. In: Proceeding of the 3rd Asian Regional Conference on Geosynthetics, pp. 661–668 (2004)
4. Cao, W.P., Chen, R.P., Chen, Y.M.: Experimental investigation on soil arching in piled reinforced embankments. Chin. J. Geotech. Eng. **29**(3), 436–441 (2007)
5. Barchard, J.M.: Centrifuge Modelling of Piled Embankments on Soft Soils. University of New Brunswick, Fredericton (1999)
6. Zhang, L., Luo, Q., Pei, F.Y., Yang, Y.: Bearing stratum effect at pile end of pile-cap-net structure subgrade based on centrifugal model tests. Chin. J. Geotech. Eng. **31**(8), 1192–1199 (2009)
7. Wang, C.D., Wang, B.L., Wang, X., Zhou, S.H.: Analysis on settlement controlling effect of pile-net composite foundation on collapsible loess by centrifugal model tests. J. Chin. Railw. Soc. **33**(4), 84–92 (2011)
8. Han, J., Gabr, M.A.: Numerical analysis of geosynthetic-reinforced and pile-supported earth platforms over soft soil. J. Geotech. Geoenvironmental Eng. **128**(1), 44–53 (2002)
9. Briançon, L., Delmas, P.H., Villard, P.: Study of the load transfer mechanisms in reinforced pile-supported embankments. In: Proceedings of 9 ICG, Brazil, pp. 1917–1920 (2010)
10. Van Eekelen, S.J.M., Benzuijen, A., Jansen, H.L.: Piled embankment using geosynthetic reinforcement in the Netherlands: design, monitoring & evaluation. In: Proceedings of the 17th International Conference on Soil Mechanics and Geotechnical Engineering, pp. 1690–1693 (2009)
11. Chai, J.C., Miura, N., Shen, S.L.: Performance of embankments with and without reinforcement on soft subsoil. Can. Geotech. J. **39**(4), 838–848 (2002)
12. Fei, K., Liu, H.L.: Field test study and numerical analysis of a geogrid-reinforced and pile-supported embankment. Rock Soil Mech. **30**(4), 1004–1012 (2009)

Comparison Test of Different Drainage in Single Well Model for Vacuum Preloading

Jing Wang[1,2(✉)] and Yongping Wang[1,2]

[1] Fourth Harbor Engineering Institute Co., Ltd., Guangzhou 510230, China
phdwangjing@163.com
[2] CCCC Key Lab of Environmental Protection and Safety in Foundation
Engineering of Transportation, Guangzhou 510230, China

Abstract. The drainage plate bending and membrane clogging often appear on vacuum preloading consolidation of dredger fill foundation. In practical, water performance and clogging application exist in plastic drainage plate. The contrast experiments of single well model carry out for three different type vertical drainage in this paper. For single drainage plate as the object, drainage seepage condition of the actual the vacuum preloading simulate in the model barrel. The strength of soil and the bending of drainage plate before and after reinforcement are compared to determine vertical drains of vacuum preloading on dredger fill foundation.

Keywords: Drainage plate · Vacuum preloading
Single well model experiment

1 Foreword

The drainage plate bending and membrane clogging problems exist in vacuum preloading consolidation of dredger fill foundation. The bending water quantity test of plastic drainage plate derived on the basis of experiments [1, 2]. The relative environmental factor is single, and less effect to water quantity. In the actual project, external factors to the equivalent drainage board is complex, such as construction, geological and design factors. These factors has a great influence on the performance of water quantity through plastic drainage plate [3, 4]. Water quantity of drainage plate should be considered with permeability decreases influence of regional performance. For the deep inserting drainage plate, the impact of drainage well resistance should be considered. The plate pressure is not fixed 350 kPa. The actual confining pressure should also consider the function of vacuum negative pressure, and confining pressure varies with depth. Influenced by various factors, the water quantity in the actual project significantly less than the indoor experiments [5, 6].

In addition, the drainage plate core and filter clogging behavior, often happen simultaneously and mutual influence in the actual project. The seepage caused by negative pressure of vacuum load is also different from the seepage caused by hydraulic pressure in hydraulic gradient.

In practical, water performance and clogging application exist in plastic drainage plate. The contrast experiments of single well model carry out for three different type

L. Li et al. (Eds.): GSIC 2018, *Proceedings of GeoShanghai 2018 International Conference: Ground Improvement and Geosynthetics*, pp. 288–296, 2018.
https://doi.org/10.1007/978-981-13-0122-3_32

vertical drainage in this paper. For single drainage plate as the object, drainage seepage condition of the actual the vacuum preloading simulate in the model barrel. The strength of soil and the bending of drainage plate before and after reinforcement are compared to determine vertical drains of vacuum preloading on dredger fill foundation.

2 Test Device

Single well test model is mainly composed of vacuum pump, gas water separator and cylinder model. These three parts are connected through a pipeline, constitute an airtight system.

The test device comprises of three sets sample models, three precision vacuum pressure regulating valves, three gas water separators and a vacuum pump.

Monitoring system mainly includes, an automatic data acquisition system, a desktop computer, pore pressure gauge, six pressure gauge hole, twelve vacuum gauges, electronic scale, settlement observation ruler and plumb.

The vacuum pump is a source of air pressure system, using the circulation technology. In the experiment continued pumping maintain a constant internal pressure system. The gas water separator is respectively connected with the cylinder model and vacuum pump. The cylinder gas water mixture pump to the separator. The gas flow is discharged through the vacuum pump, and the water is stored in the container to record water discharge. Cylinder model is the main place of seepage system in vacuum under negative pressure, as the core part in the test. Model container is high 100 cm, cylinder diameter 40 cm, internal soil height of 80 cm, central drainage board inserted in soil samples. In order to simulate the construction site conditions, soil samples located at the top by fine sand under two layers of fabric wrapped 5 cm thick cushion. Drainage plate head and vacuum tubes are buried in the sand cushion in the middle part of the transfer of vacuum loading and drainage with vacuum preloading for plastic sealing film in the sand cushion top. The adhesive through the sealant and the top edge of the cylinder container model, model of tube in the experiment can be effectively applied in the vacuum pumping tube negative pressure drainage plate.

3 Testing Program

3.1 Drainage Parameters

In order to better explore the drainage plate bending and clogging behavior of foundation reinforcement effect, the selection of small diameter conventional hot-rolled non-woven and small aperture anti-clogging nonwovens two kinds of membrane were compared. Two kinds of membrane pore size, porosity on the numerical approach while the gradient is larger than that by single well model test can verify the applicability gradient ratio test in silt soil samples on filter clogging behavior judgment; moreover, the common drain membrane mostly small aperture hot rolled non-woven fabrics, by single well model test results of two different types of non-woven fabric can be compared the results contribute to effective feedback in practical engineering. In order to better explore the drainage plate bending and clogging behavior of foundation rein-forcement effect, the selection of small aperture conventional hot rolling non-woven fabrics and small aperture anti-clogging nonwovens two membrane were compared. Two kinds of membrane in the numerical aperture, porosity and close to the gradient ratio is relatively large. Through the single well model test, the applicability is verified of gradient ratio test for silt soil samples on filter clogging behavior judgment. Further-more, the project used mostly for small aperture drainage plate membrane. Through the comparison of fruit nonwovens single well model test results of two different types of non-woven fabrics, can help the results effectively feedback in practical engineering.

Table 1. Single well model test scheme

Project	Example	Vacuum time
1	Anti-clogging integral plastic drainage board	Vacuum until settlement and water discharge stability
2	Conventional separation type plastic drainage plate	
3	100 mesh nylon and sand	

The main comparison experiments of single well model for three different vertical drainage body. Three drainage body respectively, are commonly used in engineering of conventional drainage board, new anti-clogging integral drainage plate and traditional sand wick. The horizontal drainage cushion layer from top to bottom is same for each test plan, as 200 g/m^2 woven, 200 g/m^2 non-woven fabric and thickness of 50 mm sand cushion, 200 g/m^2 layer of non-woven fabric. The sealing material is two layer thickness of $0.12 \sim 0.14$ mm sealing film, as shown in Table 1.

3.2   Test Procedure

For the sake of approaching the actual engineering, the steps of the test are in accor-dance with the actual construction procedure.

Fill the model bucket with fill to 80 cm height.

Making 3 different drainage bodies, tying three vacuum probes along the length direction in the drainage body, and closing the vacuum probe and the manometer line (see Fig. 1).

Insert a drainage plate. The drainage plate with a long rope tied to the stick. A head on the outside of the barrel and untied the rope model. Soil center stick into the circular model together with the barrel drains together. Unlock the connection, and to remove the drainage body sticks. The static model of the bucket, the silt soft soil.

A sand cushion with a thickness of 50 mm is laid on the soil surface, and a vacuum probe is embedded into the sand cushion together with a suction straw.

The sand cushion is covered with a layer of woven fabric and a layer of non-woven fabric, so as to prevent the gravel from breaking the sealing film under the action of negative pressure.

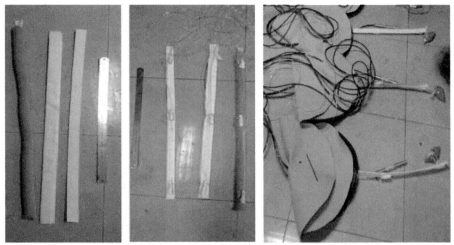

(a) Three drainage (b) Drainage and vacuum measuring heads (c) Drainage and gauges

Fig. 1. Connection of drainage, monitoring and vacuum system

The cover sealing film. The central cut open a small mouth, so that the gathering pipeline. Through the steps of connecting various instruments, sealing film and a barrel top, sealed with sealant isolated air pipeline between.

The vacuum Straw gas water separator. The other end is connected between the gas water separator and a vacuum pump for vacuum pressure regulating valve connected to adjust the vacuum pressure. After connecting various monitoring instruments and vacuum system, start the initial reading of the monitoring measurement reading content, such as pore pressure, settlement etc.

Open vacuum pump and automatic measurement instrument, to start reading.

4 Soil Performance Comparison Before and After Reinforcement

After the test, the demolition model bucket overlying the sealing membrane. Using the sampler sampling, take out the full height soil column to test water content at different depth and the soft soil wet density on surface (see Fig. 2). Shear strength index and water content compared before and after reinforcement. The unconfined compressive strength compared for 3 model bucket soil.

4.1 Water Content and Wet Density

The moisture content of this index can reflect the reinforcement effect of vacuum preloading and water before and after reinforcement rate difference is big, the more water from that row, the lower water content, the reinforcement effect is better. The wet density reflects the degree of soil compaction, soil after vacuum preloading and wet

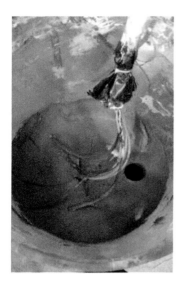

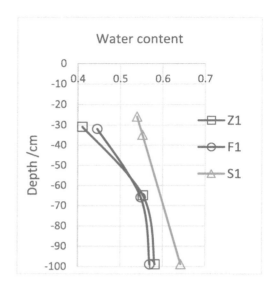

Fig. 2. Soil after reinforcement   **Fig. 3.** Curve of water content and depth

density should be increased. Each bucket model soft soil within the depth of 3 points, located at upper, middle and bottom. The determination of water content, depth water content curve, as shown in Fig. 3; Determination of soil on the surface of the wet density, the results shown in Table 2.

From Table 2, the wet density increase before and after reinforcement in varying degrees. The sequence is uniform of settlement, integral is larger than separation type and bagged sand well.

From the content point of view, it is obvious reinforcement effect. Reducing the moisture content of the most obvious is the overall drainage type and separation type of the body. The upper soft soil moisture rate were decreased by 30.5% and 28.5%, and the reinforcement effect is obvious. The upper sand wick was reduced by only 18%. In the middle, upper than soft soil, water content increased to a certain extent. The reinforcement effect has been attenuated, three were around 55% in value. The difference is not much difference between the decay rates of the reinforcement effect. The two water drain rate increased obviously, and the water bag sand well rate rises less. To the bottom, reinforcement the effect of drainage plate and sand well, the decay rate is just the opposite. Add less water integral and separate rate, similar to the reinforcement effect in the middle of the water and sand wick rate increased rapidly to 64%, has been connected with the initial moisture content of 71% close. Show that the reinforced bag sand well bottom has relatively weak decay to the point, the effective reinforcement depth as drainage plate depth. The monitoring results before comprehensive, sand well vacuum water amount, the lowest settlement minimum, reducing the moisture content is at least consistent with the law.

Table 2. Wet density of surface soil before and after reinforcement

Status	Anti-clogging integral	Conventional separation type	Bagged sand well
Before reinforcement	1.567	1.567	1.567
After reinforcement	1.767	1.719	1.698

The rate of this index from single well water model test to measure the effect of reinforcement. The reinforcement effect of drainage plate and drainage sand well drainage bodies show different patterns. The reinforcement effect of soft soil drainage plate for the upper region is better than the sand well. But for the middle part of the soft soil, the reinforcement effect quickly the middle and bottom attenuation. The reinforcement effect is basically the same. And although the upper bag sand well in soft soil reinforcement effect as drainage board. From the upper to middle reinforcement effect attenuation degree is low. Until the bottom of the reinforcement effect was attenuated more obviously.

Through the analysis of its causes, performance of initial stage of drainage plate work than sand well. The drainage plate with the rapid consolidation of soft soil layer and bending deformation, resulting in well resistance increase, water performance significantly weakened. Therefore to the middle layer, the reinforcement effect has more attenuation. The middle and bottom drainage less. The reinforcement effect of soft plate bending the soil is basically consistent. The reinforcement effect mainly restricted upper bending. For sand well, due to the factors of membrane pore size, and the upper part of the vacuum degree is relatively large. The clay soil is easy to sand, water into the upper part of the siltation. As test days in advance, eventually completely blocked therefore reinforcement effect along the depth of the law, in the upper part of the reinforcement effect, the consolidation degree is very low. This is the test to seventeenth days, the internal bag sand well without reason model bucket of vacuum.

4.2 Shear Strength and Compressive Strength

In order to verify the three model tests of reinforced soil after improve the strength degree. Through the direct shear test to determine the strength index of friction angle and cohesion, is set up in 25 kPa, 50 kPa, 100 kPa, 200 kPa and 400 kPa under the pressure of vertical consolidation fast shear test, as a contrast group (see Fig. 4a). At the same time the interception of soft soil 1 cm $\sim$ 9 cm below the surface, away from the earth the water column row 8 cm, determination of the unconfined compressive strength, horizontal comparison of reinforcement effect (see Fig. 4b).

The three model barrels of soft soil by vacuum preloading, the shear strength index were improved significantly. The extent of the increase is consistent with the pore water pressure, settlement, water content and other indicators reflect the law. The integral vacuum is the highest, the largest settlement, the shear strength increases the maximum the friction angle increases. As cohesion increases the friction 243%. Separated drainage plate angle increased 56%, cohesion increased 95%. Sand well strength index

(a) Direct sheer test   (b) Unconfined compressive strength test

Fig. 4. Soil strength test after reinforcement

Table 3. Summary of the soil strength after reinforcement

Index	Undisturbed soil	Consolidated quick direct shear test					Model test		
		20 kPa	50 kPa	100 kPa	200 kPa	400 kPa	Integral	Separation type	Sand well
Angle of friction (°)	1.6	3.2	3.9	4.7	7.2	7.4	3.4	2.5	2.2
Cohesive strength (kPa)	2.7	4.4	5.9	12.3	27.6	53.7	9.3	5.3	2.5
Water content (%)	71.6	48.3	44.5	40.2	35.8	33.2	41.2	45.7	53.3
Unconfined compressive strength (kPa)	–	–	–	–	–	–	11.0	8.9	4.3

increase minimum friction angle increased 0.6°, cohesion can be considered a basic change (Table 3).

Through the strength index and soil unconfined compressive strength has reflected the reinforcement effect, is consistent with the foregoing, the better reinforcement effect, the soil compaction, unconfined compressive strength is also higher.

From Fig. 5, the moisture content among the consolidation pressure of 0–100 kPa, especially 0–25 kPa decreased obviously. The reason lies in the soft soil initial water rate of 71% is greater than the liquid limit of 51%, soft soil in the state of flow, with more free water. At low confining force free water loss, water content decreased rapidly. In 100 kPa ~ 200 kPa, after the water content is lower than the liquid limit, low moisture content curve flattened. At this point the water body has free water, but

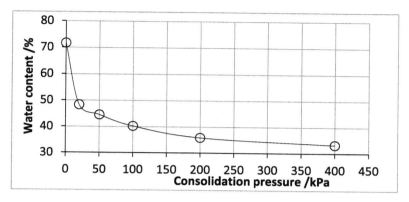

Fig. 5. Curve of consolidation pressure and water content

located in the diffusion layer, by electrostatic attraction and closely adsorbed on the surface of the soil particles weakly bound water. This part of the water for a smaller proportion of the water discharge from the soil is less so the moisture content decreased slowly. The consolidation pressure increased to 200 kPa, the indirect contact of soil particles increased. Soil skeleton formation, resist the consolidation pressure. Water content has no obvious change.

5 Conclusion

In this paper, through the indoor model test of single well, sand drains, drainage plate and conventional separation type anti-clogging integral drainage board has three vertical drainage body representative soft soil reinforcement effect. On the three indicators in the process of reinforcement, the main conclusions are as follows.

(1) Soil surface water content of three model test decreased significantly after reinforcement. Two kinds of drainage plate affected by the bending deformation of the upper part. The reinforcement effect occurs obvious attenuation in the middle. The reinforcement effect of bag sand well is obvious attenuation at the bottom.

(2) Strength index of soft soil has change a lot with water content. When the water content is greater than the liquid limit, strength index with decrease of the water content increased slowly. When the water content is less than the liquid limit, strength index with decrease of the water content increased rapidly.

(3) The water content decreased obviously as the consolidation pressure in the 0–100 kPa. In 100 kPa ~ 200 kPa, the moisture content decreased curve flattened. Increased to 200 kPa, the moisture content has no obvious change. The soil changed from more free water gradually to the weakly bound water.

References

1. Wang, Y.Z., Jing, F.H.: Effect of mineral composition on macroscopic and microscopic consolidation properties of soft soil. Soil Mech. Found. Eng. **50**(6), 232–237 (2014)
2. Song, J., Wang, Q., Zhang, P.: Laboratory research on fine particles migration of high clay dredger fill in consolidation process. J. Eng. Geology **20**(6), 1042–1049 (2012)
3. Walker, R.T.: Vertical drain consolidation analysis in one, two and three dimensions. Comput. Geotech. **38**(8), 1069–1077 (2011)
4. Rujikiatkamjorn, C., Indraratna, B., Chu, J.: 2D and 3D numerical modeling of combined surcharge and vacuum preloading with vertical drains. Int. J. Geomech. **8**(2), 144–156 (2008)
5. Wang, J., Dong, Z., Mo, H.: Fractal properties of filter membrane for silt clogging evaluation on PVD improved soft clays. KSCE J. Civil Eng. **21**(3), 636–641 (2016)
6. Rujikiatkamjorn, C., Indraratna, B.: Analysis of radial vacuum-assisted consolidation using 3d finite element method. Faculty of Engineering and Information Sciences (2007)

Design Charts for Reinforced Slopes with Turning Corner

Fei Zhang[1]([⊠]), Dov Leshchinsky[2], Yufeng Gao[1], and Guangyu Dai[1]

[1] Key Laboratory of Ministry of Education for Geomechanics
and Embankment Engineering, Hohai University, Nanjing 210098, China
[2] Department of Civil and Environmental Engineering, University of Delaware,
Newark, DE 19716, USA

Abstract. Presented are design charts for reinforced slopes with turning corner. These charts are based on results of variational limit equilibrium of three-dimensional (3D) stability of slopes which are also equivalent to rigorous upper bound in limit analysis of plasticity. Motivated by a commentary in AASHTO, the seismic effects are also included in the charts. The charts are convenient for determining the required tensile strength and embedment length of the reinforcement. As batter and seismic acceleration increase, the 3D effects of turning corner become more significant in design. Conventional design based on 2D analysis maybe unsafe in terms of reinforcement length while it is overly conservative in terms of reinforcement strength.

Keywords: Reinforced soil · Stability design · Three-dimensional analysis
Limit equilibrium · Seismicity

1 Introduction

Turning corner is commonly formed when two reinforced slopes/walls intersect. Past experiences exhibited lesser performance at the corner of reinforced walls than their straight sections, especially following an earthquake event – see commentary in AASHTO [1]. Using the conventional two-dimensional (2D) design analysis does not provide insight into the observations stated in AASHTO [1] commentary. Obviously, the corner problem has three-dimensional (3D) characteristics. Several researchers [2–4] have performed 3D analyses of slopes with variable corner angle, and indicated significant corner effects on stability of slopes. Zhang et al. [7] employed the method of 3D slope stability analysis to investigate the possible loads (e.g., pore water pressure and earthquakes) on stability of reinforced slopes with turning corner. The results demonstrated that the increase in possible pore water pressure or seismic loading can result in greater tensile force mobilized by the reinforcement. Compared to 2D analysis, the 3D one may require longer length of reinforcement. While the analysis in this paper follows the technique presented by Zhang et al. [7], the 3D results are revisited to produce charts that are insightful with respect to AASHTO's commentary. The insight is novel and has practical implications to design of reinforced corners. The design charts enable engineers to determine the required tensile strength and minimum required embedment length of the reinforcement when designing reinforced structures with turning corner.

© Springer Nature Singapore Pte Ltd. 2018
L. Li et al. (Eds.): GSIC 2018, *Proceedings of GeoShanghai 2018 International Conference: Ground Improvement and Geosynthetics*, pp. 297–306, 2018.
https://doi.org/10.1007/978-981-13-0122-3_33

2 Mechanism and Formulations

A comprehensive description of the 3D failure mechanism and relevant formulation is presented elsewhere [2, 5]. Some equations are restated here for the sake of completeness. The limit equilibrium (LE) variational extremization yields the 3D geometry of the critical slip surface. Figure 1 illustrates the two fundamental solutions for the potential 3D failure surface. Leshchinsky and Baker [4] established a combined 3D failure surface, which consists of a central cylinder and two end caps, as shown in Fig. 2a. The mechanism can be expressed in the Cartesian coordinate system as

$$\begin{cases} x_1 = x_c + A \exp(-\psi_d \beta) sin^2 \alpha \sin \beta \\ y_1 = y_c + A \exp(-\psi_d \beta) sin \alpha \cos \alpha \\ z_1 = z_c + A \exp(-\psi_d \beta) sin^2 \alpha \cos \beta \end{cases} \tag{1a}$$

$$\begin{cases} x_2 = x_c + A \exp(-\psi_d \beta) \sin \beta \\ y_2 = y_c + A \exp(-\psi_d \beta) \cos \alpha / sin \alpha \\ z_2 = z_c + A \exp(-\psi_d \beta) \cos \beta \end{cases} \tag{1b}$$

where (x_c, y_c, z_c) is the rotating center; (x_1, y_1, z_1) and (x_2, y_2, z_2) pertain to the portion of the end cap and central cylinder, respectively; A is an unknown constant; ψ_d = tan (ϕ_d) (where ϕ_d is the design internal angle of friction and can be expressed as $\phi_d = \tan^{-1}[\tan(\phi)/F_s]$, F_s = factor of safety on soil strength).

The potential 3D sliding body is rotating about the y'-axis, parallel to the plane of symmetry. The mobilized maximum tensile force at the ith reinforcement layer, T_{max-i} is assumed to act horizontally opposing sliding, as shown in Fig. 2. In addition, the

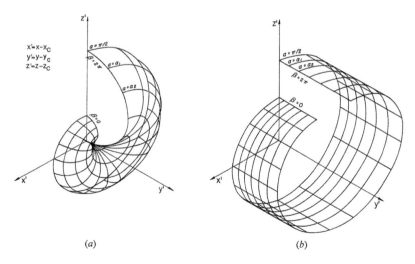

(a)   (b)

Fig. 1. 3D geometry of the fundamental components of failure surfaces [5].

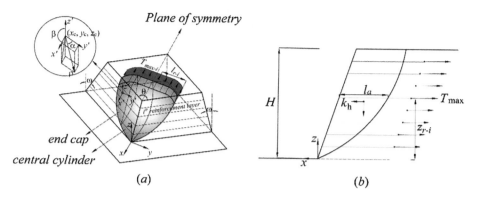

Fig. 2. Geometrical notation: (a) Coordinate system for reinforced slope with a turning corner and 3D failure surface; (b) Slip surface on plane of symmetry.

tensile force is assumed to be uniformly distributed amongst all layers, noted as T_{max}. Only one-half of the 3D sliding body needs to be considered because of symmetry. Similarly to the 2D analysis using a log spiral by Leshchinsky et al. [6], the moment limit equilibrium equation for the 3D sliding body can be written explicitly about the rotating y'-axis, without specifying the distribution of the normal stress over the slip surface. Considering earthquake impacts, the moment due to the horizontal seismic force is calculated using the traditional pseudo-static approach. It is included in the moment equilibrium equation as

$$\gamma\left(\iint_{D_1}(M_1 + k_h M_{1s})dxdy + \iint_{D_2}(M_2 + k_h M_{2s})dxdy\right) - T_{max}\sum_{i=1}^{n}l_{r-i}(z_c - z_{r-i}) = 0$$

$$(2)$$

where γ is unit weight of reinforced soil; k_h is horizontal seismic coefficient; l_{r-i} is one-half length of the 3D slip surface at its intersection with reinforcement layer i, as shown in Fig. 2a; z_{r-i} is elevation of the reinforcement layer i; n is number of reinforcement layer. D_1 and D_2 define the projected regions of the end caps and central cylinder into the x-y plane, respectively. The terms M_{12}, M_{22}, M_{1s} and M_{2s} are expressed as

$$M_1 = -\frac{\bar{Z} - z_1}{H}A\exp(-\psi_d\beta)sin^2\alpha\sin\beta \tag{3a}$$

$$M_2 = -\frac{\bar{Z} - z_2}{H}A\exp(-\psi_d\beta)\sin\beta \tag{3b}$$

$$M_{1s} = -\frac{\bar{Z} - z_1}{H}\cdot\frac{\bar{Z} - z_c + A\exp(-\psi_d\beta)sin^2\alpha\cos\beta}{2} \tag{3c}$$

$$M_{2s} = -\frac{\bar{Z} - z_2}{H} \cdot \frac{\bar{Z} - z_c + A\exp(-\psi_d\beta)\cos\beta}{2} \tag{3d}$$

where H is the slope height; $\bar{Z} = \bar{Z}(x, y)$ is the corner surface elevation and can be expressed as

$$\bar{Z} = \begin{cases} H\cot\omega\left(-x\sin\dfrac{\theta}{2} - y\cos\dfrac{\theta}{2}\right) & x \in \left[-y\cot\dfrac{\theta}{2} - H\tan\omega/\sin\dfrac{\theta}{2}, -y\cot\dfrac{\theta}{2}\right] \\[3mm] H & x \in \left(-\infty, -y\cot\dfrac{\theta}{2} - H\tan\omega/\sin\dfrac{\theta}{2}\right) \end{cases} \tag{4}$$

Leshchinsky et al. [6] defined a non-dimensional format of the maximum tensile force of the reinforcement, as $K = nT_{max}/(0.5\gamma H^2)$. Introducing this format into Eq. (2) can yield

$$K = \frac{nT_{max}}{\frac{1}{2}\gamma H^2} = \frac{\left(\displaystyle\iint_{D_1} M_{12}dxdy + \iint_{D_2} M_{22}dxdy\right) + k_h\left(\displaystyle\iint_{D_1} M_{1s}dxdy + \iint_{D_2} M_{2s}dxdy\right)}{\dfrac{H^2}{2n}\displaystyle\sum_{i=1}^{n} l_{r-i}(z_c - z_{r-i})} \tag{5}$$

To obtain the maximum tensile force of reinforcement for a given corner (i.e., batter ω, corner angle θ, design friction angle ϕ_d, number of reinforcement layer n, horizontal seismic coefficient k_h), it is necessary to maximize Eq. (5) with respect to the unknown parameters (i.e., A, x_c, y_c, z_c) describing the 3D slip surface as

$$K = \max_{z(x,y) \le \bar{Z}(x,y)} f(A, x_c, y_c, z_c | \omega, \theta, \phi_d, n, k_h) \tag{6}$$

The number of reinforcement layers $n = 500$ is adopted in the calculations. Zhang et al. [7] showed that 500 layers is a reasonably conservative approximation. Once the maximum tensile force is determined, the corresponding critical slip surface can be given. The distance between the slope face and the critical slip surface l_a in the plane of symmetry is obtained for each layer, and its maximum is defined as L_a.

3 Design Charts

To ensure stability, the required tensile resistance should be greater than the maximum tensile force. Numerous calculations were carried out to produce suitable design charts. In these charts, the ratio of the 3D to 2D results K^{3D}/K^{2D} are plotted as a function of batter ω for various corner angles θ, as shown in Fig. 3. Two design friction angles $\phi_d = 30°$ and $40°$ are considered here. When the corner angle $\theta = 180°$ the solution degenerates 2D (see – Table 1) and then $K^{3D}/K^{2D} = 1.0$. The ratio decreases with increasing batter, indicating significant impacts of the turning corner on the stability of reinforced slopes, especially as the batter increases. As the corner angle decreases, its influence becomes more profound. Under static conditions, using 2D results for reinforcement strength can yield a conservative design of reinforced slopes with turning corner. However, note that when the corner is subjected to seismic excitations, it can attract greater seismic loading than free standing slopes with straight alignments. Figures 4 and 5 illustrate K^{3D}/K_s^{2D} for corners under seismic conditions. Note that K_s^{2D} is the 2D solution for corner under *static* condition. As expected, the ratio increases with the increasing seismic acceleration coefficient k_h and corner angle θ.

Table 1. Values of K and the corresponding L_a/H obtained from 2D analyses.

		ω						
		0°	5°	10°	15°	20°	25°	30°
$\phi_d = 30°$	K	0.361	0.319	0.283	0.252	0.223	0.196	0.169
	L_a/H	0.396	0.376	0.363	0.351	0.344	0.340	0.341
$\phi_d = 40°$	K	0.230	0.196	0.167	0.141	0.117	0.094	0.073
	L_a/H	0.328	0.299	0.276	0.259	0.246	0.236	0.226

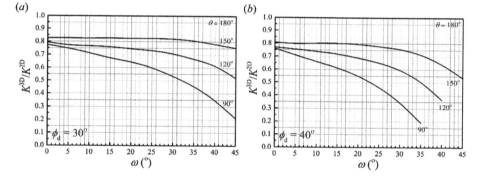

Fig. 3. Design charts for required tensile resistance under static conditions.

The required tensile resistance of reinforcement can be used to determine the anchorage length (L_e) over which sufficient pullout resistance behind the slip surface can develop. To obtain the total required length of the reinforcement in the context of classical internal stability, the distance between the slope face and the critical slip surface also needs to be calculated. Hence, the maximum of the distance L_a, is found along the slope height. Figure 6 illustrates the value of L_a obtained from 2D analyses for static case ($k_h = 0.0$) and seismic cases with $k_h = 0.1$, 0.2 and 0.3. The seismic effects obviously result in longer reinforcement. Based on the derived 3D critical slip surface, the corresponding value of L_a^{3D} in the symmetrical plane can be calculated, as shown in Figs. 7 and 8. The ratio of L_a^{3D}/ L_{a-s}^{2D} is defined as L_a^{3D} obtained from 3D analyses for a given k_h, divided by L_{a-s}^{2D} obtained from 2D analyses under the static conditions (see – Table 1).

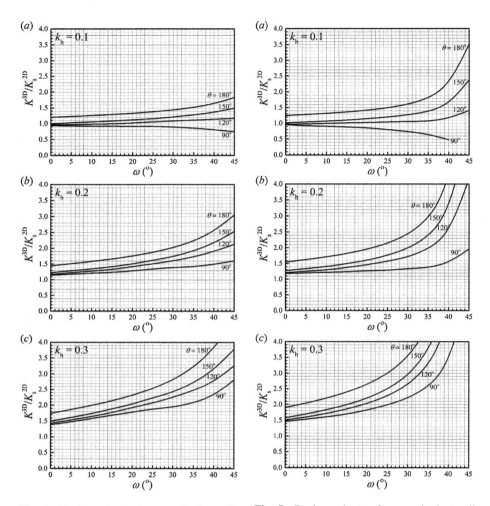

Fig. 4. Design charts for required tensile resistance under seismic loading ($\phi_d = 30°$).

Fig. 5. Design charts for required tensile resistance under seismic loading ($\phi_d = 40°$).

As the corner angle increases, the ratio gradually decreases but is always larger than the unit under both of static and seismic conditions. It indicates that using the 2D static result may underestimate the required embedment length of the reinforcement for 3D conditions. That is, while 2D analysis in lieu of 3D analysis will render conservative results with respect to strength of reinforcement, it will produce unconservative results with respect to required length. Figure 9 shows the traces of the slip surfaces for vertical corners derived from 2D and 3D analyses in the plane of symmetry. The 2D analyses result in shallower slip surface than 3D analyses thus indicating the potential importance of 3D analysis at a corner.

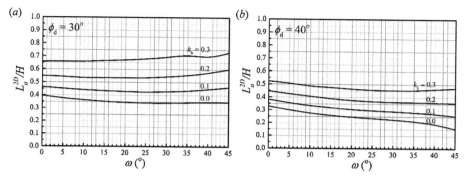

Fig. 6. Maximum distance between the slope face and the critical slip surface L_a/H obtained from 2D analyses.

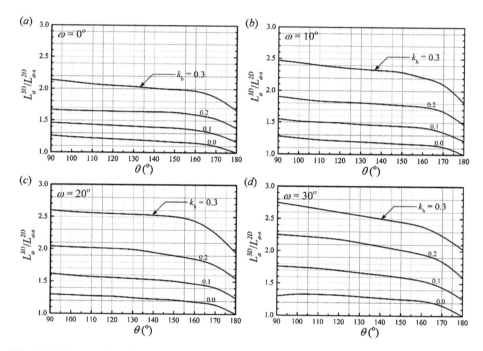

Fig. 7. Influence of turning corner on maximum distance between the slope face and the critical slip surface L_a/H ($\phi_d = 30°$).

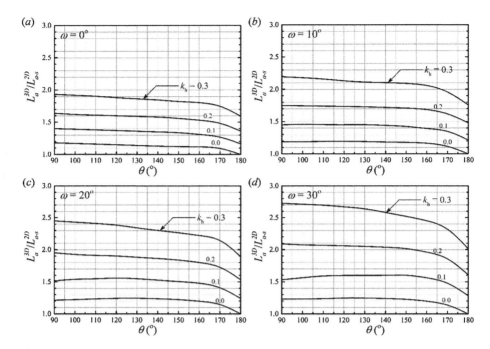

Fig. 8. Influence of turning corner on maximum distance between the slope face and the critical slip surface L_a/H ($\phi_d = 40°$).

4   Illustrative Example

The presented charts provide a convenient tool for the design of reinforced slopes with turning corner. To demonstrate the usefulness of these charts, an example is presented. Consider the following problem: batter $\omega = 0°$, height $= 6$ m, corner angle $\theta = 90°$, unit weight of reinforced soil $\gamma = 20$ kN/m^3, design friction angle $\phi_d = 30°$, number of reinforcement layer $n = 10$. For static condition, using the chart shown in Fig. 3a, one can obtain the corresponding ratio $K^{3D}/K^{2D} \approx 0.77$. As shown in Table 1, $K^{2D} = 0.361$ and K^{3D} then is calculated as 0.278. The required tensile strength of reinforcement for each layer can be determined as $T_{max} = 0.5K^{3D}\gamma H^2/n \approx 10$ kN/m. The required length of L_a between the slope face and the critical slip surface can be determined from Fig. 7a, as $L_a^{3D}/L_{a-s}^{2D} \approx 1.26$. From Table 1, the 2D result of $L_{a-s}^{2D}/H = 0.396$ leading to $L_a^{3D} \approx 3.0$ m, which is greater than the 2D result of $L_{a-s}^{2D} = 2.4$ m. If the seismic acceleration $k_h = 0.3$ is considered for seismic design, using the same procedure will yield $T_{max} \approx 18$ kN/m and $L_a^{3D} \approx 5.1$ m. Figures 10a and b illustrate the critical 3D slip surface for the example under static condition and seismic condition with $k_h = 0.3$.

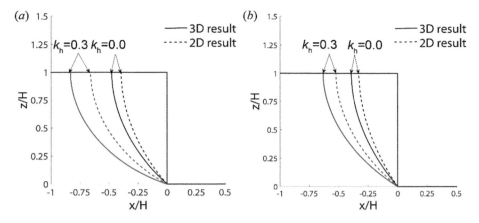

Fig. 9. Traces of 3D and 2D critical slip surfaces on symmetry plane for: (a) $\phi_d = 30°$; (b) $\phi_d = °$ and $\theta = 90°$.

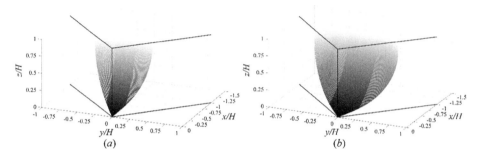

Fig. 10. 3D view of the critical slip surface for the illustrative example: (a) static condition; (b) seismic condition with $k_h = 0.3$.

5   Conclusions

Presented is an analytical approach to stability design of reinforced slopes with turning corner. The approach is based on the variational LE analyses of 3D slope stability. Motivated by the commentary in AASHTO [1] for designing a wall corner under earthquake conditions, the seismic effects are included in the analyses using pseudo-static method. The calculated results are presented in a condensed format of stability charts, which can be used to obtain the load in reinforcement as well as the required minimum length of reinforcement needed for internal stability. An example is given to demonstrate their usefulness. While the 3D effects of turning corner increase stability thus require smaller tensile strength of the reinforcement than that under 2D conditions, the required length of reinforcement for 3D stability is longer than for 2D stability. When the corners are subjected to earthquakes, seismicity significantly increases the required length of reinforcement comparing with 2D analysis. Having shorter reinforcement in a corner under seismic conditions implies that fewer layers will be utilized for internal stability as top layers might be entirely within the active wedge. It means that lower or fewer layer layers will share the load needed for stability. This further implies that some layers might be overstressed although 3D analysis yields smaller strength than needed for 2D which is employed in common design. However, since geosynthetic strength based on static conditions typically renders an over-design strength under seismic conditions, it is unlikely that the reinforcement will rupture during an earthquake. Conversely, reinforcement that is too short will result in increased deformations at the corner, as observed and commented in AASHTO [1]. Using longer reinforcement at the corner will likely remedy the problem.

References

1. AASHTO: Standard Specifications for Highway Bridges, Washington, DC (2012)
2. Giger, M.W., Krizek, R.J.: Stability analysis of vertical cut with variable corner angle. Soil Found. **15**(2), 63–71 (1975)
3. Giger, M.W., Krizek, R.J.: Stability of vertical corner cut with concentrated surcharge load. J. Geotech. Engrg. Div. **102**(1), 31–40 (1976)
4. Leshchinsky, D., Baker, R.: Three-dimensional slope stability: end effects. Soil Found. **26**(4), 98–110 (1986)
5. Leshchinsky, D., Baker, R., Silver, M.: Three dimensional analysis of slope stability. Int. J. Numer. Anal. Methods Geomech. **9**(3), 199–223 (1985)
6. Leshchinsky, D., Zhu, F., Meehan, C.L.: Required unfactored strength of geosynthetic in reinforced earth structures. J. Geotech. Geoenviron. Eng. **136**(2), 281–289 (2010)
7. Zhang, F., Gao, Y., Leshchinsky, D., Yang, S., Dai, G.: 3D Effects of turning corner on stability of geosynthetic-reinforced soil structures. Geotext. Geomembr. **46**(4), 367–376 (2017)

Experimental Study on Normalized Stress-Strain Behavior of Geogrid Reinforced Rubber Sand Mixtures

Fang-cheng Liu[1], Meng-tao Wu[1(✉)], and Jun Yang[2]

[1] School of Civil Engineering, Hunan University of Technology,
Zhuzhou 412007, Hunan, People's Republic of China
mengtao.china@gmail.com
[2] Department of Civil Engineering, The University of Hong Kong,
Pokfulam, Hong Kong S.A.R., China

Abstract. Rubber-sand mixtures (RSMs) have long been recognized as light filling back material and energy absorbing material with wide usage in civil engineering. As addition of rubber particles into sands usually decreases the strength of base sand, it is necessary to reinforce RSMs to satisfy the need of practical engineering. This paper presents studies on the behavior of geogrid reinforced RSMs. Conventional triaxial shear tests were carried out on reinforced/ un-reinforced RSMs. Four kinds of geogrid reinforcing patterns, i.e., horizontal reinforcing with one layer, two layers, three layers, and vertical reinforcing were taken into accounted, and three kinds of confining pressures, i.e., 50 kPa, 100 kPa and 200 kPa were applied. Experimental results indicate that: (1) Comparing to un-reinforced RSMs, the stress-strain curves of RSMs reinforced by geogrid are enhanced in turn for vertical reinforcing, horizontal one layer reinforcing, horizontal two layer reinforcing, horizontal three layer reinforcing respectively. (2) The stress-strain relationship of RSMs reinforced by geogrid exhibits strain hardening characteristics instead of strain softening before reinforced. (3) The reinforcement effect coefficient on the failure stress of RSM is higher than that of pure sand, and at the low confining pressure, this phenomenon is more obvious. (4) Geogrid reinforcement can restore the strength of the RSMs with less effect on the modulus. That is, reinforced RSMs can maintain the merits of low modulus and high strength simultaneously. (5) The stress-strain relationship of the reinforced RSM could be well normalized with the failure stress adopted as the normalized factor, based on the established normalization equation, the stress - strain curves could be predicted well.

Keywords: Rubber sand mixture · Geogrid reinforcement
Rubber mass content · Stress-strain relationship · Normalized behavior

1 Introduction

Scrap tires are encountered all over the world in increasing numbers. One of the main issues associated with the management of scrap tires has been their proper disposal. Granulated rubber crumbs or chips from recycled scrap tires exhibit a low unit solid

© Springer Nature Singapore Pte Ltd. 2018
L. Li et al. (Eds.): GSIC 2018, *Proceedings of GeoShanghai 2018 International Conference: Ground Improvement and Geosynthetics*, pp. 307–317, 2018.
https://doi.org/10.1007/978-981-13-0122-3_34

weight, low purchase price, low elastic modulus, and high elastic deformability. The use of granulated rubber or waste tire shreds as a new geo-material or in the form of mixtures with sand, so called Rubber Sand Mixtures or RSMs, has become a popular approach in ground improvement [1, 2] and energy dissipation material for isolated system [3, 4]. However, in general, the addition of rubber particles will result in a decrease in the strength of the mixture relative to the matrix sand. Especially, the internal friction angle decreases with increasing of rubber content [5, 6], which will cause some adverse effects, such as earth pressure increasing on retaining structures when RSMs are used as backfilling, or bearing capacity decreasing when RSMs are used as foundation cushion. Geogrid has widely been recognized as an effective flexible soil reinforcing material. So it may be a suitable method to enhance strength characteristics of RSMs by geogrid with the purpose of maintaining the merits of RSMs such as light weight and high elasticity at the same time. However, very few studies have been carried out on this to date.

On the other hand, recent researches have demonstrated that the stress-strain relationship of soil has a normalized property [7–10], i.e., the stress-strain relationship of soil under different consolidation pressures can be normalized into a straight line with a normalized stress, where the normalized stress is named normalization factor.

Therefore, in this paper, triaxial shear test was firstly carried out to study the stress-strain characteristics of geogrid reinforced rubber sand mixtures (*GGRSM*), and then the geogrid reinforcing effects on RSMs as well as the normalized stress-strain behavior of GGRSM were analyzed, in order to provide reference for engineering application of RSMs and subsequent studies on it.

2   Materials and Methods

2.1   Test Materials and Mixture Preparation

The rubber used in tests is black granular debris obtained from crushed waste tires, and the sand used here is dry ISO standard sand (Fujian Sand), as shown in Fig. 1. The grading curves of both rubber particles and sand are shown in Fig. 2.

Fig. 1. Rubber particles, sand particles and rubber sand mixtures used in tests

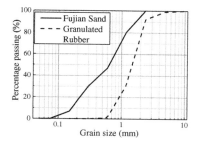

Fig. 2. Grading curves of rubber particles and sand

The sand and rubber described above were mixed by weight. As mixtures with different mix ratio exhibit different mass density, all of the test specimens were compacted to a constant relative density during specimen preparation, for the convenient of comparison. The maximum, minimum dry density, and control density during specimen preparation, of RSMs with different mix ratio are shown in Table 1.

Table 1. Mass densities of RSM with different rubber contents

Rubber content (%)	ρ_{dmin} (g/cm^3)	ρ_{dmax} (g/cm^3)	Relative density	Control density (g/cm^3)
0	1.51	1.86	0.7	1.74
10	1.38	1.71	0.7	1.60
20	1.21	1.50	0.7	1.40
30	1.04	1.29	0.7	1.20
40	0.87	1.17	0.7	1.06

2.2 Reinforcement Pattern

The geogrid used in the test is made of glass fiber, with grid size 12.7 mm × 12.7 mm. Mechanical parameters of it include elongation percentage less than 3%, both longitudinal and transverse tensile strength 60 kN/m. Four kinds of geogrid reinforcing patterns, i.e., vertical cylindrical reinforced, horizontal reinforced with one layer, two layer and three layer, are considered respectively, as described in Fig. 3, in which the shaded grid area represents geogrid.

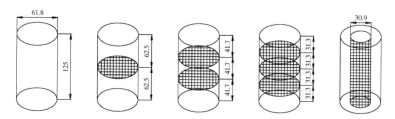

Fig. 3. Graphical representation of different reinforcing pattern in RSMs samples (unit: mm)

2.3   Test Procedures

Tests were carried out in a SLB-1 triaxial apparatus (*Nanjing soil equipment Corporation, Jiangsu, China.*) with samples size 61.8 mm in diameter and 125 mm in height. Conventional triaxial tests (*Standard for Soil Test Method: GB/T 50123-1999*) were performed on dry RSMs, and the test type is consolidated and undrained. Three confining pressures, 50 kPa, 100 kPa, and 200 kPa were taken into accounted and a constant rate of 0.32 mm/min shear loading was conducted.

3   Experimental Results

3.1   Stress - Strain Curves and Reinforcement Effect

Figure 4 presents deviatoric stress - axial strain curves of RSM (*unreinforced*) and GGRSM (*geo-grid reinforced*) with different rubber content at confining pressure of 50 kPa. It shows that geogrid reinforcing raises the stress-strain curves of RSMs obviously and turns softening stress-strain relationship into hardening stress-strain

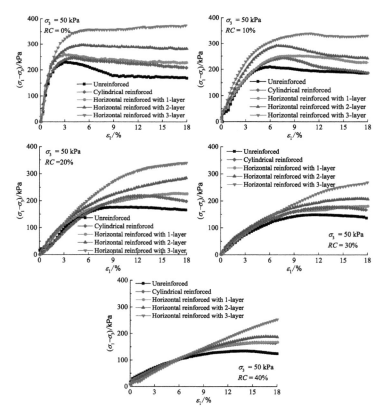

Fig. 4. Influence of reinforcing patterns on shear stress-shear strain relations of RSMs

relationship. The reinforcement effects increase with geogrid pattern in turn for cylindrical reinforcing, horizontal one layer reinforcing, horizontal two layers reinforcing, horizontal three layers reinforcing respectively. With increase of rubber content in RSMs, reinforcing effects attenuate and difference between different geogrid patterns diminishes.

Table 2 gives failure stress of reinforced/unreinforced RSMs with different confining pressure and different geogrid reinforcing pattern, in which the failure stress were determined from peak stress or the deviatoric stress at axial strain 15%. It indicates that failure stress increases monotonically with reinforcing pattern non-geogrid, cylindrical geogrid, horizontal 1 layer geogrid, horizontal 2 layers geogrid, horizontal 3 layers geogrid.

Table 2. Failure stress of RSM under different test conditions (Unit: kPa)

Rubber content (%)	σ_3 (kPa)	Reinforcing patterns				
		RSM (unreinforced)	GGRSM (cylindrical)	GGRSM (1-layer)	GGRSM (2-layer)	GGRSM (3-layer)
0	50	230.0	249.2	260.1	297.1	371.6
	100	408.6	443.6	450.0	482.2	534.5
	200	692.9	738.7	766.4	852.8	924.2
10	50	210.0	244.1	251.9	292.4	340.6
	100	371.3	406.6	404.7	453.8	526.7
	200	648.0	709.2	722.0	800.0	878.6
20	50	176.8	220.1	225.1	270.0	330.4
	100	332.8	390.0	394.5	445.7	510.8
	200	568.3	676.1	683.3	787.7	867.3
30	50	149.0	176.7	178.7	206.6	253.8
	100	281.5	310.6	323.4	355.5	397.9
	200	490.2	537.7	553.0	601.3	669.9
40	50	134.4	166.0	165.4	185.3	223.4
	100	240.0	272.9	258.6	284.6	322.8
	200	435.9	488.6	480.0	505.9	574.8

Define coefficient R_σ representing the strength reinforcement effect of geogrid on RSMs as,

$$R_\sigma = \frac{(\sigma_1 - \sigma_3)_f^R}{(\sigma_1 - \sigma_3)_f} \tag{1}$$

where $(\sigma_1 - \sigma_3)_f^R$ is the failure stress of GGRSM, and $(\sigma_1 - \sigma_3)_f$ is the failure stress of RSM with the same mix ratio and confining pressure.

Figure 5 gives variations of strength reinforcement effect coefficient with geogrid reinforcing pattern and rubber content respectively.

It could be seen that, (1) R_σ increases in turn under geogrid reinforcement pattern cylindrical loading, horizontal one layer loading, horizontal two layers loading, horizontal three layers loading respectively, and the two patterns cylindrical loading and horizontal one layer loading have close reinforcing effect. R_σ of geogrid reinforcement on RSMs are generally higher than that of pure sand. (2) RSMs with rubber content 20% seem to get the maximum R_σ. (3) For RSMs with the same rubber content, R_σ decreases with confining pressures, which may be attributed to more lateral deformation of the specimen at low confining pressure and consequently larger restraints on soil applied by geogrid.

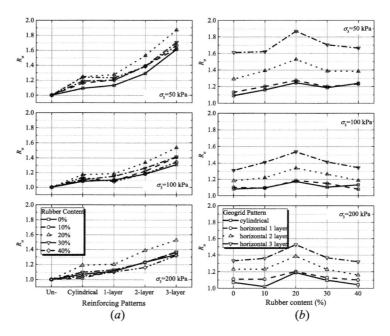

Fig. 5. Variation of reinforcing effect coefficient on RSM by geogrid with *(a)* different reinforcing patterns, *(b)* different rubber content.

3.2   The Deformation Modulus of Mixtures

As can be seen from test results, the stress-strain curve of RSM behaves as a typical hyperbolic form. Therefore, the extended Duncan-Zhang hyperbola model can be used to fit the $(\sigma_1 - \sigma_3) \sim \varepsilon_1$ relations to obtain the deformation modulus of mixtures

$$\sigma_1 - \sigma_3 = \frac{E_0 \varepsilon_1}{1 + (\varepsilon_1/\varepsilon_r)^\alpha} \tag{2}$$

where E_0 is the initial elastic modulus, ε_r is the reference strain, and α is the exponential parameter.

Figure 6 presents variation of initial elastic modulus with rubber content under different test conditions. It shows that the initial modulus of RSM decreases rapidly with the increase of rubber particle content. When the rubber content exceeds 20%, the modulus attenuation rate tends to slow down, and the lower confining pressure, the more obvious this phenomenon. However, in general, the influence of geogrid reinforcement on the initial modulus of RSM is limited, that is, the $E_0 \sim RC$ curves of RSMs reinforced by geogrid change little compared with un-reinforced RSMs.

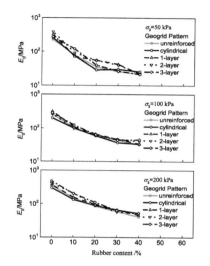

Fig. 6. Variation of the modulus of RSMs with rubber content;

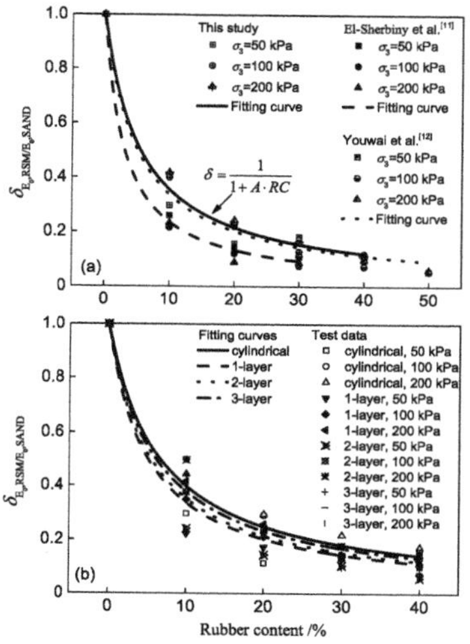

Fig. 7. Comparison of modulus attenuation laws: *(a)* this study and earlier researches for unreinforced RSM, *(b)* this study for GGRSM.

Define modulus attenuation ratio $\delta_{E_0,RSM/E_0,SAND}$ representing the attenuation degree of initial elastic modulus with rubber content as,

$$\delta_{E_0,RSM/E_0,SAND} = \frac{E_{0,RSM}}{E_{0,SAND}} = \frac{1}{1+A \cdot RC} \tag{3}$$

where A is the attenuation parameter, RC is abbrev of rubber content. The modulus attenuation ratio of RSMs for each test condition was calculated and compared with the earlier results of El-Sherbiny et al. [11] and Youwai et al. [12], shown in Fig. 7. It can be seen that, (1) For unreinforced RSM (Fig. 7(a)), the modulus attenuation curve obtained by this study is closer to 'Youwai' and higher than that of 'El-Sherbiny'. This indicates that the differences of tested sand (producing area, specifications), granulated rubber (particle size, shape, and crushing degree), test equipment and test control methods can lead to respective test results. Meanwhile, the similar laws revised from different experiments can provide reference for the mechanical studies of RSMs. (2) For GGRSM (Fig. 7(b)), reinforcing patterns have a certain impact on modulus attenuation, but the difference is not much in general. The modulus attenuation degree turns from weak to strong with reinforcing pattern as follows: vertical reinforcing, horizontal three layer reinforcing, horizontal two layer reinforcing, horizontal one layer reinforcing.

The modulus attenuation parameters A obtained under different reinforcing patterns are shown in Table 3, which can provide reference for follow-up study of the dynamic modulus attenuation characteristics.

Table 3. Fitting values of initial modulus attenuation parameter of RSM

Reinforcing patterns	RSM (unreinforced)	GGRSM (cylindrical)	GGRSM (1-layer)	GGRSM (2-layer)	GGRSM (3-layer)
A	0.182 (This study) 0.324 (El-Sherbiny et al.) 0.197 (Youwai et al.)	0.151	0.197	0.184	0.165

3.3 Normalized Stress-Strain Behavior

Normalized stress-strain behavior is important and useful for recognize characteristics of soils. Usually a normalized equation for $(\sigma_1 - \sigma_3) \sim \varepsilon_1$ relations under different normalization factors may be summarized as

$$\frac{\varepsilon_1}{(\sigma_1 - \sigma_3)} X = a\varepsilon_1 + b \tag{4}$$

where a is the slope of normalized line, b is the intercept of normalized line, and X is the normalization factors. Here for reinforced/unreinforced RSMs, the failure stress $(\sigma_1 - \sigma_3)_f$ is selected as the normalization factor. The relationships between failure stress and confining pressure are shown in Fig. 8, in which two rubber content of 0% and 20%, respectively, are given.

It can be seen from the $(\sigma_1 - \sigma_3)_f \sim \sigma_3$ curves that failure stress is nearly in perfect linear proportion with confining pressure. For example, $(\sigma_1 - \sigma_3)_f \sim \sigma_3$ relationship of RSM with $RC = 20\%$ could be expressed as

$$\begin{cases} (\sigma_1 - \sigma_3)_{f,\text{Unreinforced}} = 2.5736\sigma_3 + 59.0509 \\ (\sigma_1 - \sigma_3)_{f,\text{Cylindrical}} = 3.0144\sigma_3 + 77.0500 \\ (\sigma_1 - \sigma_3)_{f,\text{3-layer}} = 3.5773\sigma_3 + 152.1500 \end{cases} \tag{5}$$

Figure 9(a) gives the normalized stress-strain curves of 20% RSM. It could be seen that the normalized effect is acceptable under different reinforcing patterns, and the normalized equation of RSMs under different test conditions are obtained as

$$\begin{cases} (\frac{\varepsilon_1}{\sigma_1 - \sigma_3}(\sigma_1 - \sigma_3)_f)_{,\text{Unreinforced}} = 0.8189\varepsilon_1 + 2.9416 \\ (\frac{\varepsilon_1}{\sigma_1 - \sigma_3}(\sigma_1 - \sigma_3)_f)_{,\text{Cylindrical}} = 0.6630\varepsilon_1 + 5.0698 \\ (\frac{\varepsilon_1}{\sigma_1 - \sigma_3}(\sigma_1 - \sigma_3)_f)_{,\text{3-layer}} = 0.5459\varepsilon_1 + 6.7437 \end{cases} \tag{6}$$

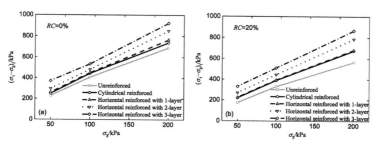

Fig. 8. Relationship between confining pressure and failure stress of pure sand and 20% RSM under different reinforcing patterns

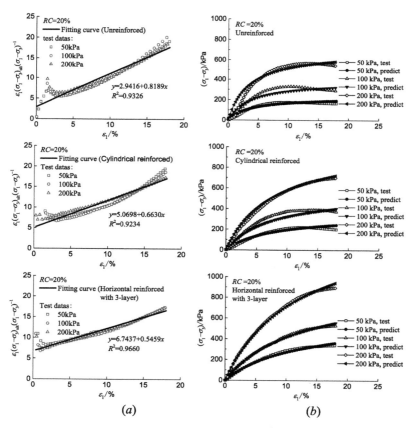

Fig. 9. (*a*) The normalized results of RSM with $RC = 20\%$ under diifferent reinforcing patterns; (*b*) Predicted stress-strain curves of geogrid reinforced RSM with $(\sigma_1 - \sigma_3)_f$ adopted as normalized factor

Substitute Eq. 5 into Eq. 6 and one can gain the following

$$
\begin{cases}
\left(\frac{\sigma_1-\sigma_3}{\varepsilon_1}\right)_{,\text{Unreinforced}} = \frac{2.5736\sigma_3+59.0509}{0.8189\varepsilon_1+2.9416} \\[2mm]
\left(\frac{\sigma_1-\sigma_3}{\varepsilon_1}\right)_{,\text{Cylindrica}} = \frac{3.0144\sigma_3+77.0500}{0.6630\varepsilon_1+5.0698} \\[2mm]
\left(\frac{\sigma_1-\sigma_3}{\varepsilon_1}\right)_{,3-\text{layer}} = \frac{3.5773\sigma_3+152.1500}{0.5459\varepsilon_1+6.7437}
\end{cases}
\tag{7}
$$

Formula (7) is the normalized stress-strain relationship of geogrid reinforced RSMs based on the normalization factor $(\sigma_1 - \sigma_3)_f$, and consequently the $\varepsilon_1/(\sigma_1 - \sigma_3) \sim \varepsilon_1$ curves of RSMs with different geogrid reinforcing pattern predicted by Eq. (7) are shown in Fig. 9(b). It can be seen that the predicted curves agree well with the tested curves, indicating Eq. (7) proposed by this paper could be used to study the normalized stress-strain behavior of geogrid reinforced RSMs.

4   Conclusions

1. Comparing to un-reinforced RSMs, the stress-strain curves of RSMs reinforced by geogrid are enhanced and turned from softening characteristic to hardening characteristic.
2. The strength reinforcing effect coefficient R_σ of geogrid on RSMs increased with geogrid reinforcing pattern in turn by cylindrical reinforcing, horizontal one layer reinforcing, horizontal two layers reinforcing, horizontal three layers reinforcing respectively. R_σ increases with rubber content of RSM first and then decreases with rubber content, and it seems the reinforcing effect tending to maximum for 20% RSM. As confining pressure increases, R_σ decreases.
3. Geogrid reinforcement can restore the strength of the RSMs with less effect on the modulus. That is, geogrid reinforced RSMs could overcome the deficiency of low strength of RSMs and maintain the merits of low modulus, great deformation capacity of them at the same time.
4. The failure stress could be adopted as the normalized factor for RSMs unreinforced/reinforced by geogrid under different rubber content and different reinforcing patterns, and the proposed equations in this paper for predict the stress-strain relationship of RSMs work well.

Acknowledgments. This research is supported by the Natural Science Foundation of China (No. 51108177), and the Innovation Foundation for Postgraduate of Hunan Province (No. CX2016B639)

References

1. Garga, V.K., O'shaughnessy, V.: Tire-reinforced earthfill. Part 1: construction of a test fill, performance, and retaining wall design. Can. Geotech. J. **37**(1), 75–96 (2000)
2. Aydilek, A.H., Madden, E.T., Demirkan, M.M.: Field evaluation of a leachate collection system constructed with scrap tires. J. Geotech. Geoenviron. Eng. **132**(8), 990–1000 (2006)

3. Reddy, S.B., Krishna, A.M.: Recycled tire chips mixed with sand as lightweight backfill material in retaining wall applications: an experimental investigation. Int. J. Geosynthetics Ground Eng. 1(4), 1–11 (2015)
4. Tsang, H.H., Lo, S., Sheikh, M.N.: Seismic isolation for low-to-medium-rise buildings using granulated rubber–soil mixtures: numerical study. Earthq. Eng. Struct. Dyn. 41(14), 2009–2024 (2012)
5. Zornberg, J.G., Cabral, A.R., Viratjandr, C.: Behaviour of tire shred sand mixtures. Can. Geotech. J. 41(2), 227–241 (2004)
6. Cabalar, A.F.: Direct shear tests on waste tires–sand mixtures. Geotech. Geol. Eng. 29(4), 411–418 (2011)
7. Li, Z.: Analysis of the normalized property of cohesive clay. Chin. J. Geotech. Eng. 9(5), 67–75 (1987). (in Chinese)
8. Zhang, Y., Kong, L., Meng, Q.: Normalized stress-strain behavior of Wuhan soft clay. Rock Soil Mech. 27(9), 1509–1513, 1518 (2006). (in Chinese)
9. Chang, D., Liu, J., Li, X.: Normalized stress-strain behavior of silty sand under freeze-thaw cycles. Rock Soil Mech. 36(12), 3500–3505 (2015). (in Chinese)
10. Ma, Q., Liu, B., Han, J.: Research on stress-strain characteristics and normalization of undisturbed saturated loess. Ind. Constr. 46(02), 68–71 (2016). (in Chinese)
11. El-Sherbiny, R., Youssef, A., Lotfy, H.: Triaxial testing on saturated mixtures of sand and granulated rubber. In: Geo-Congress 2013: Stability and Performance of Slopes and Embankments III, pp. 82–91 (2013)
12. Youwai, S., Bergado, D.T.: Strength and deformation characteristics of shredded rubber tire sand mixtures. Can. Geotech. J. 40(2), 254–264 (2003)

Experimental Study of Static Shear Strength of Geomembrane/Geotextile Interface Under High Shear Rate

Yang Shen, Ji-Yun Chang$^{(\boxtimes)}$, Shi-Jin Feng, and Qi-Teng Zheng

Key Laboratory of Geotechnical and Underground Engineering
of Ministry of Education, Department of Geotechnical Engineering,
Tongji University, Shanghai, China
1610148@tongji.edu.cn

Abstract. Due to its perfect anti-seepage performance, geomembrane is widely used in the liner system of municipal solid waste landfill together with geotextile as a protective layer. The relatively low shear strength of the interface between geomembrane and geotextile contributes to the formation of a potential slip surface within the liner system, which may result in instability of the landfill. In the present study, a new technology of direct shear test was developed, including a newly designed experimental apparatus and corresponding testing methods. A series of large-scale direct shear tests were performed to measure the interface shear strength between a smooth/textured geomembrane (GMS/GMX) and a non-woven geotextile (NW GT) under monotonic loading with three levels of normal stresses (σ_n = 100, 250, 500 kPa) under high shear rate of 100 mm/min. The experimental data indicate that the shear strength of GM/GT interface mainly depends on the normal stress, and shear distance. In general, these large-scale direct shear tests have revealed the complex intrinsic relationship among peak and residual shear strength, normal stress level and types of interface.

Keywords: Landfill liner · Geomembrane/Geotextile interface
Static shear test · Shear strength

1 Introduction

In waste disposal landfills, geomembranes (GMs) are widely used as liquid and vapor barriers because of their favorable anti-seepage performance. Furthermore, geotextiles (GTs) are usually used to protect geomembranes from damages that may occur due to angular soil particles [1–3]. However, the shear strength of the interface between GM and GT used in liner system is closely related to landfill stability, which has been extensively analyzed in previous studies [5, 6, 11, 15].

Stark et al. [12] conducted series of torsional ring shear tests on interface comprised of high-density polyethylene geomembranes (HDPE GMs) and nonwoven geotextiles (NW GTs) to study the effect of texturing and fiber type on shear strength. Similarly, direct shear tests and ring shear tests on GM/GT interfaces are performed by Jones and Dixon [10], which revealed the strain softening behavior and residual interface shear

L. Li et al. (Eds.): GSIC 2018, *Proceedings of GeoShanghai 2018 International Conference: Ground Improvement and Geosynthetics*, pp. 318–326, 2018.
https://doi.org/10.1007/978-981-13-0122-3_35

strengths of this interface. Furthermore, GM/GT interface shear strength parameters were measured using both inclined board (tilting table) and direct shear box tests by Wasti and Ozduzgun [13], and the test results indicated that testing methods have great influence on strength parameters. Although abovementioned shear tests all contribute to deep understanding of the mechanism of shearing of GM/GT interface, low shear rate is the common defect, which means these studies focus on the static shear strength only. Shaking-table tests to measure the dynamic interface shear strength properties between GTs and GMs were performed by Yegian and Lahlaf [14], but the normal stresses are inadequate to represent the true situation of liner system of landfill due to the limitation of apparatus.

Specifically, a leading and comparatively mature dynamic large-scale shear apparatus was developed by Fox et al. [7] and quantity of shear tests on geosynthetic interfaces were reported [8, 9]. However, no standard large-scale shear test on GM/GT interface has been performed, which demonstrates the necessity to carry out relevant investigation. Based on large test data, Esterhuizen et al. investigated the constitutive behavior of geosynthetic interfaces [4]. Direct shear tests for textured GM/GT interface were conducted and a constitutive model was proposed by Bacas et al. [2, 3] Akpinar et al. [1] investigated the effect of temperature on the shear strength of GM/GT interfaces, but only small changes in interface strength were observed for the considered range of temperatures. Nevertheless, the shear behavior of GM/GT interface under rapid shear rate has not been reported.

This paper presents a newly developed large-scale shearing apparatus to measure the static shear characteristics of the interface between a HDPE GMS/GMX and NW GT. Experimental investigations of the shear strength of the GM/GT interface were conducted under three normal stress levels (σ_n = 100, 250, 500 kPa), displacement rate of 100 mm/min and amplitudes of 70 mm to reach the residual shear strength, with large scale specimens (600 mm × 200 mm). The monotonic shear strength of GM/GT interface under high loading rate is concluded through analysis of test results.

2 Test Device

A geosynthetic interface dynamic shear apparatus powered by two high pressure oil pumps and controlled by a set of server systems was developed. The mainframe of the geosynthetic interface shear apparatus is exhibited in Fig. 1.

As seen in Fig. 1, the apparatus is consisted of three main subsystems: (1) the shear test system, (2) the control system and (3) the oil pump system. Two high-precision displacement sensors were directly connected with the oil-pumped actuators, both horizontally and vertically. When the bottom shear box moves during the shear tests, the shear/horizontal displacement was detected by the displacement sensor and test data were collected. The upper and bottom shear boxes were designed to model the shear process between the geomembrane and the soil, as shown in Fig. 2(a). The black gripping plates with acute 1 mm high steel teeth are fixed on the shear boxes, as shown in Fig. 2(b).

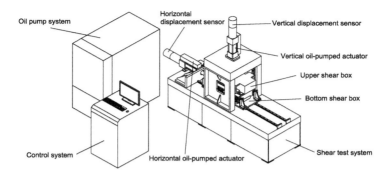

Fig. 1. Mainframe of the large-scale shear test apparatus.

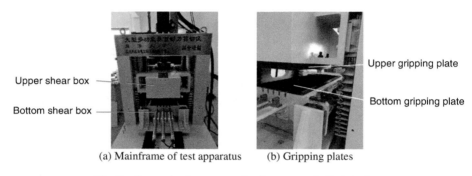

(a) Mainframe of test apparatus   (b) Gripping plates

Fig. 2. Dynamic shear apparatus for geosynthetic interface.

The size of the tested specimen is 600 mm × 200 mm, which is a quite large scale one compared with traditional apparatuses. More specifications of the newly developed apparatus are listed in Table 1. The newly developed large-scale shearing apparatus is not only capable of conducting static/dynamic shear tests of geosynthetic/geosynthetic interface but be further used for other testing procedures, such as geosynthetic/soil interface shear tests, as well. For the geosynthetic/soil interface, the steel bottom gripping plate could be removed from the bottom shear box, and the soil specimen is

Table 1. Geosynthetic interface dynamic shear apparatus specifications.

Feature	Specification
Specimen size	600 mm × 200 mm
Maximum normal force	60 kN
Maximum shear force	35 kN
Maximum specimen thickness	100 mm
Maximum horizontal displacement	120 mm
Specimen gripping system	Steel gripping plates, teeth spacing distance: 2 mm
Minimum displacement rate	0.1 mm/min
Maximum displacement rate	100 mm/min

then permissible to be filled in the shear box. The geomembrane specimen, fixed by the upper gripping plate, is in contact with the soil specimen, filled in the bottom shear box. In this study, the static shearing tests on GM/GT interface were performed.

3   Materials and Procedures

3.1   Test Materials

The geosynthetic interface investigated in this paper consists of double smooth surface geomembranes (GMS), double textured surface geomembranes (GMX) and nonwoven geotextiles (NW GT). The GM is high density polyethylene (HDPE) microspike liner. It has a constant thickness of 1.53 mm. The specification of the nonwoven geotextile (NW GT) used in this study is 700 g/m^2. The visual thickness of GT is around 6 mm, while the thickness measured with vernier caliper with slight compaction is 2 mm.

As shown in Fig. 3, the upper gripping plate was fixed to grip GM specimens and the bottom plate gripping GT specimens was movable by the pulling and pushing of oil pump actuator. Relative displacement occurred on the GM/GT interface during the shearing process.

Upper Gripping Plate (Fixed)

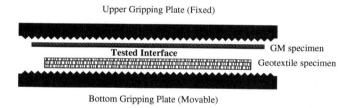

Bottom Gripping Plate (Movable)

Fig. 3.   Schematic profile of the interface tested.

3.2   Test Procedures

A consistent testing procedure was used for all shearing tests. Each test involved a multi-step process, which started with cutting the specimens from large rolls to fit appropriately in the apparatus. The specimen size of GM was 700 mm × 200 mm, longer than the GT size of 600 mm × 200 mm, to guarantee an adequate shear distance. Specifically, before each test, a preload stage was run to impose the tested normal stress on the specimen for at least 2 h to reach a steady thickness state of the specimen's thickness.

For each test, a constant normal stress was imposed on the interface for 2 h, and then the monotonic shearing process started with constant rate of 100 mm/min. The displacement amplitude is set as 70 mm, which is deemed enough to reach the residual shear strength of the interface. During the shearing process, the relative displacement of the interface, shear stress and change of thickness were documented. After the shearing test, the specimen was taken off carefully from the gripping plates and both GM and GT specimens were photographed in high-definition format. Herein, a total of 6 tests were conducted, and the shearing test program is listed in Table 2.

Table 2. Geosynthetic interface dynamic shear test program.

Test	Interface type	Normal stress/kPa	Shear distance/mm	Shear rate/mm/min
T01	GMS/NW GT	100	70	100
T02	GMS/NW GT	250	70	100
T03	GMS/NW GT	500	70	100
T04	GMX/NW GT	100	70	100
T05	GMX/NW GT	250	70	100
T06	GMX/NW GT	500	70	100

4   Experimental Results

4.1   Visual Inspection of the Tested Specimens

Figure 4 displays the GMX/GT specimens after tests. It is obvious that failure is more sever at the right side for all three normal stresses. Notably, there exists shearing layer within every geotextile because of the significant crimp and movement of the fibers, which demonstrates that failure and damage occurs within geotextile specimens.

(a) Left side of GMX/GT under 500 kPa   (b) Right side of GMX/GT under 500 kPa

Fig. 4. Pictures of the specimens after test.

4.2   Shear Strength of the GM/GT Interface

As seen in Fig. 5, three shear stress curves of static shear tests of GMS/GT interface were plotted in one coordinate system, and similar patterns were observed. The peak strengths were obtained at the horizontal displacement of about 7 mm, and the residual strengths were very close to a steady state. For the static test under normal stress of 100 kPa, the peak strength is around 30 kPa and the residual strength is around 20 kPa. For the test of 250 kPa, the peak is around 68 kPa and the residual is around 52 kPa. For the test of 500 kPa, the peak is around 139 kPa while the residual is about 116 kPa. It is very obvious that both the peak and residual strength are approximately proportional to the tested normal stresses. The peak strength of GMS/GT interface is about 28% of the normal stress and the residual strength is about 22% of the normal stress.

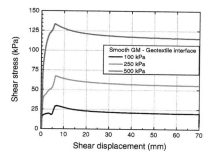

Fig. 5. Shear stress curves of Smooth Geomembrane/Geotextile interface.

For the tests utilizing the specimens of GMX, much higher shear stresses were detected compared with those utilizing GMS specimens. All three shear stress curves of GMX/GT interface obtain the peak strength at the horizontal displacement of around 11 mm, which is very remarkable in Fig. 6. For GMT/GT test of 100 kPa, the peak shear strength is about 50 kPa and the residual is about 42 kPa; for the test of 250 kPa, the peak is about 127 kPa and the residual is about 107 kPa; for the test of 500 kPa, the peak is about 249 kPa and the residual is about 198 kPa. The peak shear strength is about 50% of the tested normal stress.

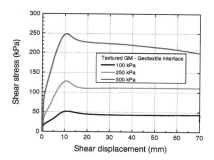

Fig. 6. Shear stress curves of Textured Geomembrane/Geotextile interface.

Remarkable is that for the situation of lower normal stresses (100 kPa and 250 kPa), the residual stress strength are obtained at the shear displacement of about 20 mm, and through the shear distance of 20 mm to 70 mm, the shear stress were nearly steady. Nevertheless, the shear stress decreased through the whole shear distance from 20 mm to 70 mm. By visual inspection of the tested specimens, serious damage and deformation of the geotextile fibers were observed. After the shear damage of the geotextile fibers, the specimen were sheared into two layers, which were captured by the GMX specimen and the bottom gripping plate respectively, as seen in Fig. 7. Hence, the inner friction occurred through the shear distance from 20 mm to 70 mm. For the situation of high normal stress of 500 kPa, both inner friction and further damaging of the geotextile fibers could occurred through the shear distance from 20 mm to 70 mm, which may be account for the reducing phenomenon of the shear stress.

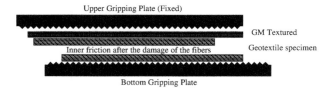

Fig. 7. The inner friction after the damage of the fibers of the geotextile specimen.

4.3   Thickness Reduction of Tested Specimens

Thickness reduction of the tested specimens are documented and displayed in Fig. 8, where the abscissa of the turning point is consistent with that of the peak shear strength in Figs. 5 and 6. The common characteristic of curves in Fig. 8 reveals thickness reduction develops during the pre-peak stage for most situations.

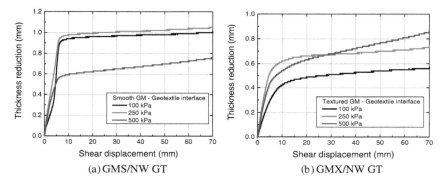

Fig. 8. Thickness reduction of the specimens during tests.

However, the thickness reduction increases steadily after the peak shear strength for specimens under normal stress of 500 kPa, which may be due to that failure occurs mainly within geotextile for such circumstances. Comparison between tests on GMS/NW GT and GMX/NW GT reveals that higher shear strength of the interface leads to lower reduction of thickness, which means minor influence on geosynthetic materials during the shearing process.

5   Conclusions

The experimental results on the static shear strength of GM/GT interface indicate that the shear strength of GM/GT interface mainly depends on the normal stress, shear distance and the interface types. The static shear strength of GMS/GT interface under the shear rate of 100 mm/min are much lower than the shear strength of GMX/GT interface. Damages and fractures of the geotextile fibers were observed in the GMX/GT interface shear tests. Minor reduction of residual shear strength through the shear distance after the peak could be attributed to the inner friction between the damaged geotextile specimens. The reported shear tests revealed the complex intrinsic relationship among peak and residual shear strength, normal stress level and types of interface.

Limitations of the shear tests of this study are obvious. The Geocomposite materials, consisting of two layers of thin geotextile and a medial layer of drainage net, are usually used upon the water resisting layer of Geomembrane and the protecting layer of thick Geotextile. Under high normal pressure, the drainage net inside the Geocomposite would unevenly impact on the Geomembrane/Geotextile interface, hence the contact interface between the Geomembrane and Geotextile is a non-ideal plane in real landfill liner system, while the shear interface of this study is an ideal plane. Recommendations, such as shear rate and water condition, are also provided to simulate the real status of the landfill liner. Hence, more detailed investigations on static shear tests of different shear rate and water-holding condition of the geotextile specimens, as well as dynamic shear tests of GM/GT interface, are needed to be conducted in the future research.

References

1. Akpinar, M.V., Benson, C.H.: Effect of temperature on shear strength of two geomembrane–geotextile interfaces. Geotext. Geomembr. **23**, 443–453 (2005)
2. Bacas, B.M., Cañizal, J., Konietzky, H.: Shear strength behavior of geotextile/geomembrane interfaces. J. Rock Mech. Geotech. Eng. **7**(6), 638–645 (2015)
3. Bacas, B.M., Konietzky, H., Berini, J.C., Sagaseta, C.: A new constitutive model for textured geomembrane/geotextile interfaces. Geotext. Geomembr. **29**(2), 137–148 (2011)
4. Esterhuizen, J.J.B., Filz, G.M., Duncan, J.M.: Constitutive behavior of geosynthetic interfaces. J. Geotech. Geoenviron. Eng. **127**(10), 834–840 (2001)
5. Feng, S.J., Shen, Y., Huang, R.Q., Li, D.P.: Seismic response and permanent displacement of landfills with liner interfaces and various foundation types. Env. Earth Sci. **74**(6), 4853–4863 (2015)
6. Fowmes, G.J., Dixon, N., Jones, D.R.V.: Validation of a numerical modelling technique for multilayered geosynthetic landfill lining systems. Geotext. Geomembr. **26**(2), 109–121 (2008)
7. Fox, P.J., Nye, C.J., Morrison, T.C., Hunter, J.G., Olsta, J.T.: Large dynamic direct shear machine for geosynthetic clay liners. Geotech. Test. J. **29**(5), 392–400 (2006)
8. Fox, P.J., Ross, J.D., Sura, J.M., Thiel, R.S.: Geomembrane damage due to static and cyclic shearing over compacted gravelly sand. Geosynthetics Int. **18**(5), 272–279 (2011)
9. Fox, P.J., Sura, J.M., Nye, C.J.: Dynamic shear strength of a needle-punched GCL for monotonic loading. J. Geotech. Geoenviron. Eng. **141** (2015). https://doi.org/10.1061/(ASCE)GT.1943-5606.0001304
10. Jones, D.R.V., Dixon, N.: Shear strength properties of geomembrane/geotextile interfaces. Geotext. Geomembr. **16**(1), 45–71 (1998)
11. Kavazanjian Jr., E., Arab, M., Matasovic, N.: Performance based design for seismic design of geosynthetics-lined waste containment systems. In: Earthquake Geotechnical Engineering Design, pp. 363–385. Springer, Cham (2014)

12. Stark, T.D., Williamson, T.A., Eid, H.T.: HDPE geomembrane/geotextile interface shear strength. J. Geotech. Eng. **122**(3), 197–203 (1996)
13. Wasti, Y., Özdüzgün, Z.B.: Geomembrane–geotextile interface shear properties as determined by inclined board and direct shear box tests. Geotext. Geomembr. **19**(1), 45–57 (2001)
14. Yegian, M.K., Lahlaf, A.M.: Dynamic interface shear strength properties of geomembranes and geotextiles. J. Geotech. Eng. **118**(5), 760–779 (1992)
15. Zania, V., Psarropoulos, P.N., Tsompanakis, Y.: Base sliding and dynamic response of landfills. Adv. Eng. Softw. **41**(2), 349–358 (2010)

Full-Scale Tests on High Narrowed Mechanically Stabilized Roadbed with Wrapped-Around Geogrid Facing

Yushan Luo[1,2(✉)], Chao Xu[2,3], and Xiang Wei[1]

[1] Shanghai Shen Yuan Geotechnical Engineering Co. Ltd.,
Shanghai 200011, China
shaluoyushan@126.com
[2] Key Laboratory of Geotechnical and Underground Engineering of Ministry
of Education, Tongji University, Shanghai 200092, China
[3] Department of Geotechnical Engineering, Tongji University,
Shanghai 200092, China

Abstract. In some mountainous areas, the construction of the geosynthetic reinforced walls/slopes has to be carried out in constrained or limited space with shorter reinforcements. While now the field tests of these "narrowed" geosynthetic reinforced walls/slopes are still lacking, therefore full-scale tests on a high narrowed mechanically stabilized roadbed with wrapped-around geogrid facing of a highway in Hubei Province were conducted. The tests include the measurements of tensile strain in geogrids, vertical earth pressure, horizontal earth pressure, layered settlement and horizontal displacement. The monitoring data showed that the tensile strain distribution around the inside rock-bench was found unusual, namely the stain was higher on top of the rock-bench while relaxing and shrinking around the rock-bench facing. And the vertical earth pressure was much smaller than theoretical earth pressure around the rock-bench facing. The settlement and lateral displacement occurred mostly during construction, and monitored accumulate displacements all met the standards, indicating that the narrowed mechanically stabilized roadbed performed well.

Keywords: High reinforced roadbed · "Narrowed" geosynthetic reinforced soil
Geogrid wrapped-around facing · Inside rock-bench · Full-scale test

1 Introduction

Due to advantages in bearing capacity requirements, seismic performance, construction techniques, application scope, cost, aesthetics and so on, the geosynthetic mechanically stabilized earth (MSE) walls have been widely designed and used all over the world. Specially, in complex topography conditions like mountainous areas, the advantages of MSE wall/steep slope are more prominent comparing to conventional retaining walls (Xu and Xing 2010; Yang et al. 2008; Jia et al. 2014).

However, the database of high MSE wall/slope in complex mountainous areas is still not enough and research about its special behavior, failure modes and reinforcing effect is not sufficient. In some cases, the construction of the MSE walls needs to

© Springer Nature Singapore Pte Ltd. 2018
L. Li et al. (Eds.): GSIC 2018, *Proceedings of GeoShanghai 2018 International Conference:
Ground Improvement and Geosynthetics*, pp. 327–337, 2018.
https://doi.org/10.1007/978-981-13-0122-3_36

construct in a constrained or limited space without excessive excavation. These "narrowed" MSE walls, typically having reinforcements shorter than conventional MSE walls, constructed in front of stable features such as a rock face, existing stable roadways or shored walls, are the newly evolved systems, "shored/narrowed mechanically stabilized earth walls" (Morrison et al. 2006).

Some researchers pointed out that due to the short reinforcement and boundary constraint, the narrowed MSE walls behaved differently with conventional MSE walls in aspects such as deformation pattern, the earth pressure, failure modes, and the distribution of the tension along the reinforcement (e.g. Leshchinsky et al. 2004; Lawson and Yee, 2005; Yang et al. 2011; Yang and Liu 2007; Morrison et al. 2006; Woodruff 2003; Xu et al. 2016a). While these conclusions were mostly drew from scaled modeling tests and numerical analysis, researches based on full-scale tests of high narrowed MSE walls/slopes are still lacking, which may restrict the application and development of MSE wall/slope technique in complex topography conditions.

Therefore, based on the high narrowed mechanically stabilized roadbed with wrapped-around geogrid facing of Yiba Highway (a highway in mountainous area from Yichang to Badong in Wuhan province, China), full-scale tests were carried out and monitored for over four years. The roadbed was fully instrumented and monitoring data including tensile strain in geogrids, vertical earth pressure, horizontal earth pressure, layered settlement and horizontal displacement in the first two years were analyzed in this paper, hoping to provide a better understanding of narrowed MSE walls using full-scale tests.

2 Description of the Narrowed Mechanically Stabilized Roadbed

The tested roadbed was the 11th section of the highway connecting Yichang and Badong cities in Hubei Province, China. It was located in complex geological and geomorphological conditions, between two tunnels, crossing a valley. The valley situates along the versant slope on top of the mountains on the Wudu River's south bank, intersecting the designed highway at a high angle. Moreover, the natural slope angles of the mountains range from 37° to 47°. According to the geotechnical investigation report, the strata from the top to the depth are mainly eluvial diluvial gravel soil of Quaternary, granitic gneiss, and hornblende schist of Proterozoic era (Table 1). In addition, the surface water, mainly coming from the surface flow after rainstorms and the transient water collected in the gully, is undeveloped to some extent. And the ground water, mostly recharged by the bedrock fracture water, is not rich as well.

Considering the difficulties in deep foundation construction, disposal of the discarded soil generated by tunnel excavation and ecological restoration, a geosynthetic reinforced slope with height from 5 m to 51.5 m was chosen to replace the previous bridge design (Xu et al. 2013). Since the global stability in many sections could not meet the national standards (NSPRC 2008), supporting concrete masonry piers were designed to embed in the base rock mass at the slope toe, acting as stable foundation to help resistant sliding. Meanwhile, in some sections, the reinforcement length was

Table 1. Physico-mechanical parameters of site rocks and soils

Material	State	Thickness (m)	Unit weight γ (kN/m^3)	Bearing capacity fa0(kPa)	Cohesion C(MPa)	Internal friction angle $\varphi(°)$
Gravel	Dense	0.8–3.0	18.7	200	/	38
Granitic gneiss/hornblende schist	Highly weathered	0–3.4	24.0	400	3	42
	Medium weathered	22.3–28.5	25.8	2000		

restricted by rock mass and was too short to meet the internal stability requirements. Anchor bolts were designed and connected to geogrids to enhance the pullout resistance. And geogrid wrapped-around facing combined with seed-nutriment-soil sacks was used to act as facing and help regional afforestation and ecological restoration. The wrapped-around length of geogrids was 1.5 m and connected to the upper layer using geosynthetic bars. The geogrids used in this project were uniaxial HDPE geogrids with ultimate tensile strength of 90 kN/m. The backfill was mostly exhumed soil from the tunnel excavation and local gravel soil, with particle size less than 15 cm and friction angle φ of 36°.

3 Instrumentation Program

The full-scale test was conducted in the section of YK73+005.5, having narrowed MSE wall, anchor bolts and concrete masonry supporting piers, which could help understand the behavior of narrowed MSE wall. The tested section is about 37 m high, with a facing batter of 1H: 2V, four offset with vertical spacing between layers of geogrids 0.3–0.5 m from the bottom up. And the length of reinforcement varies from 12–24 m, with top six layers spreading over the section to help control differential settlement (see Fig. 1).

The narrowed MSE wall system was extensively instrumented at typical locations according to numerical analyses before. The locations of the instrumentation are illustrated in Fig. 1. TXR vibrating string earth pressure cells were used to record the vertical/horizontal earth pressure (denoted as T in Fig. 1). The tensile strain was measured by HYDG-2405 strain gauges (denoted as Y in Fig. 1). Layered settlement extensometer and inclinometer tubes were embedded to help monitor the layered settlement and lateral deformation of the complex roadbed.

The full-scale test began at the same time with the start of construction around late Sep. 2011. This section was constructed in 121 days, and after construction the monitoring work went on till 2016. Since the afforestation work went well, some of the data were difficult to collect at some locations. Therefore, the data from 2014 to 2016 is not reported in this paper. Figure 2 shows variation of height of backfill with time at the test section.

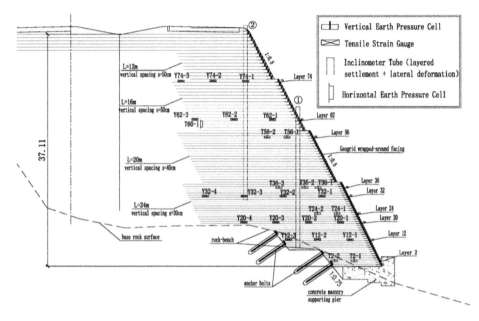

Fig. 1. Cross section and instrumentation plan of the YK73+005.5

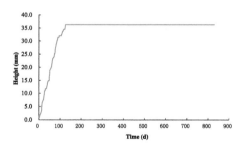

Fig. 2. Variation of height of filled earth with time in YK73+005.5

4   Results and Discussions

Full-scale test monitoring data including tensile strain of geogrids, vertical/horizontal earth pressure, lateral deformation and layered settlement will be discussed in this section.

4.1   Tensile Strain of Geogrids

Figure 3(a)–(e) presents the variation of tensile strain in geogrids with time at different depth, from the construction period to 2 years after the completion of construction.

(1) It could be seen that the tensile strain in geogrids increased with increment on the backfill thickness in construction period. And the tensile strain reacted faster and grew more quickly in the beginning of construction. The monitored data fluctuated in construction period mainly caused by the rolling machine, paving process and uneven particle size of backfill.

(2) The measurement results, performed 2 years after construction of the tested section, showed the tensile strain in reinforcement geogrids mostly between 0.1%–1.2%, namely the maximum tensile force in layers were less than 20 kN/m, much less than the ultimate tensile strength. It suggested that just 20% of ultimate tensile strength in reinforcement was motivated in the narrowed MSE wall, proving the wrapped-around high roadbed performed well.

(3) The distributions of tensile strains in different layers were not the same at different locations. At middle and higher part of the roadbed, the distribution of tensile strains were generally similar, namely increasing to a peak at about 4–5 m away from facing, and decreasing with longer distance to facing (refer to Fig. 3(d) and (e)). This distribution could be attributed to the non-linear soil-geogrid interaction and restriction of wrapped-around facing. However at lower part of the roadbed, where the reinforcement length was limited, the tensile strain distribution around the rock mass (like an inside rock-bench) was found unusual. The tensile strain was higher on top of the inside rock-bench (e.g. data recorded at gauge Y12-2, Y20-4 and Y32-4) and fluctuated strongly during construction. While around the inside rock-bench facing, the tensile strain was big at the beginning, but started relaxing quickly and became lower than testing points of the same layer (e.g. data recorded at gauge Y20-3 and Y32-3). This trend was in good agreement with the results in some scaled modeling tests (Xu et al. 2016b). Due to the stiffness difference between backfill and rock-bench, the tensile strain was higher on top of the rock-bench while relaxing and shrinking around the rock-bench facing.

(4) In the two years after construction of the experimental roadbed, the tensile strain of geogrids at most locations decreased to some extent, and some even shrink to as low as half of the peak recorded tensile strain. This could be attributed to heavy rainfall in the site. Infiltration of rain made the soil-geogrid interaction dropped and tensile force in reinforcement decreased. It is also plausible that minor soil movements occurred over time within the reinforced backfill, allowing for localized mobilization of soil strength thus reducing the need for force contribution by the geogrids, leading to relaxation. However, the data recorded at the end of the 2nd layer connected to anchor bolts was not relaxing after construction, but kept on growing (refer to Fig. 3(a), gauge Y12-3). The increase of tensile stains in reinforcements connecting to anchor bolts indicates that anchor bolts were playing a part in controlling displacement, further proving that adding anchor bolts in design was rational.

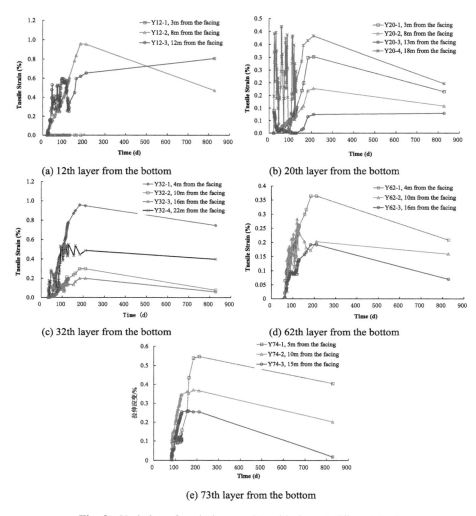

(a) 12th layer from the bottom

(b) 20th layer from the bottom

(c) 32th layer from the bottom

(d) 62th layer from the bottom

(e) 73th layer from the bottom

Fig. 3. Variation of strain in geogrids with time at different depth

4.2   Earth Pressure Distributions

Figure 4(a)–(d) presents the variation of earth pressure (including vertical and horizontal) with time/backfill thickness at different depth.

Vertical Earth Pressures:

(1)   According to Fig. 4(a)–(d), the vertical earth pressure at different depth increased nearly in linear increment on the backfill thickness during construction period. While at the same depth, the vertical pressure distribution was non-linear. The measured vertical earth pressure near the wrapped-around facing was much

smaller than that inside the backfill, which is close to the calculated vertical earth pressure. The smaller vertical earth pressure near facing helps prove that the reinforcement could obviously adjust the earth pressure distribution.

(2) Note that the measured vertical earth pressure at 24th layer near the rock-bench facing was much smaller than calculated, and even smaller than that at upper locations (refer to Fig. 4(b), cell T24-2). This special distribution was the same with numerical analysis before. With the backfill slowly moving towards unlimited zone, vertical earth pressure near the rock-bench facing was released to some extent.

(3) After completion of the backfill construction, the measured vertical earth pressures at different locations were still increasing slowly due to pavement construction and surcharge load. The arching effect and constrained boundary could probably be the reasons of the decreasing increment. Since the vertical earth pressure increment was quite small, the internal earth pressure in the high narrowed mechanically stabilized roadbed with wrapped-around geogrid facing could be considered reached stabilization.

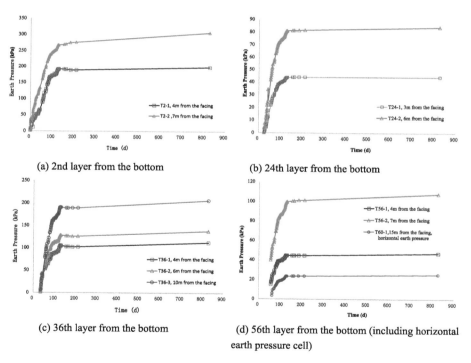

(a) 2nd layer from the bottom

(b) 24th layer from the bottom

(c) 36th layer from the bottom

(d) 56th layer from the bottom (including horizontal earth pressure cell)

Fig. 4. Variation of earth pressure with time at different depth

Horizontal Earth Pressure:

As seen in Fig. 4(d), the horizontal earth pressure increased with the backfill thickness grew. The increasing rate's slowing down probably caused by the lateral displacement nearby, which led to the pressure released. With the measured vertical pressure of cell T56-2, the horizontal earth pressure coefficients Kr (i.e., ratio of horizontal force along interface over corresponding vertical earth pressure nearby) could be proximately calculated. However the Kr is 0.53, larger than Kr calculated using Coulomb theory (about 0.26–0.33, with friction angle φ of 30°–36°). It seems that the cell T56-2 was pushed tilting/not vertical under the effect of rolling compaction and dynamic surcharge loading on top (Xu et al. 2016b).

4.3   Layered Settlement of the Roadbed

Figure 5(a) and (b) presents variation of settlement of reinforced backfill with time at different depth (the minus means settlement). It is noted that layered settlement extensometer and inclinometer shared the tubes, and 2# tube was broken during construction. In the first 2 months of construction, the measured data fluctuated seriously due to factors like rolling compaction, splicing tubes. So here we limit our discussion to the data of later phase of construction and after construction.

(1) It could be seen in Fig. 5, with the backfill thickness grew, the settlement at different layers increased, slowly and not in high rate. And the settlement at the middle and lower segment of tubes was bigger than top and bottom data. The measured maximum settlement of the roadbed at different depth were all less than 30 mm and the data began to stabilize, indicating that the backfill compaction quality was well-controlled, reinforcement connecting to anchor bolts played a crucial role in settlement control in the complex roadbed.

(2) After construction, some of the measured layered settlement were slowly decreasing with time, contrast to the increasing trend of most cells. It is possible that it stem from factors like backfill displacement, reinforcement tensile strain shrink and surcharge pressure less than compaction load.

4.4   Lateral Deformation of the Roadbed

Two inclinometer tubes were embedded over 19.5 m deep inside the full-scale test section to help monitor the later deformation of the narrowed MSE roadbed. Figure 6 (a) and (b) shows the variation of lateral deformation with depth.

(1) It could be seen that the lateral deformation grew quickly at different depth in the beginning of construction period. During construction period, the maximum measured lateral displacements of tube 1# and tube 2# were separately at 3 m and 4.5 m from top of the tube. The values were 144.18 mm and 129.01 mm, separately corresponding to 0.39% and 0.35% of wall height.

(2) From top to the bottom of each inclinometer tube, the lateral deformation increment and increasing rate both decreased with depth. It could be attributed to the special boundary zone around rock-bench and accumulation of upper lateral deformation. The middle and lower part of tube #1 shown smaller lateral displacement, showing that connecting geogrids to anchor bolts could help control

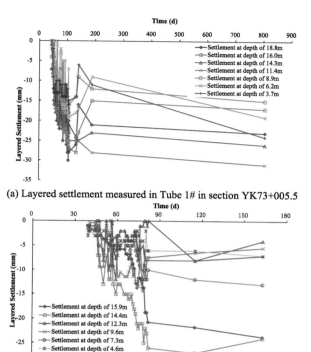

(a) Layered settlement measured in Tube 1# in section YK73+005.5

(b) Layered settlement measured in Tube 2# in section YK73+005.5

Fig. 5. Variation of settlement of reinforced backfill with time at different depth

lateral deformation around the rock-bench facing. However, the middle and lower part of tube #2 (corresponding to the part over the rock-bench) recorded larger lateral deformation and still increasing even after construction. It suggested that more measures are still needed to help control lateral deformation above the rock-bench in narrowed MSE walls.

(3) It is noteworthy that the lateral displacements recorded in two tubes both decreased to some extent after construction, consistent with the measured data of tensile strain in geogrids. Overall, the lateral deformation occurred mostly during construction, and measured lateral displacements all met the requirements in AASHTO standards (2002), less than 1.67% of the roadbed height. After construction, the lateral displacement grew at a rate of approximately 0.095 mm/d, much smaller than the rate during construction (1.965 mm/d). These measured data collectively demonstrate that the use of geogrids connecting anchor bolts and wrapped-around geogrid facing in high narrowed mechanically stabilized roadbed in mountainous areas was rational and proper.

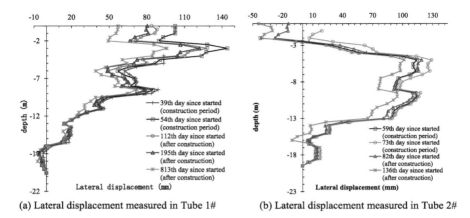

(a) Lateral displacement measured in Tube 1#   (b) Lateral displacement measured in Tube 2#

Fig. 6. Variation of deep horizontal displacement with depth

5   Conclusion Remarks

Based on a high narrowed mechanically stabilized roadbed with wrapped-around geogrid facing of a highway in Hubei Province, China, the full-scale test of complex section having inside rock-bench was extensively instrumented. The monitoring data over two years were analyzed and the behavior of the narrowed MSE wall was discussed. Conclusions could be drawn from this study as follows.

(1) The measured data shown that the geogrid reinforced roadbed performed well. Measures like wrapped-around geogrid facing, limited reinforcement connecting to anchor bolts could well control the lateral deformation of the narrowed MSE wall, adjust the earth pressure distribution and layered settlement. Moreover, spreading top layers over the backfill to stable rock mass could help control the differential settlement. Authors hope the full-scale test will contribute to the application and development of narrowed MSE wall technique in complex topography conditions, especially in mountainous areas.

(2) At upper layers, tensile strain of geogrids kept increasing to a peak at about 4–5 m away from facing, and decreasing with longer distance to facing. While around the inside rock-bench the tensile strain distribution was found unusual, namely the stain was higher on top of the rock-bench while relaxing and shrinking around the rock-bench facing. The unusual distributions may be due to the stiffness difference between backfill and rock-bench, which is a special character of the narrowed MSE wall system. Similarly, the vertical earth pressure around the rock-bench facing was found unusual too, much smaller than that at upper locations. It is probably due to the backfill moving towards unlimited zone, thus leading to a stress reduction zone.

(3) The deformation of the geogrids reinforced roadbed occurred mostly during construction. After construction, the settlement and lateral displacement generally increased slowly and began to stabilize. However, there was still increasing lateral

deformation over the rock-bench even after construction, indicating that more measures are still needed to help control lateral deformation above the bench in narrowed MSE walls.

Acknowledgement. The authors are indebted to the Transportation and Communication Department of Hubei Province for its financial supporting for this study (No. 2011-700-3-42). Also, we appreciate the kind help of colleagues of the laboratory and Nete Geosynthetics LTD in full-scale tests. Comments from reviewers and editors to improve the clarity and quality of the paper are welcome and will be appreciated.

References

American Association of State Highway and Transportation Officials (AASHTO): Standard Specifications for Highway Bridges, 17th edn., 829 pp, Washington D.C. (2002)

Jia, M.-C., Huang, W.-J., Ye, J.-Z., Xu, C.: Field tests on super-high geogrid-reinforced soil embankment without concrete panel. Chin. J. Geotech. Eng. **36**(12), 2220–2225 (2014). (in Chinese)

Lawson, C.R., Yee, T.W.: Reinforced soil retaining walls with constrained reinforced fill zones. In: Proceedings of Geo-Frontiers 2005 Congress, Austin, TX. ASCE GSP 140 (2005)

Leshchinsky, D., Hu, Y., Han, J.: Limited reinforced space in segmental retaining wall. Geotext. Geomembr. **22**(6), 543–553 (2004)

Morrison, K., Harrison, F., Collin, J., Dodds, A., Arndt, B.: FHWA-CFL/TD-06-001 Shored Mechanically Stabilized Earth (SMSE) Wall Systems Design Guidelines. FHWA-CFL/TD-06-001, Washington, DC, USA, Federal Highway Administration, Washington, DC, USA (2006)

The National Standards of People's Republic of China (NSPRC): GB/T17689-2008. Geosynthetics - Plastic Geogrids. China Planning Press, Beijing (2008). (in Chinese)

Woodruff, R.: Centrifuge modeling for MSE-shoring composite walls. MS thesis. University of Colorado, Boulder, CO, USA (2003)

Xu, C., Xing, H.-F.: Geosynthetics. Mechanical Industry Press, Beijing (2010). (in Chinese)

Xu, C., Luo, Y.-S., Chen, H., Jia, B.: Effects of interface connections on narrowed mechanically stabilized earth walls. Environ. Earth Sci. **75**(21), 1–12 (2016a)

Xu, C., Luo, Y.-S., Zhu, H., Wang, J., Yang, F.: Performance of high geosynthetic-reinforced embankments. Geotechnical Special Publication, n231 GSP, pp. 515–518 (2013)

Xu, C., Luo, Y.-S., Jia, B., Chen, H.: Effects of connection forms on shored mechanically stabilized earth walls by centrifuge modeling tests. Chin. J. Geotech. Eng. **38**(1), 180–186 (2016b). (in Chinese)

Yang, G.-Q., Lv, L., Pang, W., Zhao, Y.: Research on geogrid reinforced soil retaining wall with wrapped face by in-situ tests. Rock Soil Mech. **29**(2), 517–522 (2008). (in Chinese)

Yang, K.-H., Zornberg, J.G., Hung, W.-Y., Lawson, C.R.: Location of failure plane and design considerations for narrow geosynthetic reinforced soil wall systems. J. GeoEng. **6**(1), 27–40 (2011)

Yang, K.-H., Liu, C.-N.: Finite element analysis of earth pressures for narrow retaining wall. J. GeoEng. **2**(2), 43–52 (2007). Taiwan Geotechnical Society

Model Test on the Deformation Behavior of Geogrid Supported by Rigid-Flexible Piles Under Static Load

Kaifu Liu[1($\boxtimes$)], Linglong Cao[1], Yi Hu[2], and Jiapei Xu[1]

[1] School of Civil Engineering and Architecture, Zhejiang Sci-Tech University,
Hangzhou, China
liukaifu@zstu.edu.cn
[2] Zhejiang Yasha Decoration Co., Ltd., Hangzhou, China

Abstract. Geogrid is widely used in the cushion of composite foundation or embankment supported by rigid or rigid-flexible piles. However, the researches on the deformation behavior of geogrid are limited. Some model tests on the behavior of geogrid supported by rigid-flexible piles were designed and completed under static load. The variations of geogrid strain at different locations with time were analyzed based on the model test results. The test results showed that the deformation behavior of geogrid varies with the location. The strain of geogrid increased with the applied load. The strain changing law of geogrid on the top of pile or soils varying with time was similar to that of the static applied load. However, the strain changing law of geogrid on the pile side was not similar to that of the applied load. The strain of geogrid at the pile side was larger than that on the top of rigid or flexible pile, and it was also larger than that on the soil among the piles. The more close to the pile (cap) side, the larger the geogrid strain. The geogrid deformation may be caused by the differential settlement among the pile and soil or by the membrane tension effect of the geogrid on the pile cap. This is very useful for the application of geogrid in the geotechnical engineering.

Keywords: Model test · Geogrid · Strain · Rigid-flexible pile
Composite foundation

1 Introduction

Soft soil is not suitable for the foundation of large-scale civil engineering structures [1, 2] because of poor shear strength and high compressibility. Geogrid reinforced pile-supported composite foundation is a good method to treat soft ground and has been widely used in railway, highway, airport and other engineering projects [3–5]. In this composite foundation, piles are used as vertical reinforced material to reduce settlement and improve bearing capacity of soft ground and geogrid are used as horizontal reinforcement material to reduce differential settlement and lateral displacement of the soils [6]. A cost estimate of a high-speed railway project shows that geogrid reinforced pile-supported composite foundation can save a large amount of money compared to

© Springer Nature Singapore Pte Ltd. 2018
L. Li et al. (Eds.): GSIC 2018, *Proceedings of GeoShanghai 2018 International Conference: Ground Improvement and Geosynthetics*, pp. 338–346, 2018.
https://doi.org/10.1007/978-981-13-0122-3_37

deep mixing pile composite foundation (another frequently used foundation) to treat soft clay [7].

Many researchers had studied the mechanism and application of geogrid in the geogrid reinforced pile-supported composite foundation. The influence of the number of geosythetics (or geogrid) layers [8–10], the modulus of geosythetics [8] and the tension strength of geogrid [8, 11] to settlement and differential settlement was analyzed by using the finite element method [8–11]. The behavior of geotextile or geogrid in the geogrid reinforced pile-supported composite foundation was analyzed by using different numerical methods or softwares, such as FLAC (Han and Gabr) [12], Plaxis (Suleiman et al.) [13], Abaqus (Zhuang et al.) [14] and two-dimensional (2D) coupled mechanical and hydraulic numerical implementation method (Yapage et al.) [15] and MIDAS (Li et al.) [16].The effect of geogrid is also studied in the field test by Zheng et al. [17]. Although these research results were useful to the application, the geogrid was only supported by one type piles in the composite foundation. There is limited study on the behavior of geogrid supported by piles with different rigidity in the composite foundation.

The rigid and flexible piles had different rigidity in the composite foundation, so the deformation behavior of the geogrid supported by rigid-flexible piles was different from that supported only by single piles (reinforced concrete piles or deep mixing piles). Considering the effect of pile rigidity and applied load, some model tests were designed to analyze the deformation behavior of the geogrid supported by rigid-flexible piles under static load in the present study.

2 Materials and Model Tests

2.1 Materials

Soils. Soils used in the model tests are as follows: (1) Silty clay. The silty clay was drying out to facilitate preservations and formulated to the required soil moisture content when used in the model test. (2) Mucky clay, taken from a foundation in Hangzhou Gongshu District. The mucky clay was formulated to the required soil moisture content of 38.52% (the moisture content of undisturbed soil)when used in the model test. (3) Fine sand, used to fill embankment. The mechanical property index of the soils was measured by the direct shear test under the undrained conditions. Basic physical and mechanical property index of the soils were showed in Table 1.

Table 1. Basic physical and mechanical property index of the soils

Soil	Moisture content/%	Natural density (g/cm3)	Friction angle/(°)	Cohesive strength/kPa
Silty clay	27.80	1.783	20.53	10.20
Mucky clay	38.52	1.573	15.00	10.00
Fine sand	15.35	-	35.00	1.00

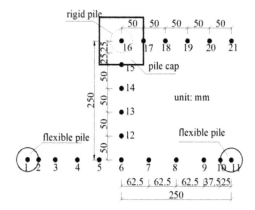

Fig. 1. The layout of strain gauges on the geogrid

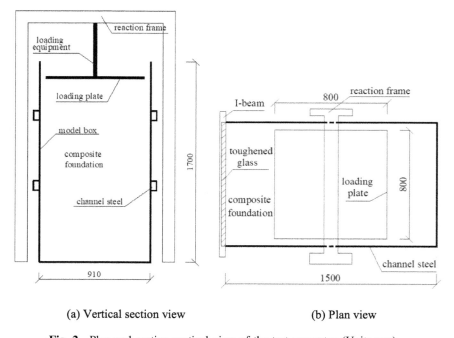

(a) Vertical section view   (b) Plan view

Fig. 2. Plan and section vertical view of the test apparatus (Unit: mm)

Geogrid. The geogrid used in the model tests was produced by Tai'an Luther Engineering Materials Co., Ltd. The type of this biaxial plastic grid was TGSG30-30. The mesh size of the geogrid was 30 mm × 30 mm, the breaking tensile strength of the geogrid was 30 kN/m and the elongation at break of the geogrid was 3%.

Some strain gauges were glued on the geogrid to investigate the strain of the geogrid, the location of strain gauges was shown in Fig. 1 (Only one half of the flexible piles and geogrid was shown in the Fig. 1). According to the symmetry rule, the strain of geogrid at the point 6 to point 10 was used in this paper. When one of the strain gauges at the points 6–10 was damaged, another one at the points 1–5 on the symmetry position could replace the damaged one.

Model Piles. Asteel pipe was used to simulate the rigid pile (prestressed concrete pipe piles in the engineering project), and 4 PVC pipe were used to simulate the flexible piles. The length of steel pipe was 100 cm, and that of the PVC pipe was 600 cm. The diameter of the rigid and the flexible piles was 50 mm. The wall thickness of steel and PVC pipe was 2 mm. One steel plate with the square side length of 100 mm and the thickness of 15 mm was put on the steel pipe to simulate the pile cap in the engineering project.

2.2   Model Tests

Test Apparatus. The model test was conducted in a test apparatus which was assembled by 1.2 cm thickness steel plates in 3 sides and 1.5 cm thickness glass plates in another side with dimensions of 1.5 m (length) × 0.91 m (width) × 1.7 m (depth). A schematic diagram is shown in Fig. 2. Two rings of channel steel were welded in the height of 0.6 m and 1.2 m to improve the stiffness of steel plate and a I-beam was soldered to the glass side to improve tensile strength and stability of model apparatus. To avoid the water leakage, every weld was sealed with glass glue.

Foundation of the Model Test. The soil layers are from the bottom to the top as the following: (1) Silty clay with the thickness of 250 mm; (2) Mucky clay with the thickness of 750 mm; (3) Silty clay with the thickness of 200 mm (pile cap is in this layer and below the geogrid); (4) Geogrid and fine sand with the thickness of 350 mm. In the model test, the distance of the flexible piles was 500 mm, the rigid pile was on the intersection of

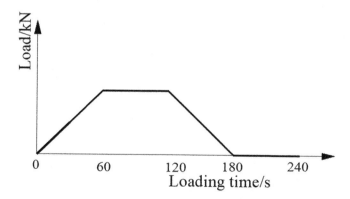

Fig. 3. The loading curve of static load

diagonal lines of the square consisted of the line from the center of four flexible piles and the thickness of one geogrid layer cushion was 50 mm.

Loading Equipment and Loading Process. The maximum load applied by the equipment was 5 kN. A 20 mm thick square steel plate was used as the loading plate with the side length of 800 mm. The static load is the only variables and the rest of parameters remained unchanged. The loading process was shown in the Fig. 3. At first, the load increased linearly from 0 to the load of 1 kN, 2 kN, 3 kN, 4 kN and 5 kN in 60 s. Then the applied load was maintained for 60 s and unloaded linearly to 0 kN in 60 s. Finally, the load of 0 kN was remained on the load plate for 60 s.

3   Results and Discussion

When the static load applied on the loading plate changed, the strain of geogrid at different location would be different. Figures 4, 5 and 6 show the strain of geogrid at different location under different static load. Obviously, higher static applied load led to bigger geogrid strain.

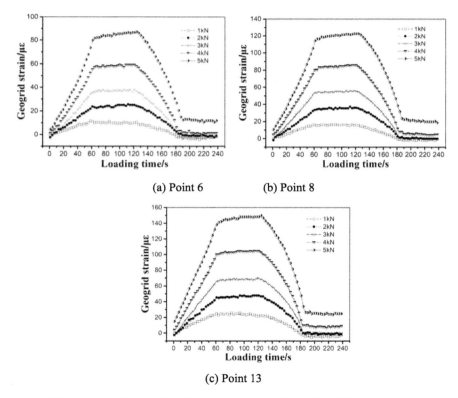

(a) Point 6   (b) Point 8

(c) Point 13

Fig. 4. Geogrid strain-loading time curves on the soil at different location

3.1   The Geogrid Strain

The Strain of Geogrid on the Soil Among the Piles. Figure 4 shows that the changing law of the geogrid strain on the soil among the piles was similar to the law of the static load applied. In the loading time of 0–60 s, the strain of geogrid increased near-linearly with the applied load. In the loading time from 60 to 120 s, the geogrid strain increased a little when the applied load was constant. In the unloading time of 120–180 s, the geogrid strain decreased near-linearly with the applied load unloading. In the next minute, the geogrid strain was almost remained unchanged as the static load did. The strain increment of geogrid in different locations slightly increased with the applied load increment of 1 kN increasing. The geogrid of could be recovered after unloading the applied load when the load was small. However, due to the large applied load, the geogrid would have some plastic deformation as shown in Fig. 4 when the time was in the stage from 185 s to 240 s. It would be drawn from Figs. 2 and 4 that the more close to the pile (cap) side, the larger the geogrid strain.

The Strain of Geogrid on the Pile Top. Figure 5 shows the strain of geogrid on the top of the flexible pile and the rigid pile varied with the loading time. The strain changing law of the geogrid on the top of the flexible pile and the rigid pile was similar to that on the soils surface. However, the strain of geogrid on the top of two different piles had a distinct difference in the numerical value. The geogrid strain on the top of the rigid pile had a much higher value while the strain of geogrid on the flexible pile head lower, compared to that on the soil surface.

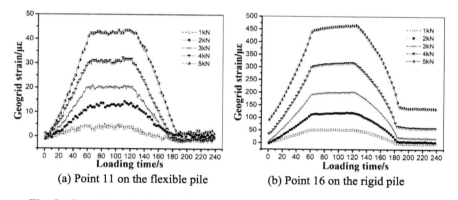

(a) Point 11 on the flexible pile   (b) Point 16 on the rigid pile

Fig. 5. Geogrid strain-loading time curves on the top of piles at different location

The Strain of Geogrid at the Pile Side. As shown in the Fig. 6, the strain of geogrid at the flexible pile side (point 10) possessed peculiar changing law under the different static load. The strain of geogrid at the flexible pile side was much bigger than that on the soil surface or the flexible pile head. In the loading time from 0 to 60 s, the strain had a near-linearly increase with the applied load. In the loading time from 60 s to 120 s, the geogrid strain kept increasing under a lower applied load and the geogrid

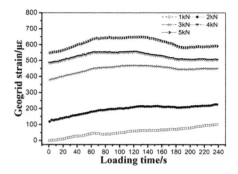

Fig. 6. Geogrid strain-loading time curves at the pile side (point 10)

strain would almost remain unchanged when a higher load applied. In the unloading stage with the time from 120 s to 180 s, the geogrid strain kept increasing under a lower applied load and the geogrid strain would decrease a little under the higher load unloading process. After the unloading process, the geogrid strain had a little increase with the time increasing.

3.2   The Geogrid Deformation Behavior Discussion

The applied load was transferred to the piles and soils, so there were a differential settlement between piles and soil because of significant difference in stiffness. This differential settlement would make the geogrid deformed. The differential settlement at the pile side was much larger than that on the soil among the piles. And the differential settlement on the soil was bigger than that on the pile top. Consequently, the geogrid strain was different at the different location. The larger differential settlement, the bigger the geogrid strain. Figures 4, 5 and 6 had verified the conclusion. However, the geogrid strain on the rigid pile was larger than that on the soil among the piles. The reason of this phenomenon may be that the differential settlement at the rigid pile side was so large that a tension in geogrid appeared because of the rigid pile cap. In other word, the large strain of geogrid on the rigid pile top was not caused by the differential settlement, but by the membrane tension effect of the geogrid on the pile cap.

4   Conclusions

In the present research, we analyzed the deformation behavior of geogrid supported by rigid-flexible piles in composite foundation under static load by the model tests. The following conclusions could be drawn from the results.

The deformation behavior of geogrid varies with the position of the geogrid. The geogrid strain increases with the applied load. The strain changing law of geogrid on the pile top or the soils varying with time is similar to that of the applied static load.

However, the strain changing law of geogrid on the pile side is not similar to that of the applied load. The strain of geogrid at the pile side is larger than that on the top of rigid or flexible pile, and it also larger than that on the soil among the piles. The more close to the pile (cap) side, the larger the geogrid strain. The geogrid deformation may be caused by the differential settlement among the pile and soil or by the membrane tension effect of the geogrid on the pile cap.

Acknowledgments. The authors would like to acknowledge the support provided by the Natural Science Foundation of Zhejiang Province, China (No. LY13E090010).

References

1. Prabakar, J., Dendorkar, N., Morchhale, R.K.: Influence of fly ash on strength behavior of typical soils. Constr. Build. Mater. **18**(4), 263–267 (2004)
2. Tang, C., Shi, B., Gao, W., et al.: Strength and mechanical behavior of short polypropylene fiber reinforced and cement stabilized clayey soil. Geotext. Geomembr. **25**(3), 194–202 (2007)
3. Wang, X., Wang, R., Liu, J.: Disposal method of unequal settlement of metro tunnel in operation in Shanghai soft ground. J. Shanghai Jiaotong Univ. **46**(1), 26–31 (2012)
4. Jiang, Z., Ping, Y., Zhou, Z.: Soft soil bank landslide analysis of Maozhou river outlet. Geotech. Invest. Surv. **2**, 18–22 (2013)
5. Shi, Q., Cao, W.: Comparison and analysis on treatment of highway soft soil foundation. Subgrade Eng. **160**(1), 111–114 (2012)
6. Jiao, D.: Model test research on geogrid-reinforced and pile-supported foundation under cyclic loading. Zhejiang University, Hangzhou (2010)
7. Lian, F.: Bearing mechanism and design method of pile-reinforced earth composite foundation. Zhejiang University, Hangzhou (2009)
8. Sa, C.T., Palmeira, E.M., Dellabiana, L.M.A., et al.: Numerical analysis of reinforced embankments on soft soils. In: Proceedings of the International Symposium on Earth Reinforcement, pp. 265–270 (2001)
9. Pham, H.T.V., Suleiman, M.T., White, D.J.: Numerical analysis of geosynthetic-rammed aggregate pier supported embankments. In: Geotechnical Engineering for Transportation Projects, pp. 657–664 (2004)
10. Shang, Y., Xu, L., Shang, L., et al.: Effect of grid strengthonpile-soilstress ratioofpile-netcomposite. J. Eng. Geol. **25**(1), 95–101 (2017)
11. Li, T., Ye, G., Zang, Q., et al.: Study on influential factors of deformation of pile-net structure reinforced bedding on high speed railway. Railw. Eng. **1**(1), 20–23 (2017)
12. Han, J., Gabr, M.A.: Numerical analysis of geosynthetic-reinforced and pile-supported earth platforms over soft soil. J. Geotech. Geoenviron. Eng. **128**(1), 44–53 (2002)
13. Suleiman, M.T., Pham, H., White, D.J.: Numerical analyses of geosynthetic-reinforced rammed aggregate pier-supported embankments. Report No. ISU-ERI, 3598 (2003)
14. Zhuang, Y., Wang, K., Liu, H.: Reinforcement performance of piled embankments. Chin. J. Geotech. Eng. **35**(S1), 436–441 (2013)

15. Yapage, N.N.S., Liyanapathirana, D.S., Poulos, H.G., et al.: Numerical modeling of geotextile-reinforced embankments over deep cement mixed columns incorporating strain-softening behavior of columns. Int. J. Geomech. **15**(2), 04014047 (2014)
16. Li, S., Ma, X., Tian, Z.: Research on influence factors of long-short pile reinforced loess foundation under embankment. J. Railw. Sci. Eng. **14**(2), 241–249 (2017)
17. Zheng, J., Zhang, J., Ma, Q., et al.: Experimental investigation of geogrid-reinforced and pile-supported embankment at bridge approach. Chin. J. Geotech. Eng. **34**(2), 355–362 (2012)

Numerical and Experimental Investigation of Tensile Behavior of Geogrids with Circular and Square Apertures

Jie Gu, Mengxi Zhang$^{(\boxtimes)}$, and Zhiheng Dai

Department of Civil Engineering, Shanghai University, Shanghai, China
mxzhang@i.shu.edu.cn

Abstract. Geogrid, one of the geosynthetics, was usually manufactured in rectangular, square or triangular apertures. In this paper, HDPE geogrids with a new shape of circular aperture was introduced. A series of laboratory pull-out tests were conducted to study on the interface properties of the geogrids. Three kinds of geogrids with different aperture shapes and different effective area ratios were prepared. The pull-out stress- displacement curves of the geogrid were depicted. Moreover, the residual resistance in all the cases was analyzed. Based on the test result, the apparent cohesion and apparent internal friction angle were determined. Furthermore, numerical modeling was conducted to demonstrate the stress distribution of the interface and the displacement of the geogrids. It was suggested that geogrids with higher effective area ratio were confined more strongly by the soil specimen and circular aperture geogrids were confined more strongly too. Circular aperture geogrids with the diameter of 48 mm leaded to the maximum coefficients of residual resistance and contributed relatively high interface parameters. The numerical modeling revealed that circular aperture geogrids leaded to general relative uniform stress distribution. Circular aperture geogrids with the diameter of 48 mm possessed the highest stiffness and enable the largest displacement region to be mobilized. The stable architecture of circular aperture geogrid resulted in the highest frictional resistance.

Keywords: Geogrid · Circular aperture · Numerical modeling
Pull-out tests

1 Introduction

Geogrid, one of the geosynthetics, is widely used in the construction of civil engineering structures (such as embankment slope, retaining wall, and foundation et al.) to improve their stability and bearing capacity [1–3]. It is understood that geogrids, a form of reinforcement buried in the soil mass of structures, can not only improve their stability but also reduce their settlement [4]. The study on interface properties of the reinforcement in the soil mass is a major research direction.

By conducting large-scale direct shear tests and pull-out tests which are commonly used in laboratory model tests, a large number of scholars at home and abroad investigated on interface behaviors of the reinforcement layers [5, 6]. For determining certain engineering design parameters, large-scale direct shear tests and laboratory

© Springer Nature Singapore Pte Ltd. 2018
L. Li et al. (Eds.): GSIC 2018, *Proceedings of GeoShanghai 2018 International Conference: Ground Improvement and Geosynthetics*, pp. 347–359, 2018.
https://doi.org/10.1007/978-981-13-0122-3_38

pull-out tests were ordinarily carried out to model the practical engineering [7]. Moraci and Recalcati [8] studied on the factors influencing the geogrid-reinforcement effect and evaluate the apparent friction, peak and residual pull-out resistance. Yang et al. [9] investigated on influence of the kinds of fills on the performance of the soil-reinforcement interface. Moraci and Gioffrè [10] developed a new theoretical method to measure the peak and the residual pull-out resistance of extruded geogrids embedded in a compacted granular soil. Zhang [11] proposed a new concept of three-dimensional reinforcement progressively. Dong et al. [12, 13] studied on the performance of geogrids with triangular apertures under static load. Mosallanezhad [14, 15] used a kind of anchoring block with dimensions of 10 mm × 10 mm × 10 mm to enhance the geogrids. Ezzein and Bathurst [16] introduced a new-type pull-out box with a transparent glass-bottom and non-contact measurement device. Zhou et al. [17] conducted pull-out tests of triangular apertures geogrids and rectangular apertures geogrids with same effective area ratio and similar mechanical behaviors. Numerical methods such as FEM, DEM and FLAC^{3D} are good method to study on the interface behavior. Zhang et al. [18] used discrete element modelling to investigate the influence of geogrid properties and soil compaction on the interface behavior. Ferellec and Mcdowell [19] used discrete element method to simulated pull-out tests considering the influence of irregular particle shapes. Wang et al. [20] studied on the influence of different top boundaries on geogrid pullout behavior. Ngoc et al. [21] study on the interface behavior of geogrid reinforced sub-ballast. Xue and Chen [22] demonstrated a more reasonable strength reduction method considering the geogrid strength reduction. Rai et al. [23] investigated stability of mine waste dump reinforced with geogrid of varying tensile strengths, permutations and combinations of spacing and aperture sizes. Review of the literature suggested that numerical modeling was widely favored (Hegde and Sitharam 2015 [24]). The frictional characteristics of soil-geogrid interface have been extensively studied. However, the influence of the opening shape, especially of circular apertures, on interface behaviors hasn't yet been well understood.

In the present paper, a series of laboratory pull-out tests of circular apertures geogrids and ordinary geogrids were carried out to discuss the influence of aperture shape on interface behaviors. The interaction behaviors of soil-geogrid interface were analyzed, as well as the residual pull-out resistance. Based on the results, the laboratory pull-out tests of geogrids with different effective area ratio were conducted, and the influence of effective area ratio on the soil-geogrid interface behaviors were studied progressively. Furthermore, a series of numerical modeling studies were conducted using FLAC^{3D}. The stress distribution of the interface and the displacement of the geogrids were examined.

2   Experimental Studies

Direct shear test and pull-out test are two major methods to study on the shear behavior of soil-grid interface properties. However there are obviously different friction mechanisms between the two tests. During the direct shear tests, geogrid is laid on the soil specimen filled in the lower shear box, and then the upper shear box is pushed parallel to the shear surface. Because of the relatively large aperture of the geogrids, the shear friction resistance between the upper and lower soil specimens contributes the most results, which cannot demonstrate the soil-geogrid interface behaviors. So, in a certain sense, conducting pull-out test to investigate on the frictional properties of the soil-geogrid interface has more practical significance.

2.1 Experimental Instrumentations

The overall arrangement of the laboratory pull-out test was shown in Fig. 1. A cube tank with inner dimensions of 300 mm × 300 mm × 300 mm (L × W × H) was fabricated from 5 aluminum plates and 1 transparent Plexiglas plate with a same thickness of 25 mm. These thick aluminum plates were rigid enough to avoid leading to a large deformation of the tank. Through the transparent Plexiglas plate, the displacement of the soil grains could be observed and obtained.

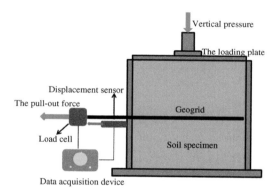

Fig. 1. Diagram of modeling box

The vertical load was applied on the loading plate by an oil pump-jack. The horizontal pull-out force applied by a motor speed reduction device was speed-controlled. The pull-out speed of 1 mm/min was kept to avoid too much influence of the rate of displacement on shear properties. The pull-out displacement was recorded by a resistance displacement sensor. The pull-out force applied on the geogrid through the clamp was monitored by a load cell.

2.2 Experimental Material

Both kinds of geogrid used in the pull-out tests were made of high-density polyethylene (HDPE), one of which is biaxial geogrid. The biaxial geogrid manufactured through cold drawn technology and enhanced with the ultrasonic welded nodes.

Two kinds of circular apertures geogrids were handmade using a simple electric knife to cut out circles from thin HDPE plate. The geometry of the geogrids was sketched and the dimension parameters were marked in Fig. 2. The main parameters of mechanic properties of the geogrid were listed in Table 1. The geogrids were hand-cut into a right size shown in Table 2. To eliminate the boundary effect, 10 mm gap

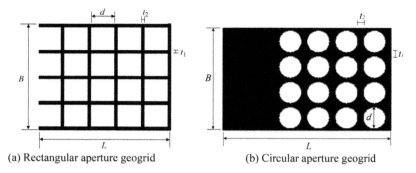

(a) Rectangular aperture geogrid   (b) Circular aperture geogrid

Fig. 2. Geometry and dimension parameters of geogrids

Table 1. The main properties of the geogrid

Tensile yield strength (kN/m)	Tensile strength at 2% strain (kN/m)	Tensile strength at 5% strain (kN/m)	Yield elongation (%)
30	10	13	13

Table 2. The dimensions of the geogrids

Aperture shape	B/mm	L/mm	d/mm	t_1/mm	t_2/mm	α
Rectangular	280	450	50	9	9	0.366
Circular	280	450	64	4	11	0.371
			48	18	27	0.596

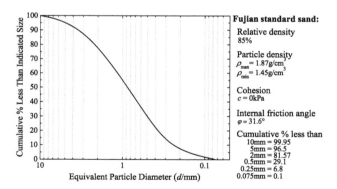

Fujian standard sand:

Relative density
85%

Particle density
$\rho_{max} = 1.87 \text{g/cm}^3$
$\rho_{min} = 1.45 \text{g/cm}^3$

Cohesion
$c = 0 \text{kPa}$

Internal friction angle
$\varphi = 31.6°$

Cumulative % less than
10mm = 99.95
5mm = 96.5
2mm = 81.57
0.5mm = 29.1
0.25mm = 6.8
0.075mm = 0.1

Fig. 3. Grain size distribution curve and main properties

remained to avoid contact between the geogrid and the sidewalls. Figure 3 depicted grain size distribution curve and listed the main property parameters of the Fujian standard sand. The tests were conducted at a uniform relative density (ID) of 85%.

Where the effective area ratio $\alpha = \frac{(B-t_1)(L-t_2)-S}{(B-t_1)(L-t_2)}$; S is the area of apertures.

2.3   Experimental Procedure and Program

Firstly, the test tank, pump-jack and motor device were fixed by bolts on a rigid test stand. Then the tank was half filled with clean dry sand. It was noted that, the shear surface must be flush with the pull-out-slot to enable the geogrids to be pulled out smoothly. Part of geogrid with the length of 2 mm was reserved to be fixed by the rigid aluminum clamp. Furthermore, sand was used to completely fill the test tank. After leveling the top-surface of soil specimen, the loading plate was centered in the tank. The vertical pressure, applied on the loading plate by the oil pump-jack, increased slowly to the required magnitude to prevent the soil mass failure under sudden high loading. After the vertical displacement had been stable, the motor speed reduction device started working. Meanwhile the horizontal displacement and pull-out stress were monitored.

Three kinds of geogrids of varying aperture shapes and effective area ratios were prepared. A series of pull-out tests were conducted under different vertical pressures, such as 100 kPa, 150 kPa and 200 kPa. Each test was repeated 3 times at least and the average values of the good results were finally taken. The results with big error would be discarded and the corresponding tests would be repeated. Table 3 summarized the program.

Table 3. Program of the pull-out test

Number	Identifier	Geogrid type	Vertical pressure (kPa)
1	R-100	Rectangle	100
2	R-150	Rectangle	150
3	R-200	Rectangle	200
4	C64-100	Circle-64	100
5	C64-150	Circle-64	150
6	C64-200	Circle-64	200
7	C48-100	Circle-48	100
8	C48-150	Circle-48	150
9	C48-200	Circle-48	200

Where Rectangle represented rectangular aperture geogrid;
Circle-64 represented circular aperture geogrid with diameter of 64mm;
Circle-48 represented circular aperture geogrid with diameter of 48mm.

3   Tests Results and Analysis

Figure 4(b–d) showed the comparison of the pull-out stress versus displacement. Obvious non-linear characteristic of the curves was observed. The pull-out stress peak value increased significantly with the increase of vertical pressure, which resulted from the soil swelling. The soil swelling was caused by shear displacement of soil-geogrid interface. However, the change of soil volume was confined by the surrounding stable soil, which exerted more additional normal stress. Furthermore, increase of normal

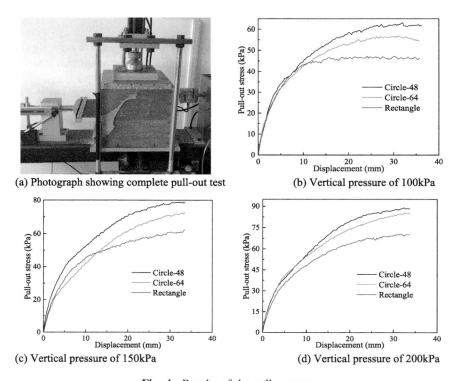

(a) Photograph showing complete pull-out test   (b) Vertical pressure of 100kPa

(c) Vertical pressure of 150kPa   (d) Vertical pressure of 200kPa

Fig. 4. Results of the pull-out tests

stress leaded to more expansion of the soil volume, which resulted in bigger pull-out stress increment.

It was shown that the sequence of pull-out stress peak values from high to low are: Circle-48, Circle-64 and Rectangle corresponding to the same applied vertical pressure. Geogrids with higher effective area ratio could contribute more constraining stress, which leaded to higher frictional resistance under same vertical pressure. Similar result was concluded that circular aperture geogrids could contribute more constraining stress compared to rectangular aperture geogrids.

During the pull-out tests, after the peak values of pull-out stress being observed, the corresponding failure would occur, and the pull-out stress would turn to the residual stress progressively. In all the cases, pull-out stress would tend to be stable and was not affected by the increase of displacement. The corresponding residual resistances were listed in Table 4.

Line chart of the coefficients of residual resistance at different vertical pressure was depicted in Fig. 5. The higher coefficients of residual resistance of circular aperture geogrids were noted comparing with rectangular aperture geogrids. The maximum coefficients of residual resistance were observed in Circle-48.

Figure 6 represented the liner fitted curves of shear-stress peak values and vertical pressure obeying Mohr-Coulomb strength criterion. Based on the fitted curves, two

Table 4. Residual resistance at different vertical pressure (kPa)

Geogrid type	Vertical pressure (kPa)		
	100	150	200
Circle-48	61	73	90
Circle-64	54	69	83
Rectangle	46	60	69

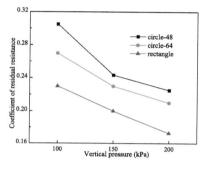

Fig. 5. Coefficients of residual resistance.

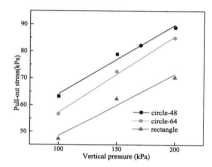

Fig. 6. Shear stress versus vertical pressure curve

interface parameters: apparent cohesion (c) and apparent internal friction angle (φ) were determined.

Where coefficients of residual resistance f was determined by following formula: $f = \tau/2\sigma$; τ represented the residual resistance; σ represented the normal stress approximately equal the vertical pressure.

Table 5 listed the interface parameters of all geogrids. It was found that the circular aperture geogrid had relatively higher interface parameters as compared to rectangular aperture geogrid. In addition, almost maximum interface parameters were observed in Circle-48.

Table 5. The interface parameters of all geogrids

Interface property	Geogrid type		
	Circle-48	Circle-64	Rectangle
c/kPa	38.499	29.049	25.378
φ/°	14.402	15.764	12.964

From Fig. 6 and Table 5, it could be seen that the interface strength properties are different from sand strength properties. The interface friction angle was about 14°, which was far smaller than the internal friction angle of sand. It could be concluded that the geogrid separated the upper and lower sand, which weaken the original interlock effect, leading to the decrease of the interface friction angle. Meanwhile, the cohesion at geogrid-soil interface was about 30 kPa, it came from the horizontal earth pressure acting on transverse rib of geogrid.

4   Numerical Modeling

FLAC³ᴰ used in the paper is a three-dimensional explicit finite-difference program offering a lot of built-in structural elements, which enabled geosynthetics and various reinforced construction to be modeled. Additionally, FLAC³ᴰ provides slip-plane model to simulate the joints and interface mechanics. It is noted that, FLAC³ᴰ also provides a powerful built-in programming language, FISH, to enable new variables and functions to be defined.

4.1   Model Consideration

In this paper, FLAC³ᴰ was used to model the pull-out tests. Using the built-in Geogrid element, a type of rectangular-aperture and two types of circular aperture geogrids with different effective area ratio were modeled (Fig. 7).

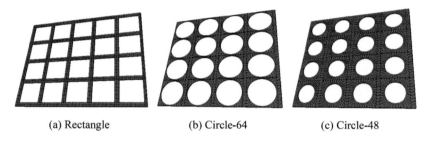

(a) Rectangle   (b) Circle-64   (c) Circle-48

Fig. 7.   Three kinds of geogrid modeled by FLAC³ᴰ

The built-in Mohr-Coulomb model was used to model the mechanical properties of the soil specimen. Linear elastic model was used to model the mechanical properties of the geogrids considering that no tensional failure occurred. By default, the soil-geogrid interface was modeled with Mohr-Coulomb criterion, whose properties were summarized in Table 6.

The side boundaries were restrained only in the horizontal direction, and the displacement along the bottom boundary was restrained in both horizontal as well as vertical directions. The geogrids were created above the soil specimen and then repositioned in the middle of soil specimen. Figure 8(a) showed the model consisting of 1416 geogridSELs and 3375 zones. The constant pressures of 100 kPa, 150 kPa and 200 kPa were respectively applied to the top surface. The pull-out test was conducted by applying a constant pull-out velocity of 1×10^{-7} m/step (Fig. 8(a)). With FISH function, the displacement and the pull-out stress were monitored. It was noted that the pull-out stress was equal to the pull-out force divided by the geogrid area.

Table 6. The properties of the geogrid in FLAC^{3D}

Property	Value
Thickness (mm)	5
Density (kg·m^{-3})	1.95
Young's modulus (Pa)	2.5×10^8
Poisson's ratio	0.4
Interface shear modulus (MPa/m)	60
Interface cohesion (kPa)	10
Interface friction angle (°)	41

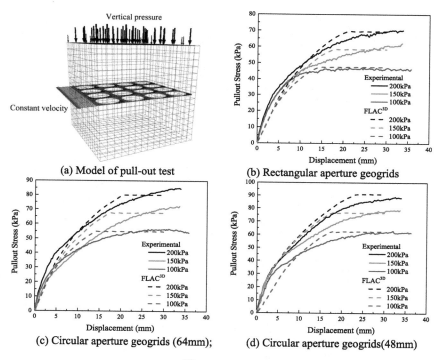

(a) Model of pull-out test   (b) Rectangular aperture geogrids

(c) Circular aperture geogrids (64mm);   (d) Circular aperture geogrids(48mm)

Fig. 8. FLAC^{3D} Modeling of pull-out tests

4.2 Model Verification

Figure 8(b–d) represented the comparison of the pull-out stress versus displacement curves of experimental and numerical studies in which 200 kPa, 150 kPa and 100 kPa are applied to the top surface respectively.

A relative good-agreement between experimental and numerical results was observed. Once the numerical modeling was validated, the stresses distributions of the interface and displacement of the geogrid at different vertical pressures were obtained to reveal the interface mechanism.

5   Numerical Results and Analysis

The contour of xx-component of stress of the geogrids at the vertical pressure of 100 kPa, 150 kPa and 200 kPa were shown in Fig. 9. The similarity of the stress distribution under different vertical pressures was noted. It could be summarized that the stress magnitudes depended on the vertical pressure while stress distribution did not. Moreover, the most uniform stress distribution was observed in Circle-48. More stress concentration was observed in the front row of ribs in Circle-64. Comparing with the case of rectangular aperture geogrid, generally uniform stress distribution was noted in the case of circular aperture geogrids.

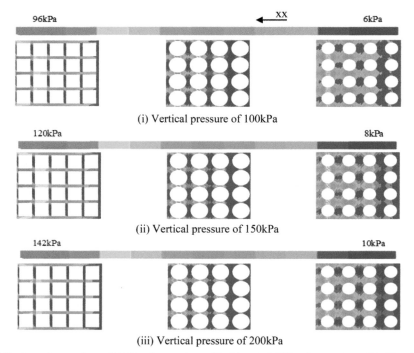

(a) Rectangular apertures; (b) Circular apertures (D=64mm); (c) Circular apertures (D=48mm)

Fig. 9. The xx-component of stress of geogrids

At the displacement of 30 mm, the displacements of all geogrids were examined. A phenomenon that the displacements of the geogrid were progressively mobilized inward from the front was noted. As it was shown in the contour plot in Fig. 10, smaller displacement region was mobilized with the vertical pressure increase. This was because the higher vertical pressure effectively confined the displacement of the geogrids. Only by applying higher pull-out stress could the geogrids be pulled out. For the same vertical pressure, the largest displacement region was mobilized in Circle-48,

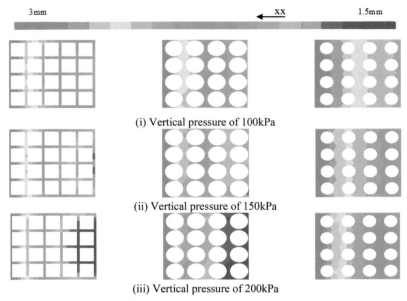

(i) Vertical pressure of 100kPa

(ii) Vertical pressure of 150kPa

(iii) Vertical pressure of 200kPa

(a) Rectangular-apertures; (b) Circular apertures (D=64mm); (c) Circular apertures (D=48mm)

Fig. 10. The displacement of geogrids

and the smallest displacement region was mobilized in Rectangular. As a result, it was concluded that Circular-48 possessed the highest rigidity and leaded to highest frictional resistance.

6 Conclusion

This paper presented a series of laboratory investigations on pull-out test. Meanwhile, the pull-out tests were modeled using FLAC3D. Had been validated by laboratory tests, the numerical model revealed the stress distribution of the interface and the displacement of the geogrids. The results indicated that:

(a) At the same applied vertical pressure, geogrids with higher effective area ratio were confined more strongly by the soil specimen. Comparing with rectangular geogrids, circular aperture geogrids were confined more strongly as it had higher effective area ratio when the aperture areas were the same.

(b) The coefficients of residual resistance had a non-linear decrease with the vertical pressure increase. Moreover, the coefficient decrease rate was lower and lower. The lowest decrease rate was observed in Circle-48, but it leaded to the maximum coefficient of residual resistance.

(c) The apparent cohesion and apparent internal friction angle were determined. It was found that the circular aperture geogrid had relatively high interface parameters as compared to rectangular aperture geogrid. In addition, almost maximum interface parameters were observed in Circle-48.

(d) Comparing with the cases of rectangular aperture geogrids, general relative uniform stress distribution was noted in the cases of circular aperture geogrids. Circle-48 geogrid, possessed the highest stiffness, enable the largest displacement region to be mobilized, which leaded to highest frictional resistance.

Acknowledgments. This study has been supported by the National Natural Science Foundation of China (NSFC) (No. 41372280). The authors would like to express their gratitude for these financial assistances.

References

1. Huang, C.C., Tatsuoka, F.: Bearing capacity of reinforced horizontal sandy ground. Geotext. Geomembr. **9**(1), 51–82 (1990)
2. Bathurst, R.J., Vlachopoulos, N., Walters, D.: The influence of facing stiffness on the performance of two geosynthetic reinforce soil retaining walls. Can. Geotech. J. **43**(12), 1225–1237 (2006)
3. Li, Z.Q., Hu, R.L., Fu, W., et al.: Study on using geogrids to reinforce embankment of expressway. Rock Soil Mech. **29**(3), 795–799 (2008)
4. Yang, Y., Liu, S.Y., Deng, Y.F.: Effect of reinforced subgrade on differential settlement by model test research. Rock Soil Mech. **30**(3), 703–706, 711 (2009)
5. Palmeira, E.M.: Bearing force mobilisation in pull-out tests on geogrids. Geotext. Geomembr. **22**(6), 481–509 (2004)
6. Juran, I., Guermazi, A., Chen, C.L., et al.: Modelling and simulation of load transfer in reinforced soils: part 1. Int. J. Num. Anal. Methods Geomech. **12**(2), 141–155 (2010)
7. Liu, W.B., Zhou, J.: Experimental research on interface friction of geogrids and soil. Rock Soil Mech. **30**(4), 965–970 (2009)
8. Moraci, N., Recalcati, P.: Factors affecting the pullout behaviour of extruded geogrids embedded in a compacted granular soil. Geotext. Geomembr. **24**(4), 220–242 (2006)
9. Yang, G.Q., Li, G.X., Zhang, B.X.: Experimental studies on interface friction characteristics of geogrids. Chin. J. Geotech. Eng. **28**(8), 948–952 (2006)
10. Moraci, N., Gioffrè, D.: A simple method to evaluate the pullout resistance of extruded geogrids embedded in a compacted granular soil. Geotext. Geomembr. **24**(2), 116–128 (2006)
11. Zhang, M.X., Zhou, H., Javadi, A.A., et al.: Experimental and theoretical investigation of strength of soil reinforced with multi-layer horizontal–vertical orthogonal elements. Geotext. Geomembr. **26**(3), 1–13 (2008)
12. Dong, Y.L., Han, J., Bai, X.H.: Numerical analysis of tensile behavior of geogrids with rectangular and triangular apertures. Geotext. Geomembr. **29**(2), 83–91 (2011)
13. Dong, Y.L., Han, J., Bai, X.H.: Behavior of triaxial geogrid-reinforced bases under static loading. In: Proceedings of the 9th International Conference on Geosynthetics, pp. 1547–1550. IFAI, Brazil (2010)

14. Mosallanezhad, M., Hataf, N., Ghahramani, A.: Experimental study of bearing capacity of granular soils, reinforced with innovative grid-anchor system. Geotext. Geomembr. **26**(3), 299–312 (2008)
15. Mosallanezhad, M., Taghavi, S.H.S., Hataf, N., et al.: Experimental and numerical studies of the performance of the new reinforcement system under pull-out conditions. Geotext. Geomembr. **44**(1), 70–80 (2016)
16. Ezzein, F.M., Bathurst, R.J.: A new approach to evaluate soil-geosynthetic interaction using a novel pullout test apparatus and transparent granular soil. Geotext. Geomembr. **42**(3), 246–255 (2014)
17. Zhou, Y.J., Zheng, J.J., Cao, W.Z., et al.: Experiment study of interface characteristics of different aperture shapes of geogrids reinforced soil. J. Southwest Jiaotong Univ. **52**(3), 482–488 (2017)
18. Zhang, J., Yasufuku, N., Ochiai, H.: Discrete element modelling of geogrid pullout test. In: Proceedings of the 4th Asian Regional Conference on Geosynthetics, pp. 11–14, Shanghai, China (2008)
19. Ferellec, J.F., Mcdowell, G.R.: A method to model realistic particle shape and inertia in DEM. Granular Matter **12**(5), 459–467 (2010)
20. Wang, Z., Jacobs, F., Ziegler, M.: Influence of rigid and flexible top boundaries on geogrid pullout behavior using DEM. In: Proceedings of the 7th International Conference on Discrete Element Methods. Springer, Singapore (2017)
21. Ngoc, N.T., Indraratna, B., Rujikiatkamjorn, C.: A study of the geogrid–subballast interface via experimental evaluation and discrete element modelling. Granular Matter **19**(3), 54 (2017)
22. Xue, J.F., Chen, J.F.: Reinforcement strength reduction in FEM for mechanically stabilized earth structures. J. Central South Univ. **22**(7), 2691–2698 (2015)
23. Rai, R., Khandelwal, M., Jaiswal, A.: Application of geogrids in waste dump stability: a numerical modeling approach. Environ. Earth Sci. **66**(5), 1459–1465 (2012)
24. Hegde, A., Sitharam, T.G.: 3-Dimensional numerical modelling of geocell reinforced sand beds. Geotext. Geomembr. **43**(2), 171–181 (2015)

Simplified Approach of Seismic Response Simulation of Geosynthetic Reinforced Slope

Sao-Jeng Chao, Te-Sheng Liu[(✉)], and Tsan-Hsuan Yu

National Ilan University, Yilan City, Taiwan
joesph5802@gmail.com

Abstract. In order to understand the seismic response of geosynthetic reinforced slope influenced by earthquake, this study uses the field recorded seismic data from an earthquake monitoring system for the geosynthetic reinforced slope built in FoGuang University with the seismic sensor monitoring instrument Palert. This study attempts to explore the effects of earthquakes on geosynthetic reinforced slope with the aid of monitoring system as well as dynamic numerical simulation. This study implements the finite element program PLAXIS to simulate the seismic responses of the geosynthetic reinforced slope with different nodal elements in order to solve the time consuming issue. The proposed simplified approach for seismic numerical simulation of the geosynthetic reinforced slope is discussed comprehensively. Simplified model provided in PLAXIS needs to be able to achieve the calculation accuracy. The seismic monitored data of this study are the actual information and make the dynamic predicted results of the numerical simulation for geosynthetic reinforced slope as a real-world matter.

Keywords: Simplified approach · PLAXIS · Geosynthetic reinforced slope
Seismic simulation · Earthquake monitoring

1 Introduction

Taiwan is located in the colliding area of Philippine plate and Eurasian Plate. Plate activities are commonly accompanied by massive earthquakes, which extremely challenge the condition of slope safety. Therefore, civil engineer often design geosynthetic reinforced structure to stabilize the slope and protect people's life for safety. Geosynthetic reinforced slope has the advantages of convenient construction, short construction period, better deformation tolerance, and seismic resisting capacity.

In this study, it is expected to recognize the response of the geosynthetic reinforced slope during earthquakes. Using the seismic data recorded by the seismic monitoring system which was set up at the FoGuang University, the predicted dynamic behavior of the geosynthetic reinforced slope in earthquake using PLAXIS finite element program can be validated. In order to understand the general response of the geosynthetic reinforced slope in the seismic simulation, this paper tries to use a simplified approach to predict the reinforced slope with geosynthetic material from the view point of time consuming.

© Springer Nature Singapore Pte Ltd. 2018
L. Li et al. (Eds.): GSIC 2018, *Proceedings of GeoShanghai 2018 International Conference: Ground Improvement and Geosynthetics*, pp. 360–367, 2018.
https://doi.org/10.1007/978-981-13-0122-3_39

In order to further understand the effect of the propagation of waves between the finite elements on the geosynthetic reinforced slope, this paper compares the simulating process using different element nodes to explore the predicted results for the geosynthetic reinforced slope. It is anticipated to use a simplified analysis to simulate the geosynthetic reinforced slope under different earthquakes in many ways, while there is no need for a lot of computing time. The seismic monitoring data used in this study is obtained from the actual condition. Therefore, the seismic simulation predicted results from the PLAXIS finite element program are quite applicable.

2   FoGuang Geosynthetic Reinforced Slope

Geosynthetic reinforced structures have developed comparatively rapidly in Taiwan in recent years. They can be used for slope stability, embankment, and highway slope protection. Considering Taiwan is located in the Pacific Rim seismic belt, the earthquake has been the serious problem which is unable to ignore in Taiwan.

In order to understand the seismic response of the geosynthetic reinforced slope under the earthquake, three seismic sensors were installed on the slope at FoGuang University. The seismic sensors (Palert) are placed on the foundation surface, the central location, and the top position of the geosynthetic reinforced slope for measuring the seismic response.

2.1   Description of the Dynamic Monitoring System Center

Ilan region is not only subjected to earthquakes a lot but also regularly attacked by typhoons. In 2008, the Xinluke typhoon landed right from Ilan area and cased the slope failed at the location of southwest side of the staff dormitory of FoGuang University. The authority of FoGuang University immediately carried out disaster restoration and use of geosynthetic reinforced structure to strengthen the slope to protect the staff dormitory building of FoGuang University.

Considering the frequent occurrences of the earthquakes in Taiwan and the location near the three seismic belt faults in Ilan area, the earthquake monitoring system is set up on the slope to achieve more understanding of the seismic response of the geosynthetic reinforced slope. The research team set up the Ilan University earthquake response monitoring center, through the remote monitoring system to do observing, and then study of the recorded data in details, so that seismic simulation can be more effective verification.

FoGuang University geosynthetic reinforced slope (Fig. 1) has three tiers, with the additional application of a 200 kN/m reinforced geogrid. The long-term allowable strength of 50 kN/m is thus accepted. The first tier (base tier) of the geosynthetic reinforced slope was 2.5 m high; the second tier (middle tier) was 15.5 m high with a reinforced slope angle of 45°. The third tier (top tier) was 12.5 high with a reinforced slope angle of 73°. Furthermore, a 2-meter-high backfill formation was additionally designed on surface of the top tier of the reinforced slope. It is noted that the designed vertical spacing of the geosynthetic reinforcement is 0.5 m for the whole structure.

Fig. 1. The geosynthetic reinforced slope at FoGuang University.

2.2 Recorded Data from Field Monitoring Program

Taiwan is in the earthquake activity frequent area, which is subjected an average of about 18,500 earthquakes per year. The seismic sensors of the earthquake monitoring system are directly set on the surfaces of the three tiers of the slope, which is named as the FoGuang University dynamic monitoring demonstration field. Specifically speaking, seismic sensor monitoring instrument Palert is set up for each tier with three-axis micro-electromechanical (MEMS) accelerometer (Fig. 2). The seismic sensors, Palerts, are used to record the acceleration time history at different tiers of the geosynthetic reinforced slope.

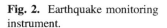

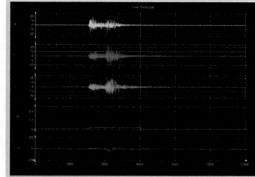

Fig. 2. Earthquake monitoring instrument.

Fig. 3. Earthquake monitoring data for three-axis.

Associated seismograph software was then used to promptly retrieve data content such as triaxial acceleration and displacement (Fig. 3), which can used by a subsequent dynamic simulation analysis. Taking advantage of the finite element numerical simulation, the current monitoring data and the numerical simulation results can be compared and provide as mutual evidence for each other. To the end, the seismic simulation can further do a good job in slope design work to reduce the slope failure caused by the earthquake hazards.

The seismic data used in this study are selected from the seismic recorded data on the geosynthetic reinforced slope. Specifically, data 0222 was adopted as the earthquake on 22th February, 2014 for the seismic simulation. The maximum acceleration of data 0222 observed in the field from the first to the third tiers were 0.092, 0.132, and 0.209 m/s^2, respectively. The amplification ratio detected from the first to the second tiers was 1.43, whereas the amplification ratio from the first to the third tiers was identified as 2.27. The seismograph (Fig. 4) illustrates the seismic activities in the three different tiers (x-axis = dynamic time, y-axis = acceleration).

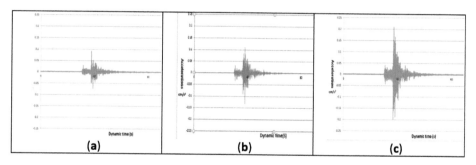

Fig. 4. Acceleration time history observed from field monitoring system: (a) base tier, (b) middle tier, (c) top tier.

3   PLAXIS FEM Analysis

3.1   Introduction of PLAXIS

PLAXIS 2D is a two-dimensional FEM analysis of deformation and stability in rock or soil that can construct a model using plane strain or axial symmetry, which is able to generate a finite element grid in the geometric model fast. Interactive behavior in various geotechnical and soil structures can be analyzed using a display window to input model conditions such as material parameters, boundary conditions, and dynamic data; furthermore, in addition to standard static analyses, earthquake data can be input for dynamic analyses by the PLAXIS finite element computer program.

3.2   Model Description

This section establishes the necessary conditions for modeling, which includes the detailed description of the model dimensions, the parameters of the soil material, and the material properties of the stiffened material. The cross-section of the adopted geosynthetic reinforced slope has three tiers, with the additional application of a 200 kN/m reinforced geogrid. The first tier (base structure) of the geosynthetic reinforced slope was designed to be 2.5 m deep, and a six-layer reinforced geogrid was applied, with 0.5 m spacing; the second tier (middle structure) was 15.5 m high with a reinforced slope angle of 45°, and a 51-layer reinforced geogrid was overlaid with 0.3 m spacing; the third tier (top structure) was 12.5 m high with a reinforced slope angle of 73° and a 41-layer reinforced geogrid with 0.3 m spacing. Furthermore, a 2-meter-high backfill formation was additionally designed on the top structure of the reinforced slope, which had been previously mixed with cement at an approximately 1.5%–2% weight ratio and processed by compaction after fill up. The design of the geosynthetic reinforced slope at FoGuang University is displayed as Fig. 5. The soil and geogrid parameters used for simulation are summarized in Table 1.

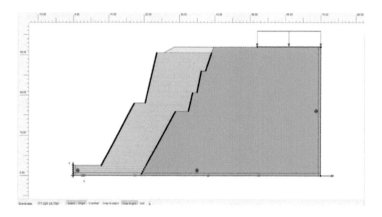

Fig. 5. Plaxis cross-section of the geometric model.

Table 1. Material properties.

Parameters	γ	γ_{sat}	c	$\emptyset$	E	υ	EA
Unit	(kN/m³)	(kN/m³)	(kN/m²)	(°)	(kN/m²)	–	(kN/m)
Backfill formation	22	22.5	7.5	29	10^5	0.4	–
Geosynthetic reinforcement	20	21	8	31	$1.5 * 10^5$	0.1	4000
Loading area	22	23.5	17	35	10^5	0.25	–

3.3   Element Node Explanation

In order to facilitate the user to automatically generate grid elements as soon as creating a complete model by PLAXIS 2D, the user can select 6 nodes or 15 nodes of the triangular element depending upon one's requirement to simulate soil mass (Fig. 6). The 15 nodes triangular element is a very accurate element, thus it can get high precision calculation results in various analysis modes; especially for the model

calculates the damage and gets a better answer. The element provides four stages of displacement interpolation, with 12 stress points to provide numerical integration calculations. Nevertheless, calculating the 15-node triangular element requires a larger memory.

The 6-node triangular element provides a 2-stage displacement interpolation with three stress points to provide numerical integration calculations. Actually, a 15-node element can be thought as a combination of four 6-node elements with the same number of stress points. However, a 15-node element calculation results are still better than the combination of four 6-node elements.

What's more, 6-node triangular elements can be calculated ideally in the case of dense meshing. But in the calculation of bearing capacity or safety analysis, the calculated damage load and safety factor are generally large, thus the best approach is to perform the analysis using a 15-node element.

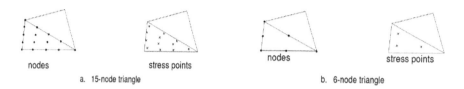

nodes   stress points   nodes   stress points

a. 15-node triangle   b. 6-node triangle

Fig. 6. The triangular elements in PLAXIS 2D with (a) 15 nodes and (b) 6 nodes.

4   Simplified Analysis of Seismic Simulation

Nowadays, contemporary structures are getting large in size and the design is getting complex and exhaustive in the design phase. The use of computer program analysis requires a lot of manpower to complete the design and simulation. After completing the design phase, it is necessary to use the computer program to repeat the calculation to ensure that the safety factor meets the design requirements.

In order to ensure that the design conforms to the relevant safety factor, the designed structure needs to be entered into the relevant program calculation stage over and over again. If the whole structure is calculated, each program calculation takes a lot of time to complete, so the simplified design seems to be essential to reduce the interval of designing time consuming.

In this study, the PLAXIS 2D program simulation is performed as an example, which is based on the analysis of the given seismic data in the numerical dynamic analysis. The recorded data form seismic sensor monitoring instrument Palert is used to simulate the seismic response of the geosynthetic reinforced slope built in the campus of FoGuang University.

The geosynthetic reinforced slope of FoGuang University is compared between the 6-node element and 15-node element simulation in PLAXIS finite program. The geosynthetic reinforced slope in the model establishment stage is set up to be 50 cm for the vertical spacing of geosynthetic material, which is a very fine simulation in a large project. In the case of element density distribution is uniform and tight enough, it can be done with a 6-node element, which can save about half of the time compared to the 15-node element. Both of the predicted results show almost negligible discrepancy.

In the dynamic calculation of the geosynthetic reinforced slope in FoGuang University, it takes a lot of time and hard disk space to calculate. Thus, the use of fewer nodes can provide more effective pre-production operations. It is recommended that the designer can use fewer nodes to run the pre-calculation and to confirm the behavior pattern of the geosynthetic reinforced structure.

In this case, the 15-node element and the 6-node element are both used to simulation the 0222 earthquake. The predicted maximum acceleration at the first tier (base tier) was 0.093 m/s^2 for both cases. The simulation results show that different nodes can obtain the same predicted maximum acceleration results. That is, the predicted maximum acceleration is 0.138 m/s^2 at the second tier (middle tier) and is 0.201 m/s^2 at the third tier (top tier). The time history of the seismic activity observed in each tier structure are shown in (Figs. 7 and 8) (x-axis = dynamic time; y-axis = acceleration).

It can be found that the simulation results are quite similar to the field recorded data. The predicted acceleration amplification ratio is 2.16 for the third tier (top tier) compared to the first tier (base tier). The monitored acceleration amplification ratio is 2.27 for the previous case. The predicted acceleration amplification ratio is 1.48 for the second tier (middle tier) compared to the first tier (base tier). The monitored acceleration amplification ratio is 1.43 for the previous case. The maximum acceleration amplification ratio error is less than 5% compared to the current monitoring value.

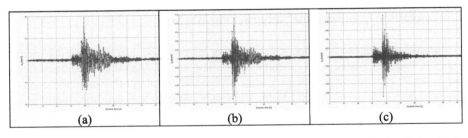

Fig. 7. Time history of seismic activity in the free-field boundary dynamic simulation in the 15-node: (a) base tier, (b) middle tier, (c) top tier.

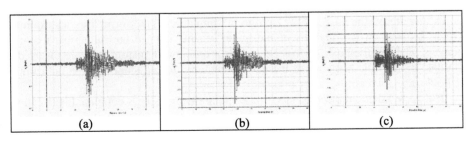

Fig. 8. Time history of seismic activity in the free-field boundary dynamic simulation in the 6-node: (a) base tier, (b) middle tier, (c) top tier.

5   Discussion

The paper uses the seismic sensor monitoring equipment Palert to obtain long-term monitoring data at the dynamic monitoring demonstration site of FoGuang University. The monitored earthquake data are used to study the acceleration time histories at the base, the middle, and the top portions of the geosynthetic reinforced slope. Following that, the PLAXIS FEM analysis software is used to predict the seismic response of the geosynthetic reinforced slope. The numerical simulation results and the current monitoring data are compared. In order to solve the time consuming issue of seismic simulation for the geosynthetic reinforced slope in the earthquake response situation, this paper pays a lot of effort on the node numbers of element to obtain the required information within a short period of time.

From the simulation results are quite similar to the field recorded data no matter with 15-node element or 6-node element. The predicted acceleration amplification ratio is 2.16 for the third tier (top tier) compared to the first tier (base tier), while the monitored acceleration amplification ratio is 2.27. The predicted acceleration amplification ratio is 1.48 for the second tier (middle tier) compared to the first tier (base tier) while the monitored acceleration amplification ratio is 1.43. It is suggested that one can use fewer node element to run the pre-calculation and to recognize the behavior pattern of the geosynthetic reinforced slope subjected to earthquake attacks.

Strength Characteristics of Glass Fiber-Reinforced Sand

Hong Sun[1(✉)], Gang Wu[2], Chun-yu Song[1], and Xiu-run Ge[1]

[1] School of Naval Architecture, Ocean and Civil Engineering,
Shanghai Jiao Tong University, Shanghai 200240, China
sunhong@sjtu.edu.cn
[2] Chinese Underwater Technology Institute, Shanghai Jiao Tong University,
Shanghai 200240, China

Abstract. The strength characteristics of glass fiber-reinforced sand composite are influenced by the controlling parameters of fiber content and confining pressure. A series of triaxial experiments were carried out to study on the behaviors of unreinforced and glass fiber-reinforced sand. The results show that glass fibers insertion can prevent the formation of the strain localization, provide an increase in strength and shearing strength parameters. Linear failure envelopes fit the data well for all the mixtures studied (unreinforced sand and the glass fiber-reinforced specimen). The deviator stress at failure and strength parameters of the sand- glass fiber composite increase as the fiber content increases. The cohesion changed more significantly than internal friction angle due to the glass fibers insertion. It is shown that the inclusion of fibers increases the contact surface between soil particles and fibers, but it has not a strong influence on the roughness and staggered arrangement of soil particles.

Keywords: Glass fiber · Reinforced sand · Triaxial loading · Strength
Cohesion · Internal friction angle

1 Introduction

Soil-fiber composite materials have been studied by a lot of experiments [1, 2]. Due to the interaction between the fibers and soil, the randomly distributed fibers can provide an increase in the strength, ductility, splitting tensile strength and flexural toughness [3, 4]. Consoli et al. [5] presented the polypropylene fiber significantly improved the unconfined strength of reinforced cement soil. Dasaka and Sumesh [6] presented coir fiber significantly increased the cohesion and the angle of internal friction by triaxial compression tests.

Glass fiber can also be a reinforcement material with high strength, good chemical properties, excellent durability, anti-leakage performance and lower price. Our group presented the stress-stain relationship and reinforcement effect of glass-fiber reinforced sand studied by triaxial tests [7]. However, the strength characteristics were not studied further. Based on the above experimental data, much more experiments were carried out. Present work describes a study of the mechanical behavior of geomaterials composed of uniform sand reinforced with randomly distributed glass fibers under triaxial

© Springer Nature Singapore Pte Ltd. 2018
L. Li et al. (Eds.): GSIC 2018, *Proceedings of GeoShanghai 2018 International Conference: Ground Improvement and Geosynthetics*, pp. 368–373, 2018.
https://doi.org/10.1007/978-981-13-0122-3_40

loading conditions. The specific objectives of the present research are to evaluate the strength behaviors of glass fiber-reinforced sand, to examine the effects of the fiber content r, confining pressure σ_3, to provide reliable data for taking glass fiber as a reinforcement material which can be applied to embankment and foundation improvement engineering.

2   Experimental Program

2.1   Materials

The sand was sampled from the region of Fujian Province, P. R. China. The sand is classified as non-plastic uniform fine sand (SP). The grain size distribution is entirely fine sand. The properties of sand are given in Table 1. Glass fibers were used throughout this investigation to reinforce the sand. The average dimension was 11 μm in diameter with density of 2.5 g/cm^3 and tensile strength of 1.25 GPa.

Table 1. Properties of sand

e_{max}	e_{min}	ρ_{max} (g·cm^{-3})	ρ_{min} (g·cm^{-3})	G_s	C_u	C_c
0.754	0.473	1.815	1.613	2.673	7.333	2.348

2.2   Sample Preparation and Testing Procedures

The compacted sand and glass fiber-reinforced specimens used in the triaxial tests were prepared by hand-mixing dry sand and glass fibers. In order to study the main interaction between soil particles and glass fibers without the influences of the other factors, the specimens are dried without water. Specimens were compacted in five layers into a 101 mm diameter by 200 mm high split mold, so that they reached the specified relative densities of 95%. For the glass fiber-reinforced specimens, fiber contents r (percentage of fiber weight divided by soil dry weight) of 0.1%, 0.2%, 0.3% and 0.5% were respectively used, 5 cm fiber length was used which is not bigger than the specimen radius.

Triaxial compression tests were performed on non-reinforced sand and glass-fiber reinforced sand specimens. Triaxial compression tests have been used in most of the experimental programs reported in the literature in order to access the influencing factors on the strength of reinforced soils, which simulated the general construction conditions of a lot of engineering projects. The test is also simple and fast, while being reliable and cheap.

All triaxial compression tests were carried out with the same triaxial apparatus until the failure of the specimen. The measurement of the axial strain was conducted with a vertical displacement sensor located on the top of pressure cell. The load was applied with a constant displacement rate (0.08 mm/min), the axial deviatoric force was measured by axial load sensor. Four confining pressures σ_3 (50 kPa, 100 kPa, 150 kPa, and 200 kPa) were applied.

3 Test Results and Analysis

3.1 Failure Shapes of the Glass Fiber-Reinforced Sand Specimens

The glass fiber-reinforced sand specimens failed on the triaxial compression tests were much different from the unreinforced sand specimens. The unreinforced sand specimens and a few glass fiber-reinforced sand specimens with the low confining pressure presented a failure composed of bulging failure followed by a failure plane, but the glass fiber-reinforced sand specimens with the high confining pressure presented a drum shape with no shear band identified, typical of ductile materials. The addition of fibers can adjust the internal structure of specimen, resulting in the more uniform change of specimen. It indicates that the strain localization is prevented by the presence of the fibers, and the glass fiber-reinforced sand composite is manifested by more uniform deformations.

3.2 Maximum Deviatoric Principal Stress - Axial Strain Curves

The maximum deviatoric principal stress ($\sigma_1 - \sigma_3$) - axial strain (ε_1) curves of the non - reinforced sand and glass-fiber reinforced sand are non-linear from the onset of loading. Figure 1 presents the stress - strain curves for glass-fiber reinforced sand for the various confining pressures at r = 0.1%. As the axial strain increases, the maximum deviatoric principal stress increases obviously at axial strain less than about 2%, and then the maximum deviator principal stress increases gradually at smaller degree until a peak stress. Finally it decreases as the axial strain increase. With the increase in confining pressure, the deviator principal stress increases. Due to the inclusion of fibers, the strength is higher.

3.3 Failure Envelopes Characteristics Influenced by Main Controlling Factors

The stresses at failure of unreinforced sand and glass fiber-reinforced sand are influenced by the confining pressures and fiber content. Figure 2 presents a series of failure envelopes for the glass fiber-reinforced sand of all cases, as well as that for the unreinforced sand in which p is equal to $(\sigma_1 + \sigma_3)/2$, the deviator stress q is $(\sigma_1 - \sigma_3)/2$. The failure envelopes for all the mixtures (unreinforced sand and the glass fiber-reinforced samples) with various fiber contents are approximately parallel straight lines. The similar results for fiber-reinforced soil were also observed by the other authors [8]. The deviator stress at failure increases with the increasing in fiber content. For example, when the confining pressure is 200 kPa, the deviator stress at failure of the fiber-reinforced soil of 0.5% fiber content increases to 463 kPa, compared with strength of 341 kPa for unreinforced sand. Chen et al. (2015) [9] reported that fiber contents lower than 0.5% would be considered optimal for engineering practice.

The inclusion of glass fiber provides an increase in strength of glass fiber-reinforced sand, because fibers and soil particles are intertwined to form the spatial frame structure, in which the friction force and spatial constraining force between soil particles and

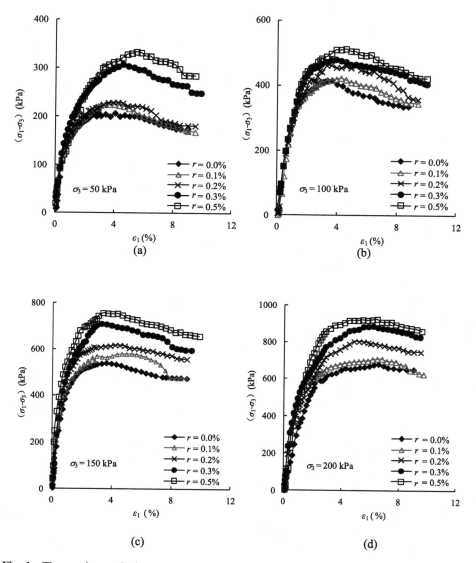

Fig. 1. The maximum deviator principal stress – axial strain curves for glass fiber-reinforced sand with various confining pressures and fiber contents: (a) Confining pressure of 50 kPa; (b) Confining pressure of 100 kPa; (c) Confining pressure of 150 kPa; (d) Confining pressure of 200 kPa.

fibers are appeared to jointly bear the external force and coordinate deformation. The interaction between the fiber and the soil could be raised by increasing the surface area of fibers, which is found to be in good agreement with Plé and Lê [10]. Tang et al. [11] reported that there is increased resistance against the applied force as the contact area between fibers and soil particles increase.

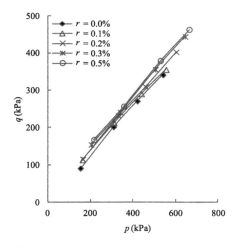

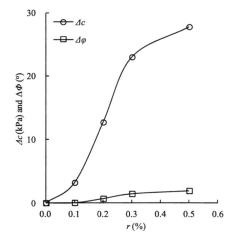

Fig. 2. Failure envelops for all specimens   **Fig. 3.** Increments of strength parameters vs. r

3.4   Strength Parameters

The Mohr-Coulomb parameters were determined by triaxial compression tests. The increments of cohesion and internal friction angle influenced by the controlling parameters are shown in Fig. 3, in which Δc is the increment of cohesion, $\Delta\varphi$ is the increment of internal friction angle. The strength parameters would be a cohesion intercept of zero for the unreinforced sand, increasing to 27.8 kPa for the 0.5% fiber content inclusion, and the internal friction angle φ increases by 1.9°. With the increase in the fiber content, the cohesion c significantly increases, especially that for the fiber content of 0.2%–0.3%. It is shown that the cohesion is nonlinear to the fiber content. While the internal friction angle φ increases slightly. However, Ahmad et al. [12] evaluated the response of fiber on the strength of reinforced silty sand. They reported that the cohesion increases linearly with the increase in fiber content.

The inclusion of glass fiber provides an increase in strength parameters of glass fiber-reinforced sand, because the inclusion of fibers increases the contact surface between soil particles and fiber so that the cohesion of glass-fiber reinforced sand increased. But it has not a strong influence on the roughness and staggered arrangement of soil particles, the internal friction angle is not relatively obvious to change. Higher content fibers increase number and surface of fiber-sand particle contact, so that strength parameters of the composite material are larger.

4   Conclusions

The triaxial experiments of unreinforced sand and reinforced sand in which glass fiber was a reinforcement material were carried out to study the strength and strength parameters. The conclusions can be drawn as follows:

(1) The addition of glass fibers can prevent the formation of the strain localization. The glass fiber insertion provides an increase in the strength and shearing strength parameters of the glass fiber-reinforced sand composite. Linear failure envelopes fit the data well for all the mixtures studied (unreinforced sand and the glass fiber-reinforced samples).

(2) The cohesion significantly increases, but internal friction angle does not increase obviously due to the inclusion of fibers. Because the inclusion of fibers increases the contact surface between soil particles and fibers, it has not a strong influence on the roughness and staggered arrangement of soil particles.

(3) The deviator stress at failure, and cohesion of the sand composite increase with the increase in fiber content, due to increasing the number and surface of fiber-sand particle contact, while the increment of cohesion is not linear to the fiber content.

Acknowledgement. The research described in this paper was financially supported by the National Natural Science Foundation of China under Grant No. 41572255 and Grant No. 51678361.

References

1. Prabakar, J., Sridhar, R.S.: Effect of random inclusion of sisal fibre on strength behaviour of soil. Constr. Build. Mater. **16**(2), 123–131 (2002)
2. Kumar, A., Gupta, D.: Behavior of cement-stabilized fiber-reinforced pond ash, rice husk ash-soil mixtures. Geotext. Geomembr. **44**, 466–474 (2016)
3. Diambra, A., Ibraim, E.: Fibre-reinforced sand: interaction at the fibre and grain scale. Geotechnique **65**, 296–308 (2015)
4. Festugato, L., Menger, E., Benezra, F., Kipper, E.A., Consoli, N.C.: Fibre-reinforced cemented soils compressive and tensile strength assessment as a function of filament length. Geotext. Geomembr. **45**, 77–82 (2017)
5. Consoli, N.C., Bassani, A., Festugato, L.: Effect of fiber-reinforcement on the strength of cemented soils. Geotext. Geomembr. **28**(4), 344–351 (2010)
6. Dasaka, S.M., Sumesh, K.S.: Effect of coir fiber on the stress-strain behavior of a reconstituted fine-grained soil. J. Nat. Fibers **8**(3), 189–204 (2011)
7. Liu, F., Sun, H., Ge, X.R.: Glass fiber-reinforced sand studied by triaxial experiments. J. Shanghai Jiaotong Univ. **45**(5), 762–766 (2011). (in Chinese)
8. Sivakumar, G.L., Vasudevan, A.K., Sumanta, H.: Numerical simulation of fiber-reinforced sand behavior. Geotext. Geomembr. **26**(5), 181–188 (2008)
9. Chen, M., Shen, S.L., Arulrajah, A., Wu, H.N., Hou, D.W., Xu, Y.S.: Laboratory evaluation on the effectiveness of polypropylene fibers on the strength of fiber-reinforced and cement-stabilized Shanghai soft clay. Geotext. Geomembr. **43**, 515–523 (2015)
10. Plé, O., Lê, T.N.H.: Effect of polypropylene fiber-reinforcement on the mechanical behavior of silty clay. Geotext. Geomembr. **32**, 111–116 (2012)
11. Tang, C.S., Shi, B., Zhao, L.Z.: Interfacial shear strength of fiber reinforced soil. Geotext. Geomembr. **28** (1), 54–62 (2010)
12. Ahmad, G., Bateni, F., Aami, M.: Performance evaluation of silty sand reinforced with fibres. Geotext. Geomembr. **28**(1), 93–99 (2010)

Three-Dimensional Numerical Analysis of Performance of a Geosynthetic-Reinforced Soil Pier

Panpan Shen[1](✉), Jie Han[2], Jorge G. Zornberg[3], Burak F. Tanyu[4],
and Dov Leshchinsky[5]

[1] Tongji University, Shanghai 200092, China
hermit_shpp@hotmail.com
[2] The University of Kansas, Lawrence, KS 66045, USA
jiehan@ku.edu
[3] The University of Texas at Austin, Austin, TX 78712, USA
zornberg@mail.utexas.edu
[4] George Mason University, Fairfax, VA 22030, USA
btanyu@gmu.edu
[5] ADAMA Engineering, Inc., Clackamas, OR 97015, USA
adama@geoprograms.com

Abstract. Geosynthetic reinforced soil (GRS) consisting of closely spaced geosynthetic layers has been increasingly used to support bridges due to its rapid construction, low construction cost, and effectiveness in eliminating bump at the end of the bridges. GRS performance tests (also referred to as GRS mini-pier tests) were used to evaluate load-deformation behavior of a frictionally connected GRS mass to support a vertical load. Numerical analysis is an alternative approach to study the behavior of the GRS mass if the numerical model is well calibrated and/or validated. In this study, a three-dimensional numerical model was developed using FLAC3D, a finite difference method-based program. The Mohr-Coulomb model was used to describe the behavior of backfill soil. Geosynthetic reinforcement was modeled as a linearly elastic material using "geogrid" structural elements. Different interfaces were considered in the numerical model to simulate the interaction between different components. The numerical model was calibrated and verified against the measured load-deformation curve and the lateral displacement profile along the height of the GRS pier. This paper also presents the calculated lateral earth pressure profile behind facing blocks along the height of the GRS pier. Comparisons show that the numerical model was able to reasonably capture the behavior of the GRS pier in the experimental test.

Keywords: Deformation · Geosynthetic · Numerical analysis
Pier · Stress

© Springer Nature Singapore Pte Ltd. 2018
L. Li et al. (Eds.): GSIC 2018, *Proceedings of GeoShanghai 2018 International Conference: Ground Improvement and Geosynthetics*, pp. 374–381, 2018.
https://doi.org/10.1007/978-981-13-0122-3_41

1 Introduction

Geosynthetic-reinforced soil (GRS) structures consisting of closely spaced geosynthetic layers (with vertical spacing smaller than 0.3 m) have been increasingly used to support bridges due to their rapid construction, low construction cost, and effectiveness in eliminating bump at the end of the bridges. Adams et al. [1] pointed out that closely spaced geosynthetic layers and compacted granular fill could provide direct bearing support for structural bridge components if designed and constructed properly.

The US Federal Highway Administration (FHWA) recommends GRS performance tests (also called GRS mini-pier tests) to evaluate the load-deformation behavior of a frictionally connected GRS mass. However, these large-scale GRS performance tests require specialized equipment and knowledge to perform properly [2]. Numerical analysis is an alternative approach to study the behavior of the GRS mass if the numerical model is well calibrated and/or validated.

In this study, a numerical analysis was carried out to simulate the performance of the GRS mini-pier. A three-dimensional (3D) numerical model was developed using FLAC3D, a finite difference method-based program. The Mohr-Coulomb model was used to describe the behavior of backfill soil. Geosynthetic reinforcement was modeled as a linearly elastic material using "geogrid" structural elements. The numerical model used different interfaces to simulate the interaction between different components. The numerical model was calibrated and verified against test results to prove its validity.

2 Overview of GRS Pier Tests

Nicks et al. [3] conducted 19 GRS pier tests to investigate the axial load-deformation relationships of the GRS mass with different reinforcement spacing. For each test, the height to width ratio of the GRS pier was kept at approximately 2 to simulate triaxial test relative dimensions. Concrete Masonry Units (CMU) were used as facing blocks in these tests. In all the tests, layers of woven polypropylene geotextile were used as the reinforcement. When a vertical load was applied on the pier, CMU blocks were removed in 8 of 19 tests, leaving the remaining 11 tests to be loaded with CMU blocks. One of the tests, TF6 with vertical geosynthetic spacing $S_v = 0.2$ m, was chosen for numerical analysis in this study since the 0.2-m reinforcement spacing is most commonly used in actual GRS structures in the field. Virginia Department of Transportation (VDOT) 21A aggregate was used as the backfill soil in TF6. Geotextile with wide-width tensile strength of 70 kN/m at 10% elongation in the machine direction and wide-width tensile strength of 70 kN/m at 8% elongation in the cross-machine direction was used in TF6 as reinforcement material.

3 Numerical Modeling

The finite difference method-based program, FLAC3D (Fast Lagrangian Analysis of Continua), was used to simulate the behavior of the GRS pier. The numerical model contained backfill soil, CMU facing blocks, concrete slab, and geosynthetic layers. Figure 1 shows the numerical model mesh.

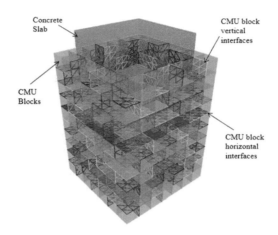

Fig. 1. Numerical model mesh

The soil was modeled as a linearly-elastic perfectly-plastic material with the Mohr-Coulomb failure criterion. An elastic modulus of 25 MPa and Poisson's ratio of 0.25 were used. A friction angle of 53° and a cohesion of 5.5 kPa were adopted in the numerical analysis based on large-scale direct shear test results [3]. A dilation angle of 23° was used based on the empirical relationship of $\psi = \phi - 30°$ from Bolton [4]. The soil was compacted to the density of 2500 kg/m^3 [3]. The geotextile reinforcement was modeled as an isotropic elastic material using "geogrid" structural elements available in the software. The tensile stiffness of the geotextile was assumed to be 700 kN/m based on the wide-width tensile stress of the geotextile at 2% elongation used in the GRS pier tests. The thickness of the geotextile was assumed to be 1 mm and its Poisson's ratio was assumed to be 0.33. Both the CMU facing blocks and the concrete slab on top of the GRS pier were modeled as elastic materials with an elastic modulus E = 2 GPa and Poisson's ratio v = 0.15. A density of 2500 kg/m^3 was used for the concrete slab and an equivalent density of 1230 kg/m^3 was used for the CMU facing blocks to account for two hollow voids in the actual blocks.

Different types of interfaces were used in the numerical analysis. They were represented by interface elements embedded in the software, which are linearly-elastic perfectly-plastic springs with the Mohr - Coulomb failure criterion. An interaction coefficient C_i of 0.6 and zero cohesion were assigned to the geotextile – soil interface considering the smooth surface of the geotextile and the interface friction angle was therefore equal to $\delta = \arctan(0.6 \cdot \tan 53°) = 38.5°$. An interaction coefficient C_i of 0.4

(i.e., $\delta = \arctan(0.4 \cdot \tan 53°) = 28.0°$) and zero cohesion were used for the vertical interfaces between soil and CMU facing blocks. Awad and Tanyu [5] conducted both direct shear and pullout tests to investigate the connection strength between geotextile and facing blocks. Based on their results, a friction angle δ of 16.2° and a cohesion of 9.5 kPa were assigned to the interface between CMU facing blocks and geotextile. A cohesion of 9.5 kPa was used here because some irregularities existing on rough surfaces of concrete blocks could behave like cohesion by preventing the geotextile from moving. A friction coefficient $\mu = 0.6$ was assigned to the interface between CMU facing blocks in the numerical analysis based on the recommendation [6]. In other words, the interface friction angle δ between CMU blocks was equal to $\arctan(\mu) = \arctan(0.6) = 31.0°$. Zero cohesion was assigned to the interface between CMU blocks since no mechanical connection was used in the tests. The horizontal and vertical surfaces of the CMU blocks were assumed to have the same roughness. The soil – slab interface was also considered in this study by assuming that the interface friction angle was equal to the friction angle of the soil.

Apart from the friction angle and the cohesion, the relative interface movement was also controlled by interface normal stiffness k_n and shear stiffness k_s. Itasca Consulting Group (2009) [7] recommended an equation for estimating the maximum interface stiffness values as follows:

$$k_n = k_s \approx 10 \times \max\left(\frac{K + \frac{4}{3}G}{\Delta z_{min}}\right) \tag{1}$$

where the parameters Δz_{min}, K, and G are the smallest dimension in the normal direction, the bulk modulus, and the shear modulus of the continuum zone adjacent to the interface, respectively. This equation was used to estimate the initial input values for the interface stiffness, and these values were subsequently adjusted by comparing the numerical results to the test data and minimizing computation time. Table 1 summarizes the interface properties used in the numerical model. In all cases, the shear stiffness was refined as 1/100 of the normal stiffness to prevent the interface from penetrating into neighboring zones.

Numerical model activation was carried out layer by layer to simulate the construction process. During the construction, CMU facing blocks separated by interfaces were used to provide lateral confinement to the inside GRS mass. Geotextile layers were placed over the backfill soil according to the reinforcement plan. In the numerical model, the reinforcement overlapped 100% of the block depth instead of 85% used in the model test to simplify the interface connection between "geogrid" structural elements and CMU blocks. After the GRS pier was constructed, a concrete slab was put on the top of the pier and a vertical load was applied on the concrete slab according to the loading plan. During the construction and the loading stages, the bottom of the numerical model was fixed in all directions and no fixity was applied to the left, right, front, and back sides of the pier.

Table 1. Interface properties

Interface	Normal stiffness k_n (MPa/m)	Shear stiffness k_s (MPa/m)	Interaction coefficient C_i	Friction coefficient μ	Friction angle δ (°)	Cohesion (kPa)
Soil and geogrid elements	/	3.4	0.6	/	38.5	0
CMU blocks and geogrid elements	/	210	/	/	16.2	9.5
Soil and CMU blocks	340	3.4	0.4	/	28.0	0
CMU blocks interfaces (vertical and horizontal)	21000	210	/	0.6	31.0	0
Soil and concrete slab	21000	210	1.0	/	53.0	0

4   Numerical Results

Figure 2 shows the applied pressure-vertical strain curves from the numerical model and test, indicating that the numerical result agreed with the test result until the applied pressure reached approximately 1500 kPa. The average vertical strain, defined as the total vertical deformation of the concrete slab divided by the height of the pier. The average vertical strain calculated by the numerical model was smaller than that measured in the test when the applied pressure reached comparatively high values. One possible explanation for this difference is that the "geogrid" structural elements in the numerical model had a linearly elastic behavior without a failure limit. Without the tensile failure, the "geogrid" structural element could continue providing tensile resistance to the surrounding backfill soil no matter whether a high applied pressure was applied. However, the experimental tests [3] showed that tensile failure of the geotextile happened when the applied pressure reached to high values. The contribution of the geotextile no longer existed after the tensile failure, causing the larger deformation of the GRS pier.

Figure 3 shows the lateral displacement profile of the CMU blocks along the height of the GRS pier at two different applied pressures. Figure 3 shows that numerical results match the test data reasonably well. As expected, a higher applied pressure resulted in a larger lateral displacement. Both the numerical and test results show that the maximum lateral displacement happened within the upper part to mid-height of the

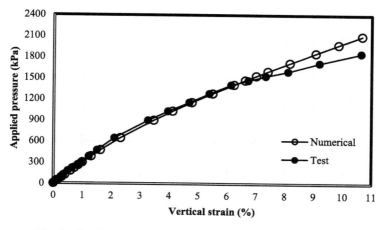

Fig. 2. Applied pressure-vertical strain curves for the GRS pier

GRS pier. The lateral displacement at the top of the GRS pier was smaller than that at the deeper depth because the surface of the concrete slab on top of the GRS pier was relatively rough as compared to the backfill soil – geotextile interface. The bottom of the GRS pier had the smallest lateral displacement due to the restraint at the bottom of the pier from moving laterally.

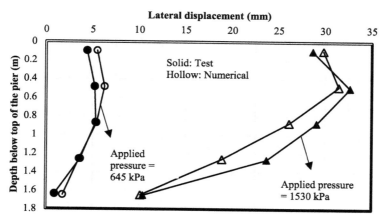

Fig. 3. Lateral displacement profiles of the GRS pier under two applied pressures

Compared to the photo taken in the model test after loading, the numerical model successfully captured the gaps and slippage developing between CMU blocks, indicating that different interfaces used in the numerical model as mentioned in the last section are necessary in order to simulate differential movement between different components (e.g., slippage between the upper and lower layers of CMU facing blocks or gaps between two CMU facing blocks at the same layer).

Figure 4 shows the calculated lateral earth pressure profile behind the CMU facing blocks along the height of the GRS pier under the applied pressure of 180 kPa. Figure 4 also shows the theoretically calculated lateral earth pressure profiles using the Rankine active earth pressure coefficient K_a and the coefficient at rest K_0. At the top of the GRS pier, the lateral earth pressure behind the CMU facing blocks was slightly higher than K_a because the interaction between the concrete slab and the backfill soil reduced the lateral movement of the backfill soil on top of the pier. In the mid-part of the GRS pier, the lateral earth pressure behind the CMU facing blocks was obviously lower than K_a. Since the GRS pier was three-dimensional, the lateral earth pressure was lower than the two-dimensional earth pressure calculated by the Rankine and Coulomb methods, due to the arching effect behind the facing [8]. Another possible reason for this phenomenon is that the backfill soil behind the CMU facing blocks was limited. The theoretical methods commonly used for estimating active lateral earth pressure are inappropriate since the thrust wedge (i.e., failure plane) cannot fully develop in the shape and size predicted by these methods [9, 10]. In other words, the GRS pier in this study was similar to a limited-space retaining wall and the failure plane was not able to fully develop due to lack of space. As a result, lower lateral earth pressure behind the CMU blocks developed. At the bottom of the GRS pier, the lateral earth pressure was close to and yet higher than K_0 since the bottom the GRS pier was fully fixed and no movement was allowed in the numerical model.

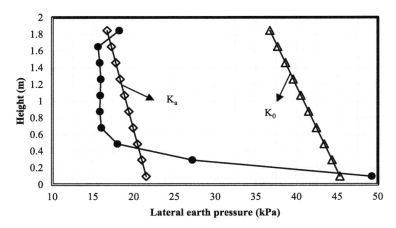

Fig. 4. Calculated lateral earth pressure behind CMU facing blocks along the height of the GRS pier under the applied pressure of 180 kPa

5   Conclusions

This study used a three-dimensional finite difference method-based program, FLAC3D, to simulate the performance of a geosynthetic-reinforced soil (GRS) pier under vertical loading. The numerical model was calibrated and verified against the test results using the applied pressure-vertical strain curve and the lateral displacement profile along the height of the GRS pier. This paper also presents the calculated lateral earth pressure profile behind facing blocks along the height of the GRS pier. The lateral earth pressure behind the CMU facing blocks was mostly lower than the active earth pressure coefficient calculated using the Rankine method due to both the three-dimensional and limited space effects.

Acknowledgements. This paper was completed while the first author visited the University of Kansas, USA as a visiting scholar, which was supported by the China Scholarship Council (No. 201506260117). Dr. Barry R. Christopher provided some guidance in the numerical analysis. The authors would like to appreciate the financial support and technical help.

References

1. Adams, M., Nicks, J., Stabile, T., Wu, J., Schlatter, W., Hartmann, J.: Geosynthetic Reinforced Soil Integrated Bridge System, Synthesis report. FHWA-HRT-11-027 (2011)
2. Nicks, J.E., Esmaili, D., Adams, M.T.: Deformations of geosynthetic reinforced soil under bridge service loads. Geotext. Geomembr. **44**(4), 641–653 (2016)
3. Nicks, J.E., Adams, M., Ooi, P., Stabile, T.: Geosynthetic reinforced soil performance testing - axial load deformation relationships. FHWA-HRT-13-066 (2013)
4. Bolton, M.: The strength and dilatancy of sands. Geotechnique **36**(1), 65–78 (1986)
5. Awad, M.I., Tanyu, B.F.: Laboratory evaluation of governing mechanism of frictionally connected MSEW face and implications on design. Geotext. Geomembr. **42**(5), 468–478 (2014)
6. British Standards Institution: Eurocode 2: Design of Concrete Structures: Part 1-1: General Rules and Rules for Buildings. British Standards Institution, London (2004)
7. Itasca Consulting Group: FLAC3D Theory and Background. Itasca Consulting Group Inc. (2009)
8. tom Wörden, F., Achmus, M.: Numerical modeling of three-dimensional active earth pressure acting on rigid walls. Comput. Geotech. **51**, 83–90 (2013)
9. Leshchinsky, D., Hu, Y.H., Han, J.: Limited reinforced space in segmental retaining walls. Geotext. Geomembr. **22**(6), 543–553 (2004)
10. Greco, V.: Active thrust on retaining walls of narrow backfill width. Comput. Geotech. **50**, 66–78 (2013)

Numerical Simulation Analysis
of Geogrid-Reinforced Embankment
on Soft Clay

Yuanyuan Zhou, Zhenming Shi, Qingzhao Zhang, and Songbo Yu[✉]

Department of Geotechnical Engineering, Tongji University,
Shanghai 200092, China
xsun@lsu.edu

Abstract. Based on a previous centrifugal test, the finite-difference program FLAC3D was employed to simulate the centrifugal test and the prototype embankment corresponding to this centrifugal test. The variations of embankment settlements, reinforcement tensile forces and horizontal displacements were analysed. The results show that the computed displacements and reinforcement tensile forces are in good agreement with the measured ones, indicating the appropriate numerical model. Reinforcement can reduce the horizontal displacement of foundation. The settlement and the difference of the prototype embankment, away from the center, are less than those of centrifugal model. It is shown that the numerical model for the prototype can fully consider the reinforcement effect, which will help to diffuse the embankment load and homogenize the shallow stress of foundation.

Keywords: Soft clay foundation · Geogrid · Centrifugal test
Prototype embankment · Numerical simulation

1 Introduction

Due to the high compressibility, high water content, high sensitivity, low strength and low-permeability of soft clays, analysis of deformation characteristics and stability computation of foundation have been tough problems in design and construction. Therefore, great importance has been attached to them in construction projects. Owing to its tensile properties, geogrid transfers stress through friction and interlocking in order to increase the strength and toughness of soil. Its application to the construction of embankment will not only mitigate the lateral displacement and differential settlement, but also enhance its integrity and stability. Consequently, geogrid has been widely applied in construction projects at home and abroad.

As for the reinforced embankments over soft foundations, many researchers [1–4] have conducted studies on the impact of foundation depth, reinforcement types, different drainage conditions and various reinforcement arrangements on characters, deformation and stability of embankment through centrifugal model test. Based on the numerical simulation of centrifugal test, some scholars analyzed embankment displacements, earth pressures, pore water pressures, reinforcement tensile forces and their variations with

© Springer Nature Singapore Pte Ltd. 2018
L. Li et al. (Eds.): GSIC 2018, *Proceedings of GeoShanghai 2018 International Conference: Ground Improvement and Geosynthetics*, pp. 382–389, 2018.
https://doi.org/10.1007/978-981-13-0122-3_42

time. They also investigated the performance and stability of reinforced embankment on soft clay foundations. The results show that reinforcement will not only improves stability but also efficiently reduces the lateral displacement of embankment [5, 6].

In this paper, a finite difference software FLAC3D which is suitable for solving nonlinear large deformation problems was employed to simulate the centrifugal test based on the test dimensions and conditions. Meanwhile, in order to investigate the correlation between variable acceleration loading of centrifugal test and step loading of prototype embankment, the numerical simulation of prototype embankment corresponding to centrifugal test has also been carried out to conduct contrastive analysis on the differences of embankment displacement, earth pressure, reinforcement tensile force between numerical calculation results and those obtained from tests, as well as the differences of numerical results between model tests and prototype embankment.

2 Numerical Calculation Model

2.1 Geometric and Constitutive Model

The prototype for centrifugal model test are embankment made of lime-stabilized soil (with the height of 6 m, the crest width of 13 m and the slope ratio of 1:1); soft silty clay foundation with the thickness of 25 m; sand cushion with the thickness of 50 cm; prefabricated vertical drains arranged in quincunx pattern at a center-to-center spacing of 2 m. According to the dimensions in centrifugal model test [7] (Fig. 1), geometric model is built on the basis of one half of the embankment (Fig. 2).

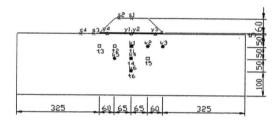

Fig. 1. Model dimensions (unit: mm) [7]   **Fig. 2.** Geometric model of calculation

Mohr-Coulomb Model has been used to model the embankment and sand cushion in this paper and basic parameters are shown in Table 1. As for the foundation soil, modulus expression of Duncan-Chang Nonlinear Model [8] substitutes for constant modulus value, that is

$$E_t = [1 - R_f \frac{(1 - \sin \varphi')(\sigma_1 - \sigma_3)}{2c' \cos \varphi' + 2\sigma_3 \sin \varphi'}]^2 kp_a(\frac{\sigma_3}{p_a})^n$$

where c' and φ' are effective cohesion and effective internal friction angle respectively; R_f is failure ratio; k and n are dimensionless test constants; p_a is atmospheric pressure.

Table 1. Properties of the embankment and sand

Material	E/MPa	v	γ/(kN $\cdot$ m^{-3})	C/kPa	$\varphi/°$	K_v/($\times 10^{-4}$cm . s^{-1})	K_h/($\times 10^{-4}$cm . s^{-1})	n
Embankment	425	0.2	19.1	109	35	-	-	-
Sand	3	0.3	12.4	0	30	1.16	1.16	0.4

Note: E = Young's modulus; v = water content; γ = unit weight; C = cohesion; φ = friction angle; K_v = vertical permeability; K_h = horizontal permeability; n = porosity.

Table 2. Material parameters of the foundation soil

Material	n	k	R_f	γ(kN $\cdot$ m^{-3})	c'/ kPa	$\varphi'/°$	K_v/($\times 10^{-6}$cm . s^{-1})	K_h/($\times 10^{-6}$cm . s^{-1})	n^*
Foundation soil	1.49	69.5	1	13.0	8.3	23.4	2.35	2.35	0.43

Note: n^* = porosity, *indicates the difference between the parameter n in Duncan-Chang Model.

Table 3. Material parameters of the geogrid

Object	E/Pa	v	t/mm	c/kPa	$\varphi/°$	k/Pa
Model	1.42e+07	0.33	1	0	27.1	3.60e+06
Prototype	3.55e+08	0.33	4	0	27.1	3.60e+06

Note: E = Young's modulus; v = Poisson's ratio; t = thickness; c = coupling spring cohesion; φ = coupling spring friction angle; k = stiffness of unit area of coupling spring.

Material parameters can be obtained through consolidated-undrained triaxial compression and permeability tests, which are shown in Table 2. The geogrid element inherent in FLAC3D was used to simulate the model geogrid. Table 3 summarizes the values of the geogrid parameters.

2.2   Calculating Process

The process of increasing centrifugal accelerations can be achieved through FISH language programming at three acceleration stages (33.3, 66.7 and 100.0 g) as illustrated in Fig. 3 [6].

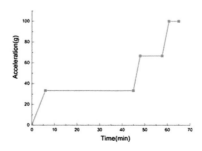

Fig. 3. Relationship between acceleration and time [6]

Since the affinity constant of permeability coefficient k is N, during the numerical simulation of centrifugal model test, the permeability coefficient k should be adjusted according to instant acceleration. Besides, the seepage coupling between water and soil skeleton should be taken into consideration during loading and resting phases.

3   Analysis of Numerical Results

3.1   Numerical Simulation of Centrifugal Model Test

The model should be modified by means of subtracting the physical quantity of each point in inconstant acceleration of foundation with no embankment from numerical calculation results. The modified settlements are shown in Fig. 4. The numerical results basically resembled those in centrifugal model test.

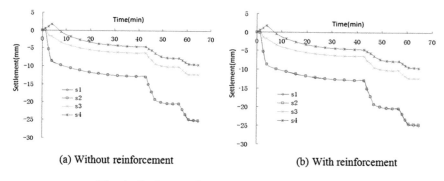

(a) Without reinforcement   (b) With reinforcement

Fig. 4.   Settlements in variable acceleration loading

Figure 5 shows the variation of reinforcement tensile forces collected from three representative positions: the center of embankment, beneath the road shoulder and midpoint of slope. Tensile forces increased with increasing acceleration and also increased during resting phases due to the continuous deformation of foundation.

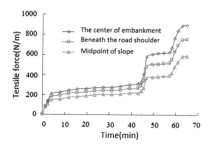

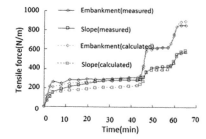

Fig. 5.   Reinforcement tensile forces in variable acceleration loading

Fig. 6.   Comparison of calculated and measured reinforcement tensile forces

Figure 6 has compared the calculated values of tensile force in the center of embankment and at the midpoint of slope with measured values of tensile force at point y2 and y3 in centrifugal test. Every phases of the curve in Fig. 6 are ideally matched except one: the first resting phase.

Four sections have been selected to perform a horizontal displacement analysis: road shoulder, midpoint of slope, 1 cm and 8 cm outside slope toe (see Fig. 7). The soil of different depths outside the slope toe all have moved outwards and the displacement value reached the maximum at depth of 5 cm. The soil above the road shoulder and midpoint of slope have moved towards the center of embankment due to differential settlement while the soil beneath have moved outwards. The maximum displacement is at depth of 15 cm. The displacement outside slope toe can be slightly mitigated through the reinforcements.

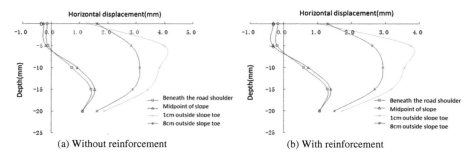

(a) Without reinforcement   (b) With reinforcement

Fig. 7. Horizontal displacements in variable acceleration loading

Four horizontal planes have been selected to perform stress increment analysis (see Fig. 8): 1.25 cm beneath earth surface, 6.25 cm beneath earth surface, 11.25 cm beneath earth surface and 16.25 cm beneath earth surface. Stress concentration appeared in road shoulder of shallow soil layer, which can be mitigated through reinforcements. Stress increments decreased with depth near the center while farther from the center stress increments appeared relatively small and increased with depth.

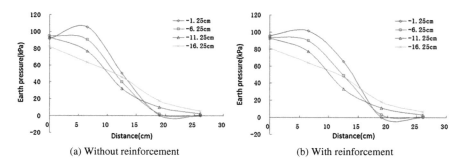

(a) Without reinforcement   (b) With reinforcement

Fig. 8. Stress increments in variable acceleration loading

3.2   Numerical Simulation of Prototype Embankment

Numerical model, which have been validated in numerical simulation of centrifugal model test, should be applied. Besides, the constitutive model and soil parameters remain unchanged. Embankment heaped load is in correspondence with staged load curve in Fig. 3, the height of which should be divided into three phases: 2 m, 4 m and 6 m. Each heaped load required 1 month and the resting phase lasted 9 months, 4 months and 1 month respectively.

Figure 9 showed the correlation between prototype settlement and time. The location of each point should be in correspondence with displacement measured points of centrifugal test in Fig. 1. The curve pattern nearly resembled that of model test [7]. The maximum settlements without and with reinforcement reached 1.44 m and 1.38 m respectively. The decreasing ratio was confined to 4% due to the reinforcements, smaller than measured decreasing ratio (8%) and larger than simulative decreasing ratio (1%). Figure 10 showed the variation of reinforcement tensile forces, which basically resembled the patterns in Fig. 5. However, the rapid increase of tensile forces at midpoint of embankment during the first resting phase strengthened the discreteness of tension distribution. Consequently, the tensile forces were larger than the measurements and calculated values in centrifugal test.

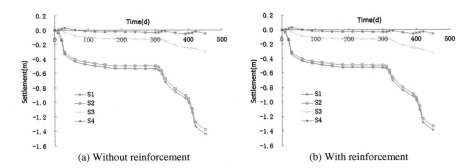

(a) Without reinforcement   (b) With reinforcement

Fig. 9. Settlements in step loading

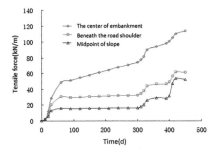

Fig. 10. Reinforcement tensile forces in step loading

Figures 11 and 12 present horizontal displacements and stress increments, which are in correspondence with Figs. 7 and 8 obtained on the basis of centrifugal model test. The curve pattern in Fig. 12 is basically in accordance with that in Fig. 8. However, near the center, the stress increment of shallow soil layer increased while the stress increment of deep soil layer decreased. Prominent difference occurred in the horizontal displacement curves. The surface soil and deep soil at road shoulder and midpoint of slope moved outwards while the interlayer soil moved inwards. As for the sections 1 m and 8 m outside the slope toe, their horizontal displacement curves intersected, proving that soil compression occurred between two sections 10 m or more beneath earth surface. After conversion, the maximum horizontal displacements were smaller than the measurements obtained from the model test. Considering the homogenization of stress distribution, reinforcements impacted more on prototype rather than the model test.

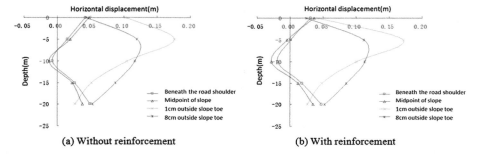

(a) Without reinforcement   (b) With reinforcement

Fig. 11. Horizontal displacements in step loading

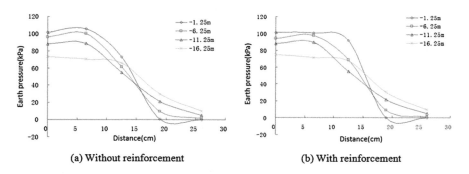

(a) Without reinforcement   (b) With reinforcement

Fig. 12. Stress increments in step loading

4   Conclusion

In order to study the impact of reinforcement on embankment performance on soft clay foundations, the centrifugal model test and prototype embankment are numerically simulated in this paper. The main conclusions are as follows:

(1) The calculated results of settlements, soil pressures and tensile forces of rein-forcements, obtained from numerical simulation with the adoption of FLAC3D have been in accordance with the results obtained from centrifugal test, proving the rationality of numerical model and parameter selection.

(2) The tensile force of reinforcement obtained through calculation displayed a trend of homogeneity first and differentiation later. Reinforcement can reduce the horizontal displacements of foundation and the stress concentration of road shoulder of shallow soil layer. Stress increments decreased with depth near the center while farther from the center stress increments appeared relatively small and increased with depth.

(3) Remarkable differences occurred between the calculated results of horizontal displacement for prototype and that for model test. Reinforcement impacted more on prototype rather than the model by promoting the transmission of embankment load and homogenizing stress on shallow foundation.

Acknowledgements. This work is supported by National Key R&D Program of China (NO. 2017YFC0806003), Fundamental Research Funds for the Central Universities, National Natural Science Foundation of China (NO. 41602287), Key Project of Natural Science Foundation of China (NO. 41731283) and National Natural Science Foundation of China (NO. 41372272).

References

1. Mandal, J.N., Joshi, A.A.: Centrifuge modelling of geosynthetic reinforced embankments on soft ground. Geotext. Geomembr. **14**(2), 147–155 (1996)
2. Viswanadham, B.V.S., Mahajan, R.: Modeling of geotextile reinforced highway slopes in a geotechnical centrifuge. In: Geotechnical Engineering for Transportation Projects, pp. 637–646 (2004)
3. Chen, J.F., Yu, S.B.: Centrifuge modeling of a geogrid-reinforced embankment with lime-stabilized soil as backfill on soft soil. Bull. Eng. Geol. Environ. **68**(4), 511–516 (2009)
4. Zhang, J., Huang, X.: Centrifuge model tests and numerical simulation of geosynthetic-reinforced embankment for expressway widening. J. Highw. Transp. Res. Dev. **28**(4), 1–5 (2011)
5. Sharma, J.S., Bolton, M.D.: Centrifugal and numerical modelling of reinforced embankments on soft clay installed with wick drains. Geotext. Geomembr. **19**(1), 23–44 (2001)
6. Chen, J.F., Yu, S.B.: Centrifugal and numerical modeling of a reinforced lime-stabilized soil embankment on soft clay with wick drains. Int. J. Geomech. **11**(3), 167–173 (2011)
7. Chen, J.F., Yu, S.B., Ye, T.F., et al.: Centrifuge test on reinforced embankment with lime-stabilized soil as backfill on soft clay. Chin. J. Rock Mech. Eng. **02**, 287–293 (2008). (in Chinese)
8. Zhang, Y., Xue, Y.Q., Wu, J.C., et al.: Parameters of Duncan-Chang model for hydrostratigraphic units in Shanghai City. Hydrogeol. Eng. Geol. **35**(1), 19–22 (2008). (in Chinese)

Experimental Study on Geocell-Stabilized Unpaved Shoulders

Jun Guo[1(✉)], Jie Han[2], Steven D. Schrock[2], Robert L. Parsons[2], and Xiaohui Sun[3]

[1] Shenzhen University, Shenzhen 518048, Guangdong, China
guoddx@live.com
[2] The University of Kansas, Lawrence, KS 66045, USA
{jiehan, rparsons, schrock}@ku.edu
[3] Louisiana Transportation Research Center, Baton Rouge, LA 70808, USA
lsun@lsu.edu

Abstract. Two-lane highways often consist of aggregate or turf shoulders, which require maintenance on a recurring basis. Rutting and edge drop-offs are the most common performance problems. The maintenance of unpaved shoulders is typically done by placing and compacting more geomaterial. This practice is considered temporary and does not address the cause of the problem; therefore, the problem often recurs. On the other hand, the typical materials used for constructing shoulders contain little organic matter, which makes vegetation growth on these shoulders difficult. Vegetation on the shoulder can prevent erosion due to water runoff and wind, thus it is desirable to maintain a healthy vegetation cover on the shoulders. The previous study showed that an aggregate and topsoil mixture had the same ability to sustain vegetation as a topsoil. The addition of the topsoil into the aggregate lowered the strength of the aggregate, thus geocell could be used to improve the performance of the mixture. This study aimed to evaluate the performance of aggregate, aggregate-topsoil mixture, and geocell-stabilized aggregate-topsoil mixture as the geomaterials for constructing shoulders. Three large-scale cyclic plate loading tests were conducted on sections of 200 mm thick base courses over 5% CBR subgrades (moderate subgrade). The base courses consisted of: (a) 200 mm thick aggregate, (b) 200 mm thick aggregate-topsoil mixture, and (c) 50 mm thick aggregate-topsoil mixture over 150 mm thick geocell-stabilized aggregate-topsoil mixture. The total and permanent deformations of the base course surface under the loading plate were monitored by the actuator. The permanent deformations were used to evaluate the performance of different test sections.

Keywords: Aggregate · Base course · Geocell · Performance
Unpaved shoulder

1 Introduction

Shoulders of highways provide space for vehicles to stop and structural support for roadways. In rural areas, unpaved shoulders are commonly constructed with aggregate or turf (i.e., native topsoil) along two-lane highways for a design traffic volume of 0 to

© Springer Nature Singapore Pte Ltd. 2018
L. Li et al. (Eds.): GSIC 2018, *Proceedings of GeoShanghai 2018 International Conference: Ground Improvement and Geosynthetics*, pp. 390–398, 2018.
https://doi.org/10.1007/978-981-13-0122-3_43

874 Annual Average Daily Traffic (AADT) according to the Kansas Department of Transportation. Typically, aggregate-surfaced shoulders have a uniform aggregate layer of 150 mm thick. Turf shoulders are constructed by bringing the grade up to the pavement edge. In Kansas, unpaved shoulders are typically constructed on compacted natural soil subgrade of a California Bearing Ratio (CBR) value at 3% to 4%. Sometimes unpaved shoulders are also constructed on lime-stabilized subgrade of a CBR value ranging from 5% to 10%. In rural areas, heavy vehicles, such as combines, tractors, and grain wagons, often stop on unpaved shoulders to allow following, faster-moving vehicles to pass. These heavy vehicles can induce deformations and excessive rutting on the unpaved shoulders. The primary performance problems of unpaved shoulders are rutting and edge drop-offs. The typical practice of unpaved shoulder maintenance is placing and compacting more material, which does not address the cause of the problems and is considered temporary [1]. The same performance problem often recurs.

On the other hand, vegetation on unpaved shoulders is desirable for their esthetic and environmental value. Moosmuller et al. [2] found vehicles of large size or poor aerodynamics traveling at speeds ranging from 80 km/h to 100 km/h could result in high turbulent kinetic densities (approximately 10 N/m^2 or higher). The high turbulent kinetic energy density can cause significant fugitive dust re-suspension and make the unpaved shoulders a major contributor to PM10. Youssef et al. [3] found vegetation cover significantly reduced windblown mass loss from soil. Vegetation cover can also effectively reduce soil erosion caused by water runoff [4, 5]. Traditionally aggregate that is used for constructing shoulders contains little organic matter and is difficult to sustain vegetation. A simple solution to improve the vegetation growth on a shoulder is to construct the shoulder with a mixture of high organic content soil and aggregate. Guo et al. [6] compared vegetation growth on compacted topsoil, aggregate, and aggregate-topsoil mixture (1:1 by dry weight) and found the compacted aggregate surface was a harsh environment for the establishment of vegetation and the aggregate-topsoil mixture showed equal ability in sustaining vegetation as the topsoil. However, previous research suggested that adding a high organic content topsoil into the aggregates lowered the strength of the aggregate. The organic matters in a geo-material would result in high compressibility, high plasticity, high shrinkage, low strength, and large creep deformation [7]. Mitchell and Soga [8] concluded that decomposed organic matters could reduce the stiffness and undrained strength as a result of the higher water content and plasticity contributed by the organic matter. The conflict between the necessity of organic matter in a geomaterial to maintain vegetation and the strength requirement for the material to support traffics makes the aggregate-topsoil mixture as a shoulder construction material difficult.

It was proposed to construct the shoulders with geocell-stabilized aggregate-topsoil mixture to mitigate the shoulder performance problems while maintaining a vegetation cover on the shoulders. Geocell is a three-dimensional geosynthetic product, which has been widely used for erosion control on steep slopes, soil stabilization of earth retention structures, revetment and flexible channel lining, and roadway load support [9]. Giroud and Noriay [10] stated the major reinforcing mechanisms of geosynthetics for unpaved roads are lateral confinement, increased bearing capacity, and tensioned membrane effect. Pokharel [11] pointed out that geocells could provide confinement by (a) the friction between geocell wall and infill material and (b) restraining of the upward and lateral movement of infill material under loading. In the past, experimental and numerical studies investigated bearing capacity increase and deformation reduction by geocells with different infill materials. Han et al. [12] found a 65% increase in the bearing capacity of sand as a result of geocell stabilization. Pokharel et al. [13] found a 170-mm thick geocell-stabilized crushed-stone section outperformed a 300-mm thick section with the same material in accelerated moving-wheel tests. Thakur et al. [14] showed the inclusion of geocell significantly increased the strength and stiffness of the recycled asphalt pavement (RAP) base. They concluded that the geocell-stabilized RAP base distributed the load to a wider area and reduced the permanent deformation under cyclic loading.

This paper presents the test results of three large-scale cyclic plate loading tests on a geocell-stabilized aggregate-topsoil mixture shoulder versus a traditional aggregate shoulder and an aggregate-topsoil mixture shoulder. The traditional aggregate test section consisted of a 200-mm aggregate base course. The control test section consisted of a 200-mm aggregate-topsoil mixture base. The geocell-stabilized test section consisted of a 50-mm thick aggregate-topsoil mixture layer over a 150-mm thick geocell-stabilized mixture base. All three shoulders were constructed over a 5% CBR subgrade. The permanent deformations of all three test sections under cyclic loading were measured and used for evaluating the performance of these test sections.

2 Test Material and Setup

2.1 Materials

Subgrade. The subgrade material used in the cyclic loading tests were a mixture of 25% kaolin and 75% Kansas River sand (by dry weight). This subgrade material had a maximum dry density of 2.0 Mg/m^3 at the optimum water content of 10.9% based on the modified Proctor method (ASTM D1557) [16].

AB-3 Aggregate. The aggregate used in the aggregate-topsoil mixture was a well-graded aggregate. This aggregate commonly used in the state of Kansas is named the AB-3 aggregate. The AB-3 aggregate contained less than 10% fine. The optimum water content was approximately 11.5% and its corresponding maximum dry density was 2050 kg/m^3 based on the modified Proctor method.

Topsoil. The topsoil was an organic silt excavated from the University of Kansas west campus, which appeared in a black color. The topsoil had a plastic limit of 34 and a plasticity index of 15. The topsoil had an optimum water content of 21% and its corresponding maximum dry density was approximately 1570 kg/m^3 based on the modified Proctor tests. Based on the ASTM standard for soil volatile organic compounds (ASTM D4547), the topsoil had an organic content of 19%.

Aggregate-Topsoil Mixture. The aggregate-topsoil mixture consisted of 50% AB-3 aggregate and 50% topsoil by dry weight. This material will be referred to as a "mixture" in the following sections. This mixture had an optimum water content of

15% and its maximum dry density was approximately 1830 kg/m^3 based on the modified Proctor method.

Geocell. The geocell used in this study was manufactured by Presto Geosystems. The geocell was made of polyethylene with a density between 0.935 and 0.965 g/cm^3. The height of the cell was 150 mm. Figure 1 presents the detailed schematic of the geocell including holes on the cell wall.

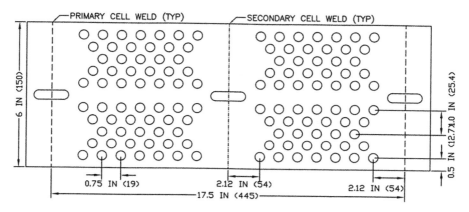

Fig. 1. Schematic of the geocell [15]

3 Test Setup

The large-scale cyclic loading tests were conducted in the large geotechnical box at the University of Kansas. The box has a dimension of 2.2 m × 2 m × 2 m (L × W × H). The load was applied by a hydraulic actuator in a trapezoidal waveform as shown in Fig. 2 to simulate the traffic loading by heavy vehicles. The maximum load was 40 kN and the minimum load was 0.5 kN. The load was applied to the test sections through a 300-mm diameter steel loading plate with a 12.5-mm thick rubber base. At the peak load

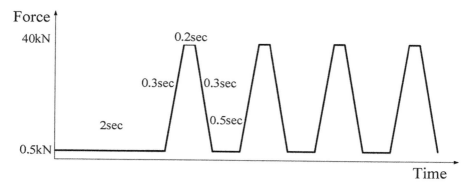

Fig. 2. Cyclic loading waveform

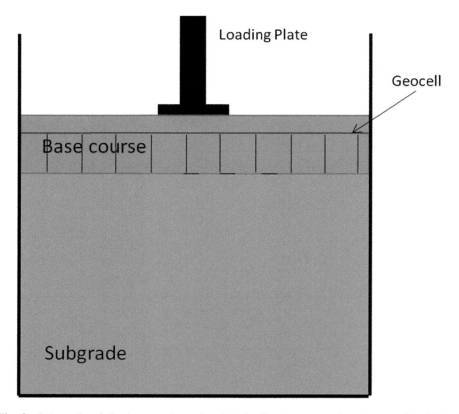

Fig. 3. Schematic of the large-scale cyclic plate loading test setup for the geocell-stabilized section

of 40 kN, the loading plate exerted a pressure of 550 kPa on the surface of the test section. Figure 3 presents a schematic of the test setup.

The subgrade of the test sections was constructed in six 150-mm thick lifts. After the subgrade material was placed and leveled in the box, an electrical vibratory compactor was used to compact the subgrade. After each lift was compacted, vane shear tests were conducted to verify the subgrade at the target CBR value of 5%.

After the preparation of the subgrade, the base materials were adjusted to their optimum water content so that a 90% or greater relative compaction could be achieved. The base courses of the non-stabilized sections were constructed in two 100-mm lifts. For each lift, the base material was first leveled and then compacted with the electrical vibratory compactor. For the geocell-stabilized section, a layer of woven geotextile was first placed over the subgrade. Then, the geocell was streched and placed over the geotextile. The edges of the geocell were fixed with rebars inserted into the subgrade. Then, the first 100 mm of mixture was placed over the geocell. The first lift of mixture was compacted with a pneumatic compactor so that the mixture in each cell was compacted evenly. The second 100 mm lift of mixture was compacted with the electrical vibratory compactor.

The constructed test sections were left overnight and covered with a plastic sheet before the cyclic plate loading tests. Each cyclic loading test was terminated if (a) the permanent deformation exceeded 75 mm or (b) the number of loading cycles exceeded 25,000. None of the three test sections resulted in permanent deformation exceeding 75 mm within 25,000 loading cycles. Due to the seating problem, the first loading cycle generated large permanent deformations on all the test sections. Thus, the deformations resulted from the first loading cycle were considered as preconditioning of the test section and were not counted into the permanent deformation calculation.

4   Test Results and Discussions

As presented earlier, the aggregate-topsoil mixture had an optimum water content and maximum dry density between those of the aggregate and the topsoil. Based on the dynamic cone penetrator results, the compacted aggregate base course had a CBR value of 11%, while the mixture base course had a CBR value of 6%. The geocell-stabilized base course had a CBR value of 7.3% which was between the values of the aggregate and mixture base courses. The different CBR values between the non-stabilized and geocell-stabilized base course may be contributed by the different compaction methods and the confinement effect of geocell during compaction. Figure 4 presents the permanent deformations of these three test sections.

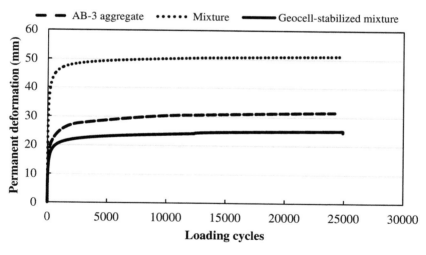

Fig. 4. Permanent deformations of the test sections

Figure 4 shows that the aggregate-topsoil mixture section had the largest amount of permanent deformation, while the geocell-stabilized mixture section had the least amount of permanent deformation. Because the aggregate-topsoil mixture had a higher fine content, organic content, and water content, the permanent deformation observed on the mixture section without geocell was almost twice that of the aggregate section.

However, the permanent deformation of the geocell-stabilized mixture section was less than half of that of the non-stabilized mixture section. The difference in permanent deformations between the non-stabilized and geocell-stabilized mixture base course indicates the effectiveness of geocell in limiting the development of permanent deformation. Based on Pokharel [17] the geocell can provide confinement in two ways: (1) friction between cell wall and soil, and (2) the restraint of the infill soil upwards and lateral movement under loading. In the cyclic loading tests, the geocell effectively restraint the soil movement thus limited the development of permanent deformation. The permanent deformation of the AB-3 aggregate section was slightly larger than that of the geocell-stabilized mixture section. At 10,000 cycles, the permanent deformation of the aggregate-topsoil mixture section was over twice that of the geocell-stabilized mixture section while the permanent deformation of the AB-3 aggregate section was 26% larger than the geocell-stabilized mixture section. The permanent deformation development of the geocell-stabilized aggregate-topsoil mixture showed the effect of geocell confinement in improving the performance of the mixture. Since the geocell-stabilized aggregate-topsoil mixture showed slightly less permanent deformation than the aggregate section, the geocell-stabilized aggregate-topsoil mixture can replace the aggregate material as a shoulder base material that can sustain vegetation as well as satisfy the requirement of traffic loading.

To evaluate the benefit of the geocell stabilization, an improvement factor is defined as the ratio of the number of cycles for the geocell-stabilized test section to that without geocell (i.e., the aggregate or mixture section) at a specific permanent deformation:

$$\text{Improvement factor}, I = \frac{N_{\text{stabilized}}}{N_{\text{non-stabilized}}} \tag{1}$$

In this study, the improvement factors were calculated at several permanent deformations of the geocell-stabilized mixture section to that of the aggregate section (I_a) or the non-stabilized mixture section (I_m). Table 1 presents the improvement factors (I_a and I_m) at the permanent deformations of 5, 10, 15, and 20 mm.

Table 1. Numbers of load cycles and improvement factors at different permanent deformations

Permanent deformation (mm)	5	10	15	20
N (Geocell-stabilized mixture)	8	32	124	709
N (Non-stabilized aggregate)	7	25	79	252
I_a		1.1	1.3	1.6	2.8
N (Non-stabilized mixture)	3	5	8	15
I_m		2.7	6.4	15.5	47.3

Table 1 shows that the geocell-stabilized aggregate-topsoil mixture section performed slightly better than the aggregate section at small permanent deformations (i.e., less than 15 mm permanent deformation), where the improvement factor I_a was less than 2. The improvement factor increased with the increase of the permanent deformation. At the 20 mm permanent deformation, the improvement factor of the geocell-stabilized mixture section to the AB-3 aggregate section was 2.8. The geocell-stabilized mixture significantly limited the permanent deformation compared to the non-stabilized mixture base course. At the permanent deformation of 10 mm, the improvement factor of the geocell-stabilized section to the non-stabilized mixture section was 6.4. At the 20 mm permanent deformation, the improvement factor increased to 47.3 relative to the non-stabilized mixture section. For both cases, the improvement factors increased as the permanent deformation increased.

5 Conclusions

This study investigated three base courses: (a) 200 mm thick aggregate, (b) 200 mm thick aggregate-topsoil mixture, and (c) 50 mm thick aggregate-topsoil mixture over 150 mm thick geocell-stabilized aggregate-topsoil mixture, as possible base materials for unpaved shoulders. Three large-scale cyclic plate loading tests were conducted in a large test box. Based on the evaluation of the permanent deformations of these test sections, the following conclusions can be made:

1. The non-stabilized aggregate-topsoil mixture section had significantly larger permanent deformations under cyclic loading than the aggregate and geocell-stabilized mixture sections.
2. The geocell stabilization significantly improved the performance of the aggregate-topsoil mixture. The inclusion of geocell in the mixture greatly reduced the permanent deformation.
3. The geocell-stabilized aggregate-topsoil mixture had smaller permanent deformations under cyclic loading than the aggregate base.
4. The geocell was more effective at a larger permanent deformation. The improvement factors of the geocell-stabilized section over other two non-stabilized sections increased as the permanent deformation increased.
5. The geocell-stabilized aggregate-topsoil mixture can be an alternative to the aggregate base for unpaved shoulder construction.

Acknowledgments. This research was sponsored by the Kansas Department of Transportation (KDOT). Mr. Jonathan Marburger was the project monitor at KDOT. Presto Geosystems supplied the geocell used in this study. Hamm Inc. from Perry, Kansas supplied the AB-3 aggregate material. Mr. Byron Whitted and Mr. Lee Crippen, the undergraduate research assistants at the University of Kansas, provided great assistance to the experimental tests. All the above support and help are greatly appreciated.

References

1. Mekkawy, M.M., White, D.J., Jahren, C.T., Suleiman, M.T.: Performance problems and stabilization techniques for granular shoulders. J. Perform. Constr. Facil. ASCE **24**(2), 159–169 (2010)
2. Moosmüller, H., Gillies, J.A., Rogers, C.F., DuBois, D.W., Chow, J.C., Watson, J.G., Langston, R.: Particulate emission rates for unpaved shoulders along a paved road. J. Air Waste Manag. Assoc. **48**(5), 398–407 (1998)
3. Youssef, F., Visser, S.M., Karssenberg, D., Erpul, G., Cornelis, W.M., Gabriels, D., Poortinga, A.: The effect of vegetation patterns on wind-blown mass transport at the regional scale: a wind tunnel experiment. Geomorphology **159**, 178–188 (2012)
4. Zhou, Z.C., Shangguan, Z.P., Zhao, D.: Modeling vegetation coverage and soil erosion in the Loess Plateau area of China. Ecol. Model. **198**(1), 263–268 (2006)
5. Istanbulluoglu, E., Bras, R.L.: Vegetation-modulated landscape evolution: effects of vegetation on landscape processes, drainage density, and topography. J. Geophys. Res.: Earth Surf. **110**(F2) (2005). https://doi.org/10.1029/2004JF000249
6. Guo, J., Han, J., Schrock, S.D., Parsons, R.L.: Field evaluation of vegetation growth in geocell-reinforced unpaved shoulders. Geotext. Geomembr. **43**(5), 403–411 (2015)
7. Han, J.: Principles and Practice of Ground Improvement. Wiley, Hoboken (2015)
8. Mitchell, J.K., Soga, K.: Fundamentals of Soil Behavior, 3rd edn. Wiley, Hoboken (2005)
9. Yuu, J., Han, J., Rosen, A., Parsons, R.L., Leshchinsky, D.: Technical review of geocell-reinforced base courses over weak subgrade. In: First Pan American Geosynthetics Conference, Cancun, Mexico, pp. 2–5 (2008)
10. Giroud, J.P., Noiray, L.: Geotextile-reinforced unpaved road design. J. Geotech. Div. ASCE **107**(GT9), 1233–1254 (1981)
11. Pokharel, S.K.: Experimental study on geocell-reinforced bases under static and dynamic loading. Ph.D. dissertation, The University of Kansas (2010)
12. Han, J., Yang, X., Leshchinsky, D., Parsons, R.: Behavior of geocell-reinforced sand under a vertical load. Transp. Res. Rec.: J. Transp. Res. Board **2045**, 95–101 (2008)
13. Pokharel, S., Han, J., Manandhar, C., Yang, X., Leshchinsky, D., Halahmi, I., Parsons, R.: Accelerated pavement testing of geocell-reinforced unpaved roads over weak subgrade. Transp. Res. Rec.: J. Transp. Res. Board **2204**, 67–75 (2011)
14. Thakur, J.K., Han, J., Pokharel, S.K., Parsons, R.L.: Performance of geocell-reinforced recycled asphalt pavement (RAP) bases over weak subgrade under cyclic plate loading. Geotext. Geomembr. **35**, 14–24 (2012)
15. Presto Geosystems: Genuine Geoweb® GW30 V - 150 mm (6 in) depth: performance & material specification summary (2011). http://www.prestogeo.com/downloads/YNjJj6EykMeI87iE9Sx4FxrNzFrvYrwx6W7fRBT3NxBLukPt3D/Geoweb%20GW30v6_summary.pdf. Accessed 10 Sept 2017
16. ASTM D1557: Standard Test Methods for Laboratory Compaction Characteristics of Soil Using Modified Effort (56,000 ft-lbf/ft^3 (2,700 kN-m/m^3)) (2012)
17. Pokharel, S.K.: Experimental study on geocell-reinforced base under static and dynamic loading. Doctoral dissertation in Civil Engineering, The University of Kansas, Kansas, USA (2010)

Model Tests on Performance of Embankment Reinforced with Geocell Under Static and Cyclic Loading

Zhiheng Dai, Mengxi Zhang$^{(\boxtimes)}$, Lei Yang, and Huachao Zhu

Department of Civil Engineering, Shanghai University, Shanghai, China
mxzhang@i.shu.edu.cn

Abstract. Geocell, a honeycomb geosynthetics, has been widely used in geotechnical engineering to improve the performances of embankments. A useful method, particle image velocimetry (PIV) technique which could track motion of soil grains, was used to determine the displacement field and the slip surface progressively. A series of experimental and analytical investigations on geocell-reinforced embankments were presented. This study focused on revealing the mechanics response and failure process of embankment under static and cyclic loading. The embankments were reinforced with geocells. And the experiments were scheduled to demonstrate the effect of the burial depth of the geocell layer, such as 0.83B, 1.67B and 2.50B (B, width of loading plate). Based on the sand displacement field, the deformation mechanisms of embankments were analyzed. Under the static loading, the steel plate would be forced down into the embankment, which made the large lateral deformation develop and progressively caused the collapse of embankments. Comparing the failure modes of different embankments, it could be found that the slip region area of reinforced embankments was wider than unreinforced embankments. Moreover, the similar phenomenon occurred in stress distribution. Under the cyclic loading condition, the results suggested that geocell-reinforcements provided more lateral confinement and reduced the cumulative plastic settlement. Analyzing the influence of burial depths, it could be concluded that shallower the geocell was embedded, the better stability of the embankment was.

Keywords: Geocell · Model test · Failure surface · PIV
Reinforced embankment

1 Introduction

Polymeric materials, with high tensile strength and reasonable structure, had a rapid development and got expansive application in civil engineering. Over the past two decades, geocell had emerged as a promising 3-dimensional reinforcement material. It was noted that geocell was easily folded and convenient in transit. When unfolded, geocell could be flexibly filled with dense sand, gravel and other fillers in hand, forming a great reinforcement layer. It was identified that geocell could confine the lateral deformation of fillers. Meanwhile, the fillers could improve the stiffness of the

© Springer Nature Singapore Pte Ltd. 2018
L. Li et al. (Eds.): GSIC 2018, *Proceedings of GeoShanghai 2018 International Conference: Ground Improvement and Geosynthetics*, pp. 399–410, 2018.
https://doi.org/10.1007/978-981-13-0122-3_44

geocell. The applications of geocell in highway and railway engineering had raised great concern among the scholars. To reveal the load-carrying mechanism and effect of geocell-reinforcement, a lot of experiments were conducted.

El-Nagger and Kennedy [1] carried out laboratory model tests to investigate on the mechanical behavior of embankments reinforced with geocells. Several factors in geocell reinforcement, such as the number of reinforcement layer, the length of reinforcement, the height of the embankment, the slope of embankment et al., were considered. It was found that the ultimate bearing capacity increased significantly, when the number of reinforcement layers increased. Gao et al. [2, 3] conducted laboratory model tests to study on the mechanical behaviors of embankment reinforced with geocells of varying geocell layers, height, opening size et al. The performance of reinforced embankment under static loading was analyzed. Latha [4] revealed the influences of the height, tensional strength and the length of geocells on deformation and bearing capacity of embankments. Through laboratory model tests and numerical modelling method, Yoo [5] demonstrated the efficacy of geocells with different burial forms, lengths, multiple layers et al. for improving the stability of reinforced embankment.

Different from static load, traffic load is instantaneous and dynamic. In addition, the amplitude of dynamic loading will change with the pavement rising and falling. So some simplification should be made for theoretical analysis. Hyodo and Yasuhara [6] measured the waveform of vertical earth pressures under the reciprocating movement of a truck with different speeds. Progressively, the result was simplified to half-sine waveform. Saad et al. [7] and Ling et al. [8] simply described the stress at the top of the subgrade as triangle wave. He [9] simplified the traffic load to square wave. Hanazato et al. [10] expressed the traffic load in the form of Fourier series. On the whole, the traffic load was always simplified into half-sinusoidal wave and triangle wave. Monitoring practical application of embankment engineering, Luo [11] found that the performance of roadbed under cyclic loading was improved effectively by the geocell-reinforcement. Geocell mattress could resist axial tension and bending moment. As a result, geocell reinforcement could not only increase the ultimate bearing capacity but also reduce the displacement of highway embankment.

Literatures available indicated that the reinforcement effect of geocell mattress of varying layer number, length, height et al. had caught great attentions. The mechanical behaviors of geocell reinforced embankment under static loading had been understood well. However, the deformation procedure and dynamic performance of the geocell reinforced embankment had not been widely reported. A series of experimental and analytical investigations on geocell-reinforced embankment were presented. This study focused on revealing the mechanics response and failure process of embankment under static and cyclic loading.

2 Experimental Program and Equipment

To develop an understanding of the influence of the burial depth of the geocell layer, a 1/12 scale model of sandy embankment was built. A loading system was used to apply quasi-static axial and half-sinusoidal cyclic loading. The induced settlement was monitored in real time. And the normal displacement of the embankment slope was recorded too. The additional earth pressure was measured by several pressure cells. The displacement field and the slip surface were obtained by the particle image velocimetry (PIV) technique. Subject to technology constraints, the laboratory model study could only achieve geometric similarity and kinematic similarity. The dynamic behaviors had not been simulated. However, the work of investigations on failure process and mechanics response still had some engineering sense.

2.1   Axial Loading System

The axial loading system consists of a loading device, a force sensor and a linear vertical displacement transducer (hereinafter referred to as LVDT). This loading system was produced by the American GCTS Corporation. The apparatus was installed right above the model and fixed on two pillars, which were symmetrical to the box. The loading device could apply quasi-static axial and half-sinusoidal cyclic loading. The maximum axial load was 10 kN. The maximum frequency was 5 Hz. The axial force was monitored by a load cell, which was screwed and fixed behind the rigid loading cap. The LVDT with a maximum range of 50 mm was used to record the vertical displacement of the cap (Fig. 1).

Fig. 1. Model box and loading system

2.2   Experimental Material and Measurement Equipment

A box with interior dimensions of 600 mm × 290 mm × 400 mm (L × W × H) was fabricated from 2 steel plates and 2 transparent Plexiglas plates (Fig. 1).

All the thickness of the plates is 25 mm, which were rigid enough to avoid the large deformation. The transparent Plexiglas plates were used to enable digital images to be taken during testing. A steel tube with a hole was fixed on the test tank. Through the hole, a T-type supporting bar was installed to fix 3 LVDTs (Fig. 2). The normal displacement of the slope was measured by the LVDTs. Along the slope from the top to the bottom, 3 LVDTs were labeled with A, B and C in sequence (Fig. 3).

According to the Design Specification for Highway Alignment (JTGD20-2006), a certain second-class highway was modeled in the test. Limited by the size of the tank, a 1/20 scale model was built by dry and clean sand with slope of 1:1.5, height of 6 m, and width of 8.8 mm at the top. Because of symmetry, simplification was made by

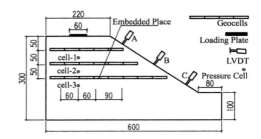

Fig. 2. Installation of LVDTs   **Fig. 3.** Model sketch and test program (mm)

simulating only half of the embankment (Fig. 3). The test chose the standard river sand from Fujian Province. The sand was dry and clean, with properties $C_u = 2.07$, $C_c = 0.87$, $G_s = 2.62$, $\omega = 0.08\%$, $\rho_{max} = 1.87$ g/cm^3 and $\rho_{min} = 1.45$ g/cm^3. The grain size distribution curve was depicted and the property parameters of the standard sand were listed (Fig. 4).

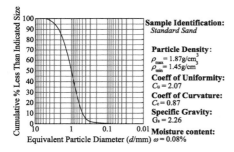

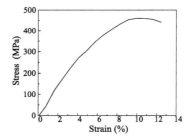

Fig. 4. Grain size distribution curve   **Fig. 5.** Strain-stress curve of the geocell sheet

Limited by the size of the embankment model, some mini-cells with height of 5 mm and weld spacing of 25 mm was manufactured. The mini-cells were made of high-density polypropylene. The axial tension strength of the high-density polypropylene sheet was beyond 450 MPa while the elongation at break was lower than 15%. The stress-strain curve of the polypropylene sheet was depicted in Fig. 5. The picture and sketch of the hand-made geocell was shown in Fig. 6.

A useful method, particle image velocimetry (PIV) technique which could track motion of soil grains, was used to determine the displacement field and the slip surface progressively. The PIV consisted of a camera (model number B5M16) with charge coupled device and image process software (namely MicroVec V3). The digital camera with 5 mega-pixels could capture and record 11.3 pictures per second. The motion of soil grains was accurately tracked by PIV, progressively analyzed by MicroVec V3 and finally presented by post processing software Tecplot (showed in Fig. 7).

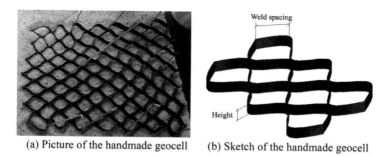

(a) Picture of the handmade geocell   (b) Sketch of the handmade geocell

Fig. 6. Handmade geocell used in the test

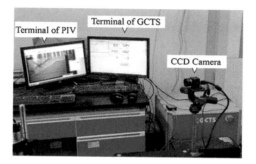

Fig. 7. Monitor and control center

Three kinds of burial depth of geocell such as 0.83B, 1.67B and 2.50B (B, width of loading plate) were prepared. The model embankments were built with the relative density of sand of 97%. The experimental programs were summarized in Table 1.

Table 1. Experimental program

Number	Identifier	Burial depth	Relative density %	Load type
1	Unreinforced-S	–	97	Static
2	0.83B-S	0.83B	97	Static
3	1.67B-S	1.67B	97	Static
4	2.50B-S	2.50B	97	Static
5	Unreinforced-C	–	97	Cyclic
6	0.83B-C	0.83B	97	Cyclic
7	1.67B-C	1.67B	97	Cyclic
8	2.50B-C	2.50B	97	Cyclic

3   Results and Analysis

3.1   Static Loading Test

Distinct from step-by-step load applied in traditional laboratory static test, the load applied in this model experiment was stress-controlled at a constant increment of 4 N/s.

Pressure-Settlement Curve of Embankment

Figure 8 showed the comparison of the bearing pressure-settlement curve of reinforced and unreinforced experimental studies. At the same settlement of 3 mm, the bearing capacities of reinforced embankment with embedded depth of 0.83B, 1.67B and 2.5B were 39.33%, 25.06%, 15.08% greater than the unreinforced respectively. While, the settlements were 25.38%, 20.45%, and 14.77% less than the unreinforced respectively at the same pressure of 218 kPa, which was ultimate bearing capacity of unreinforced embankment. It was verified that geocells-reinforcement could not only improve the ultimate bearing capacity but also reduce the displacement of embankment. Comparing with unreinforced embankment, geocells with the burial depth of 0.83B, 1.67B and 2.5B improved the ultimate bearing capacity by 75.84%, 45.71%, and 18.98% respectively. These results indicated that with the increase of embedded depth, the settlement of the embankment increased. On the contrary, the ultimate bearing capacity would reduce significantly.

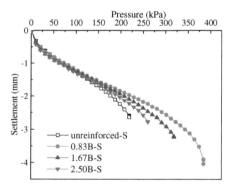

Fig. 8. Bearing pressure-settlement curve

Normal Displacement of the Slope

Figure 9 showed the comparison of the normal displacement of the slope. By observation on position A, the normal displacement of unreinforced-S (Table 1) was 1.34 mm which was 54.97% greater than 0.60 mm of 0.83B-S, and 31.79% greater than 0.91 mm of 1.67B-S, and 15.89% greater than 1.13 mm of 2.50B-S. Focusing on position B or position C, similar laws could be found. At the case of unreinforced-S, the displacement, along the slope to the bottom, decreased obviously. Moreover, similarities occurred at the cases of 0.83B-S, 1.67B-S, and 2.50B-S. The minimum displacement at position A was observed in 0.83B-S. The minimum displacement at

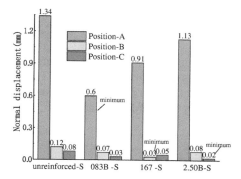

Fig. 9. Comparison of the normal displacement

position B was noted in 1.67B-S. Regularly the minimum displacement at position C was found in 2.50B-S.

The results also showed that the displacement of soil, close to geocells, was strongly confined. In general, failure surface occurred at the top of embankment, which made the effect of the geocells of 2.50B-S suppressed. With the burial depth increased, the reinforced embankment performed worse behaviors and had a trend to be same as unreinforced embankment. The normal displacement of 2.50B-S was found pretty similar to unreinforced-S. In other words, the improvement of the geocells gradually died down. It was suggested that geocells should be embedded at a right shallow place.

Displacement Vectors and Contour

Figure 10(a–b) showed the displacement vector fields of embankments at the corresponding ultimate limit state. The most displacement vectors pointing to top-left were observed, indicating that the failure modes of embankments resulted from the lateral extrusion of granules. Failures primarily occurred at the side of the loading plate, gradually developed through the embankments to its profile, and finally formed the circular sliding surfaces as it was sketched by red lines. Comparing with unreinforced embankment, larger magnitude of displacement vectors was observed at the case of reinforced embankment. It was concluded that the membrane effect and moment-resisting ability enabled the applied load to be dispersed, which resulted in the

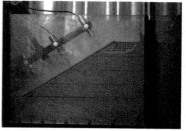

(a) embankment without reinforcement

(b) embankment with reinforcement (u=0.83B)

Fig. 10. Displacement vector diagram of the embankment

improvement of ultimate bearing capacity and deformability. With wider slip region, better stability and more evenly stress distribution were deduced in the case of reinforced embankments.

Figure 11 showed the contours of displacement at pressure of 204 kPa. An evident displacement concentration was observed in unreinforced-S (Fig. 11a), while smaller displacement magnitude and relative uniform displacement distribution was noted in 0.83B-S (Fig. 11a). The results verified former deduction.

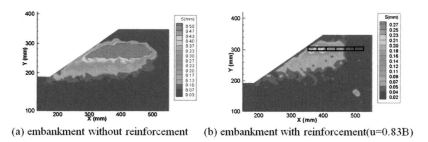

(a) embankment without reinforcement   (b) embankment with reinforcement(u=0.83B)

Fig. 11. The displacement contour (p = 204 kPa)

Figure 12(a–c) showed the displacement contours of embankments reinforced with geocell of three burial depths. The contours demonstrated a right shallow depth about 0.83B, in which the geocell should be embedded, enabled larger displacement region to be motivated and leaded to smaller value of displacement. By observation on Fig. 12 (a–c), displacement concentration mainly emerged at the top of embankments. The reinforcement effect was fully exerted in the large-deformation region, which leaded to greatest improvement of elastic modulus and strength of soil around geocell. Further, the applied load and inductive displacement were successfully transmitted to deeper field. Geocells embedded in 1.67B and 2.50B exerted obviously weak effect on transmitting the load and displacement. Because only soil close to geocells was strongly confined. The mechanical properties of soil in the large-deformation region were not enhanced gradually, with the embedded depth increasing.

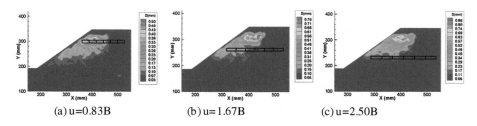

(a) u=0.83B   (b) u=1.67B   (c) u=2.50B

Fig. 12. The displacement contour of reinforced embankments (p = 256 kPa)

3.2   Cyclic Loading Test

Through loading experiment, Terrel et al. [12] concluded that the total strain and rebound strain had little difference under the half-sine wave or triangle wave. Thus, half-sinusoidal wave was adopted to simulate the traffic load in this paper. In order to facilitate the comparison, the amplitudes of cyclic loading were determined based on bearing capacities of the unreinforced embankment under static loading. Thus, a half-sinusoidal cyclic loading with frequency of 1 Hz, vibration times of 10000 and amplitude of 131.34 kPa (60% of ultimate bearing capacity of unreinforced embankment 218.9 kPa) was applied in the cyclic loading test (Fig. 13).

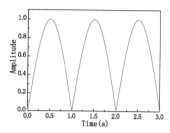

Fig. 13. A kind of half-sinusoidal wave

Vertical Cumulative Displacement

Figure 14 showed the comparison of vertical cumulative displacement versus time curves between unreinforced-C, 0.83B-C, 1.67B-C and 2.50B-C in first 20 s. In the beginning, all the curves were plumb, indicating that the soil was rapidly compacting. About 1 s later, the curves stopped declining and tended to be horizontal. An obvious branch of unreinforced-C was noted and the value of cumulative settlement of unreinforced embankment became bigger than reinforced. It was suggested that geocells started to exert their reinforcement effect.

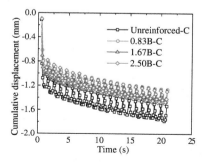

Fig. 14. Cumulative displacement at the first 20 s

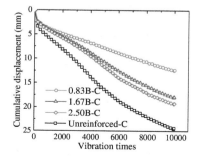

Fig. 15. Cumulative displacement versus vibration times curves

Figure 15 showed the vertical cumulative displacement versus vibration times. After 10000 vibration times, the vertical cumulative displacement of unreinforced-C was 24.38 mm, 49.71% higher than the 0.83-C of 12.26 mm, 26.87% higher than 1.67B-C of 17.83 mm and 21.45% higher than 2.50B-C of 19.15 mm. It was concluded that the improvement of elastic modulus and strength could reduce the plasticity of soil, effectively remising the vertical cumulative displacement. Similar suggestion to static loading test was provided that right shallow depth could exert the reinforcement effect more.

Earth Pressure in the Embankment

Figure 16 showed the earth pressure versus time curve of pressure cell 2# of 0.83B-C in the first 20 s. Generally, the curve fluctuation was in keeping with the excitation of the applied half-sinusoidal cyclic loading. In the first 10 s, the soil was in the state of be rapidly compacting, which contributed a great change of the amplitude. After then, the soil was softening slowly. In this time, the settlement cumulated constantly, the earth pressure reduced gradually. (The overall arrangement of earth pressure cells was shown in Fig. 7.)

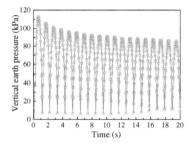

Fig. 16. The earth pressure versus time curve

Figure 17 presented the earth pressure versus time curve of unreinforced-C and 0.83B-C. Observing on the earth pressure measured by cell-1, relative high amplitude of 0.83B-C curve was noted. Impacted by the same action of displacement, higher reaction force was exerted in 0.83B-C than that unreinforced-C. In other words, the embankment of 0.83B-C performed higher rigidity. It was concluded that the geocell reinforcement had improved the rigidity and leaded to larger stress magnitude at the top field of embankment. Further, comparing the earth pressure measured by cell-2 and cell-3, similar phenomenon was observed. The larger magnitude of the stress filed at the lower region of geocell-reinforced embankment was caused. It was indicated that geocell reinforcement enabled the applied force to be transmitted into the lower stable region. Comparing with unreinforced embankment, more uniform and reasonable stress distribution was noted.

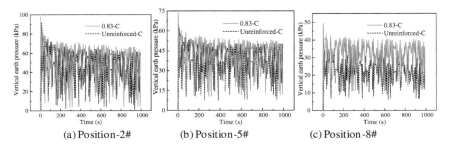

(a) Position-2#   (b) Position-5#   (c) Position-8#

Fig. 17. Earth pressure versus time curve for different depths

4   Conclusions

1. With the increase of embedded depth, the improvement effect of the geocells gradually died down. Geocell, embedded at a right shallow place, not only reduced the settlement of the embankment but also improved the ultimate bearing capacity much better.
2. Under static loading, geocell reinforcement could strongly confine the displacement of its ambient soil, which suggested that geocell mattress should be buried at the large deformation region of the embankment.
3. Comparing with the unreinforced embankment, the cumulative displacement was reduced obviously by the geocell mattress. Under cyclic loading, geocell reinforcement could greatly improve the elasticity and strength of the embankment.
4. It was concluded that the geocell reinforcement could improve the rigidity of the embankment as well as lead to larger stress magnitude at the top field. Moreover, geocell reinforcement enabled the applied force to be transmitted into the deeper stable region, which caused a more uniform and reasonable stress distribution in embankment.

Acknowledgments. This study has been supported by the National Natural Science Foundation of China (NSFC) (No. 41372280). The authors would like to express their gratitude for these financial assistances.

References

1. El-Naggar, M.E., Kennedy, J.B.: New design method for reinforced sloped embankments. Eng. Struct. **19**(1), 28–36 (1997)
2. Gao, A., Zhang, M.X., Zhu, H.C., et al.: Model tests on geocell-reinforced embankment under cyclic and static loadings. Rock Soil Mech. **37**(7), 1921–1928 (2016)
3. Gao, A., Zhang, M.X., Liu, F., et al.: Model experimental study of embankment reinforced with geocells under stepped cyclic loading. Rock Soil Mech. **37**(8), 2213–2221 (2016)
4. Latha, G.M., Rajagopal, K., Krishnaswamy, N.R.: Experimental and theoretical investigations on geocell-supported embankments. Int. J. Geomech. **6**(1), 30–35 (2006)

5. Yoo, C.: Laboratory investigation of bearing capacity behavior of strip footing on geogrid-reinforced sand slope. Geotext. Geomembr. **19**(5), 279–298 (2001)
6. Hyodo, M., Yasuhara, K.: Analytical procedure for evaluating pore-water pressure and deformation of saturated clay ground subjected to traffic load. In: Proceedings of the 6th International Conference on Numerical Methods in Geomechanics, pp. 653–658. A. A. Balkema, Rotterdam (1988)
7. Saad, B., Mitri, H., Poorooshasb, H.: Three-dimensional dynamic analysis of flexible conventional pavement foundation. J. Transp. Eng. **131**(6), 460–469 (2005)
8. Ling, J.M., Wang, W., Wu, H.B.: On residual deformation of saturated clay subgrade under vehicle load. J. Tongji Univ. **30**(11), 1315–1320 (2002)
9. He, G.J.: Laboratory test and research on the settlement of soft foundation under low embankment considering the influence of traffic load. M.A. thesis, Hehai University, Nanjing, China (2015)
10. Hanazato, T., Ugai, K., Mori, M., et al.: Three-dimensional analysis of traffic-induced ground vibrations. J. Geotech. Eng. **117**(8), 1133–1151 (1991)
11. Luo, Q., Zhou, H., Wang, Z.J.: Test study on geocell-stabilized railway subgrades. J. China Railway Soc. **26**(3), 98–102 (2004)
12. Terrel, R.L., Awad, I.S., Foss, L.R.: Techniques for characterizing bituminous materials using a versatile triaxial testing system. In: American Society for Testing and Materials, Philadelphia, ASTM STP 561, pp. 47–66 (1974)

Comparison Analysis on Behavior of Geosynthetic Reinforcement in Piled Embankments Under Plane Strain and Three-Dimensional Conditions: Numerical Study

Zhen Zhang[1(✉)], Meng Wang[1], Guan-Bao Ye[1], and Jie Han[2]

[1] Department of Geotechnical Engineering, Tongji University,
Shanghai 200092, China
zhenzhang@tongji.edu.cn

[2] Civil, Environmental and Architectural Engineering (CEAE) Department,
The University of Kansas, Lawrence, KS 66045, USA

Abstract. In the system of Geosynthetic-Reinforced Pile-Supported (GRPS) embankment, geosynthetic reinforcement contributes to load transfer from surrounding soil to pile through tensioned membrane effect. Among the design methods of GRPS embankments, tension in geosynthetic reinforcement were mostly developed based on a plane strain (2-D) condition or a strip between two supports. However, piles are typically installed in a square or triangular pattern in practice. It has not been quite clear whether the calculated tension in a 2-D model can represent the real 3-D condition well or not. In this paper, plane strain and three-dimensional numerical analyses were conducted to investigate the behavior of geosynthetic reinforcement in 2-D and 3-D conditions. Pile caps were simplified as square rigid stationary parts arranged in a square or triangular pattern and the surrounding soil between the piles were modeled using modulus of subgrade reaction. Geosynthetic reinforcement was modeled using an orthotropic shell element. The results show that the maximum tension and the sag in the geosynthetic reinforcement were smaller in the models with pile cap in a triangular pattern than those in the models with pile cap in a square pattern. The maximum strains in the geosynthetic reinforcement in a 3-D condition were larger than those in a 2-D condition. A series of conversion factor were proposed to correct the maximum tension in geosynthetic reinforcement in a 2-D condition to that in a 3-D condition considering the pile cap shape and arrangement, and the surrounding soil support.

Keywords: Geosynthetic reinforcement · GRPS embankment
Plane-strain condition · Three-dimensional condition · Membrane effect

© Springer Nature Singapore Pte Ltd. 2018
L. Li et al. (Eds.): GSIC 2018, *Proceedings of GeoShanghai 2018 International Conference: Ground Improvement and Geosynthetics*, pp. 411–419, 2018.
https://doi.org/10.1007/978-981-13-0122-3_45

1 Introduction

Geosynthetic reinforcement acts to enhance load transfer from surrounding soil to pile and reduce differential settlement between pile and surrounding soil in the system of Geosynthetic-Reinforced Pile-Supported (GRPS) embankment. Prediction of maximum strain and vertical deflection in geosynthetic reinforcement is an important issue in design of GRPS embankment. Various studies have been focused on the behavior of geosynthetic reinforcement in GRPS embankments (Jones 2007; Filz et al. 2009; McGuire et al. 2009; Halvordson et al. 2010). However, the current commonly-used design methods around the world, such as Rogbeck et al. (2003), GB/T 50783-2012 (2012) and BS8006 (2010), were fundamentally developed under a plane strain (2-D) condition, were only available for piles installed in a square grid and did not consider subsoil support. Filz and Smith (2007) and Nunez (2013) indicated that there existed great differences between the predictions and the field measurements of the strains in geosynthetic reinforcements. Almeida et al. (2007) performed a series of field tests of GRPS embankment with biaxial geogrids and presented that the maximum deflections of geogrids under a 3-D condition were 14% larger than those under a 2-D condition.

In this paper, 2-D and 3-D finite-element methods were used to compare the behaviors of geosynthetic reinforcement in pile-supported embankments in terms of load transfer, vertical deflection and strain of geosynthetic reinforcement. Since EBGEO (2010) takes account of both the subgrade support and the geosynthetic deflection, it was applied for comparison with the numerical results. In the numerical model, the piles were simplified as square stationary parts arranged in a square or triangular pattern and the supports from the surrounding soil between piles were considered. Finally, based on the numerical results, conversion factors were proposed to account for the difference in the maximum strain in geosynthetic reinforcement under the 2-D and 3-D conditions.

2 Numerical Modeling

The software ABAQUS was adopted to establish the numerical models of GRPS embankment over soft soils. The numerical models similar to the study by Filz et al. (2009) were adopted in this study. For simplicity, the pile caps were modeled as rigid stationary parts with square or round shape. The surrounding soils were modeled by a linear elastic foundation system with modulus of subgrade reaction, and the embankment load were modeled as pressure acting up on the geogrids (Jones 2007; Halvordson 2007, Filz et al. 2009). Referring to the study by Filz et al. (2009), the stress concentration ratio, which is defined as a ratio between the average stresses acting up on geogrids in the area underlain by pile caps and subsoil, was taken as 6 in all cases. The geosynthetic reinforcement layer was modeled as orthotropic linearly elastic material using S4R element, i.e., a four-node doubly curved shell element with reduced integration (Seay 1998; Park et al. 2007). The tensile stiffness of geogrids in both orthotropic directions was $J = 1200$ kN/m.

Figure 1 shows the plan views of the analysis models with caps in square and triangular patterns under 2-D and 3-D conditions. In both the 2-D and 3-D models, the caps had a same spacing. In the 2-D model with round cap, the width of the cap was equal to the side length of the square cap which had an equivalent cross sectional area with the round cap under the 3-D condition. The 3-D models with caps in square and triangular patterns had a constant area replacement ratio of pile caps of 16%. The area replacement ratio of pile caps is consistently less than 20% in practice (Han 1999). The bottom boundaries of the pile caps were fixed in all three directions (i.e., x, y, and z directions). The four-sided boundaries of the numerical models were set as symmetrical boundaries. The geosynthetic reinforcements were tied with the top surfaces of the pile caps. Table 1 tabulates all the numerical models in this study. The selected moduli of subgrade reaction are referred to study by Filz et al. (2009).

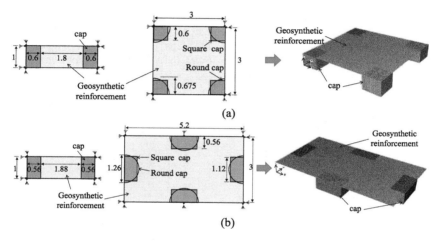

Fig. 1. Schematic of numerical models under 2-D and 3-D conditions: (a) in square pattern; (b) in triangular pattern (Unit: m)

Table 1. Specific information of each numerical case

Case no.	Analysis method	Pile grids	Pile cap			k/kN/m^3	Embankment load	
			Shape	Size/m	Spacing/m		σ_{cap}/ kPa	σ_{soil}/ kPa
1	3-D	Sq.	Sq.	1.2	3	100/200/300/400/500	108	18
2	3-D	Tri.	Sq.	1.12	3	100/200/300/400/500	108	18
3	3-D	Sq.	R.	1.35	3	100/200/300/400/500	108	18
4	3-D	Tri.	R.	1.26	3	100/200/300/400/500	108	18
5	2-D	Sq.	Sq.	1.2	3	100/200/300/400/500	108	18
6	2-D	Tri.	Sq.	1.12	3	100/200/300/400/500	108	18

Note: Tri. and Sq. in column Pile grids represent the pile caps in triangular and square patterns, respectively; Sq. and R. in column Pile cap represent the pile caps with square and round shape, respectively; k is the modulus of subgrade reaction σ_{cap} and σ_{soil} are the average stresses acting up on geosynthetic reinforcement in the area underlain by pile caps and subsoil, respectively.

3 Results and Discussion

3.1 Effect on Efficiency of Geosynthetic Reinforcement

Figure 2 shows the variation of geosynthetic efficiency with the modulus of subgrade reaction. The geosynthetic efficiency of load transfer is defined as the ratio of the net average stress acting on geosynthetic layer to the average stress acting up on geosynthetic layer. It can be seen that under the 2-D condition the geosynthetic efficiencies were greater in a range from 1.7% to 3.2% than those under the 3-D condition irrespective of the cap arrangement pattern, but the difference between the 2-D model and the 3-D model reduced with an increase of modulus of subgrade reaction. The difference in geosynthetic efficiency between 2-D and 3-D model might be the reason that the conversion from 3-D model to 2-D model would increase the equivalent area replacement ratio in the 2-D model. The geosynthetic efficiency and the difference of the geosynthetic efficiency between the 2-D and 3-D models were decreased with the increase of modulus of subgrade reaction. Filz et al. (2009) also confirmed the similar findings. In addition, it is noted that the pile caps in a triangular pattern helped geosynthetic reinforcement transfer slightly more load to the pile caps than the pile caps in a square pattern.

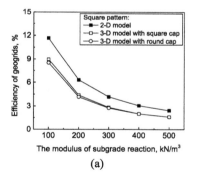
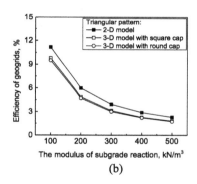

Fig. 2. Efficiency of geogrids: (a) square pattern; (b) triangular pattern

3.2 Effect on Vertical Deflection of Geosynthetic Reinforcement

Figure 3 shows the maximum vertical deflections of geosynthetic reinforcement obtained from the numerical results under the 2-D and 3-D conditions and the results by EBGEO (2010). It can be seen that the calculated results by EBGEO (2010) were significantly larger than the numerical results. EBGEO (2010) is conservative for the calculation of vertical deflection of reinforcement. This can be explained by the reason that EBGEO (2010) assumes a triangular-shaped line load on a reinforcement strip between two adjacent piles.

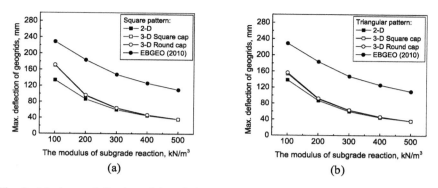

Fig. 3. Maximum deflection of the reinforcement: (a) square pattern; (b) triangular pattern

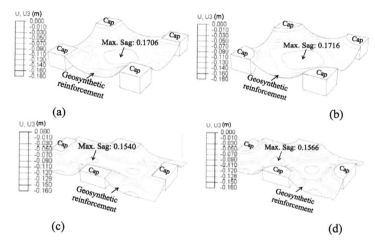

Fig. 4. Vertical deflection contours in 3-D models: (a) square caps in a square pattern; (b) round caps in a square pattern; (c) square caps in a triangular pattern; (d) round caps in a triangular pattern

The 3-D condition posed a larger maximum vertical deflection than the 2-D condition. Almeida et al. (2007) found that maximum deflection under 3-D condition was 14% larger than that under a 2-D condition through the field tests without subgrade support. The numerical results show that the dimensional effect (i.e. 2-D and 3-D conditions) had significant effect on the vertical deflections of geosynthetic reinforcement when the subsoil had low modulus of subgrade reaction. However, the difference in the vertical deflections of geosynthetic reinforcement due to the dimensional effect was small when the modulus of subgrade reaction was greater than 400 kN/m^3.

In order to investigate the influence of pile cap shape and pile arrangement pattern on vertical deflection of geosynthetic reinforcement, the vertical deflections contours in the 3-D models with a modulus of subgrade reaction of 100 kN/m^3 were extracted from the numerical results (see Fig. 4). The pile cap shape (i.e., square or round cap) had minor influence on the maximum vertical deflections of the geosynthetic reinforcement. The maximum vertical deflection of the geosynthetic reinforcement occurred at the midway between the pile caps and the models with caps in a triangular pattern reduced the maximum deflection as compared with the model with caps in a square pattern.

3.3 Effect on Strain in Geosynthetic Reinforcement

Figure 5 shows the maximum strains in the geosynthetic reinforcement obtained from the numerical analysis and the results by EBGEO (2010). It can be seen that the results by EBGEO (2010) had good agreements with the numerical results of the 3-D models with round pile caps. However, EBGEO (2010) underestimated the maximum strains in the geosynthetic reinforcement in the models with square caps under a 3-D condition. In addition, the maximum geosynthetic strains in the 2-D condition were smaller than those in the 3-D condition irrespective of the cap arrangements (i.e., in square and triangular patterns). The softer subsoil would lead to a significant increase of maximum strain in geosynthetic reinforcement.

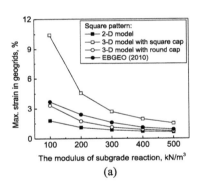 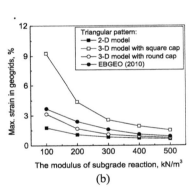

Fig. 5. Maximum principal strain calculated by numerical model and EBGEO (2010): (a) square pattern; (b) triangular pattern

3.4 Conversion Factor of Strain Between 2-D and 3-D Condition

To rectify this underestimation of maximum strain in geosynthetic reinforcement using the 2-D model, a conversion factor is proposed herein, which is defined as the ratio of the maximum strain in geosynthetic reinforcement by the 3-D model to that by the 2-D model:

$$\varepsilon_{max}^{3D} = F\varepsilon_{max}^{2D} \tag{1}$$

in which, ε_{max}^{3D} and ε_{max}^{2D} are the maximum strains in geosynthetic reinforcement under the 2-D and 3-D conditions, respectively; and F is the conversion factor. Figure 6 shows the conversion factor based on the numerical results. It can be seen that the cap shape and subsoil stiffness had significant influences on the conversion factor of maximum strain in geosynthetic reinforcement, while the pile cap arrangement pattern had a slight influence on the conversion factor. The conversion factors tended to 1.0 with an increasing in the subsoil stiffness. Due to the severe stress concentration effect of the square cap, the conversion factors under the square cap condition were larger than those under the round cap condition. Rogbeck et al. (2003) proposed a simple method to account for the conversion factor of the strain under the plain strain condition to that under the 3-D condition for the case of piles in a square pattern, in which the conversion factor was developed based on the relative coverage area of the strip as compared with the area of strip under the 3-D condition. Accordingly, the conversion factor based on the Rogbeck method (2003) was included in Fig. 6. Since the Rogbeck method (2003) did not account for the subsoil stiffness, the calculated conversion factor was a horizontal line. The conversion factor by Rogbeck method (2003) was similar with the model with round cap when the subsoil was soft (i.e., modulus of subgrade reaction less than 200 kN/m³). However, the Rogbeck method (2003) underestimated the 3-D dimensional effect in the models with square pile caps.

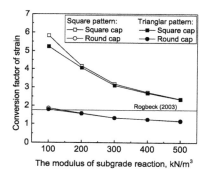

Fig. 6. Conversion factor of maximum strain in geosynthetic between 2-D and 3-D condition.

4 Conclusions

In this study, the plane-strain (2-D) and 3-D numerical analyses were conducted to investigate the 3-D dimensional effect on the behavior of geosynthetic reinforcement in the GRPS embankment system. According to the numerical results and analysis, the following conclusions can be drawn:

1. The 2-D condition overestimated the geosynthetic efficiency compared to the 3-D condition. The geosynthetic efficiency increased with the decrease of modulus of subgrade reaction. The pile cap shape (i.e., square or round cap) and arrangement pattern (i.e., piles in a square or triangular pattern) had minor effect on the geosynthetic efficiency.
2. The 3-D condition would lead to a larger maximum vertical deflection of geosynthetic reinforcement than the 2-D condition. The subsoil stiffness had a great influence on the vertical deflections of geosynthetic reinforcement.

3. The maximum strain in geosynthetic reinforcement predicted by EBGEO (2010) agreed with the results of 3-D models with round pile caps, but EBGEO (2010) underestimated the maximum strain in geosynthetic reinforcement for case with square caps. The 2-D model had smaller maximum geosynthetic strains than the 3-D models irrespective of the cap arrangements.
4. The pile cap shape and subsoil stiffness had significant influence on the conversion factor. The conversion factor by Rogbeck method (2003) was similar with the model with round cap when the subsoil was soft (i.e., modulus of subgrade reaction less than 200 kN/m^3). However, the Rogbeck method (2003) underestimated the 3-D dimensional effect in the models with square pile caps.

Acknowledgements. The authors appreciate the financial supports provided by the National Natural Science Foundation of China (NSFC) (Grant No. 51508408 & No. 51478349) and the Pujiang Talents Scheme (No. 15PJ1408800) for this research.

References

Almeida, M.S.S., Ehrlich, M., Spotti, A.P., Marques, M.E.S.: Embankment supported on piles with biaxial geogrids. Proc. Inst. Civil Eng.-Geotech. Eng. **160**(4), 185–192 (2007)

BS8006-1: Code of Practice for Strengthened/Reinforced Soils and Other Fills. British Standard Institution, British (2010)

EBGEO: Recommendations for Design and Analysis of Earth Structures using Geosynthetic Reinforcements. German Geotechnical Society, German (2010)

Filz, G.M., Plaut, R.H.: Practical implications of numerical analyses of geosynthetic reinforcement in column-supported embankments. In: Advances in Ground Improvement: Research to Practice in the United States and China, pp. 55–62 (2009)

Filz, G.M., Smith, M.E.: Net vertical loads on geosynthetic reinforcement in column-supported embankments. Soil Improvement, pp. 1–10 (2007)

GB/T 50783-2012: Technical code for composite foundation. Zhejiang Provincial Department of Housing and Urban-Rural Development, China Plan Press, China (2012). (in Chinese)

Halvordson, K.A., Plaut, R.H., Filz, G.M.: Analysis of geosynthetic reinforcement in pile-supported embankments. Part II: 3D cable-net model. Geosynth. Int. **17**(2), 68–76 (2010)

Halvordson, K.A.: Three-dimensional analysis of geogrid reinforcement used in pile-supported embankments. MS thesis, Virginia Polytechnic Institute and State University, Blacksburg (2007)

Han, J.: Design and construction of embankments on geosynthetic reinforced platforms supported by piles. In: Proceedings of 1999 ASCE/PaDOT Geotechnical Seminar, pp. 66–84 (1999)

Jones, B.M.: Three-dimensional finite difference analysis of geosynthetic reinforcement used in column-supported embankments (2007)

McGuire, M.P., Filz, G.M., Almeida, M.S.: Load-displacement compatibility analysis of a low-height column-supported embankment. In: Contemporary Topics in Ground Modification, Problem Soils, and Geo-Support, pp. 225–232 (2009)

Nunez, M.A., Briançon, L., Dias, D.: Analyses of a pile-supported embankment over soft clay: full-scale experiment, analytical and numerical approaches. Eng. Geol. **153**, 53–67 (2013)

Park, S., Yoo, C., Lee, D.: A study on the geogrid reinforced stone column system for settlement reduction effect. In: Seventeenth International Offshore and Polar Engineering Conference. International Society of Offshore and Polar Engineers (2007)

Rogbeck, Y., Alén, C., Franzén, G., Kjeld, A., Oden, K., Rathmayer, H., Oiseth, E.: Nordic guidelines for reinforced soils and fills. Nordic Geosynthetic Group of the Nordic Geotechnical Societies, Nordic Industrial Fund (2003)

Seay, P.A.: Finite element analysis of geotextile tubes (1998)

Review of Effects of Poor Gripping Systems in Geosynthetic Shear Strength Testing

Charles Sikwanda[1]([⊠]), Sanelisiwe Buthelezi[1], and Denis Kalumba[2]

[1] University of Cape Town, Cape Town, South Africa
SKWCHA001@myuct.ac.za
[2] Department of Civil Engineering, University of Cape Town,
Cape Town, South Africa

Abstract. The use of geosynthetic materials in geotechnical engineering projects has rapidly increased over the past several years. These materials have resulted in improved performance and cost reduction of geotechnical structures as compared to the use of conventional materials. However, working with geosynthetics requires knowledge of interface parameters for design. These parameters are typically determined by the large direct shear device in accordance with ASTM-D5321 and ASTM-D6243 standards. Although these laboratory tests are standardized, the quality of the results can be largely affected by several factors that include; the shearing rate, applied normal stress, gripping mechanism and type of the geosynthetic specimens tested. Amongst these factors, poor surface gripping of a specimen is the major source of the discrepancy. If the specimen is inadequately secured to the shearing blocks, it experiences progressive failure and shear strength that deviates from the true field performance of the tested material. This leads to inaccurate, unsafe and cost ineffective designs of projects. Currently, the ASTM-D5321 and ASTM-D6243 standards does not provide a standardized gripping system for geosynthetic shear strength testing. Over the years, researchers have come up with different gripping systems that can be used such as; glue, metal textured surface, sand-blasting, and sandpaper. However, these gripping systems are regularly not adequate to sufficiently secure the tested specimens to the shearing device. This has led to large variability in test results and difficulties in results interpretation. Thus, it is recommended that the future direction of geosynthetic shear strength testing should consider understanding the effects of gripping systems, which will contribute to easy data interpretation and increase result accuracy and reproducibility.

Keywords: Geosynthetics · Shear strength parameters · Gripping systems

1 Introduction

Geosynthetics are man-made products which are manufactured from polymeric materials. They can be used for reinforcement, separation, filtration, drainage, or other geotechnical-related functions of a civil engineering project, structure, or system [1]. Geosynthetics can be classified into different groups based on manufacturing methods and their intended use. These groups include geotextiles, geogrids, geocomposites,

© Springer Nature Singapore Pte Ltd. 2018
L. Li et al. (Eds.): GSIC 2018, *Proceedings of GeoShanghai 2018 International Conference: Ground Improvement and Geosynthetics*, pp. 420–429, 2018.
https://doi.org/10.1007/978-981-13-0122-3_46

geomembranes, geonets, geosynthetic clay liners, and geofoam. Although geosynthetics are used in other civil engineering projects, their primary field of application is in geotechnical and geoenvironmental engineering [2]. The use of geosynthetic products has been increasing since they were introduced on the market and according to [3], the world's demand for geosynthetics is expected to be more than 5.0 and 6.5 billion square meters by the end of 2017 and 2019, respectively. However, when geosynthetics are used in field applications (e.g. landfill slopes), the interface interaction between two geosynthetics or geosynthetics and soil becomes the most critical section where shear failure is likely to occur. In order to evaluate the interfaces, shear strength parameters are determined in the laboratory mainly using a large direct shear device in accordance with ASTM-D5321 [4] and ASTM-D6243 [5] standards. But, the direct shear device has the disadvantage of requiring a proper gripping system for shear to take place in the desired interface (which is achieved when the intended failure surface has the lowest shear resistance of all possible sliding surfaces) [6]. If unintended slippage occurs, tensile failure within the tested specimens may be generated which may result in shear not taking place in the desired interface. This leads to obtaining inaccurate results of the true field performance of the tested geosynthetics. This leads to inaccurate, unsafe and cost ineffective designs of projects. Thus, the use of proper gripping systems improves the accuracy and reproducibility of test results and reduces difficulties encountered in test data interpretation [6].

In recent years, concerns have been raised over the effects of gripping systems in geosynthetic shear strength testing on large direct shear devices. A great effort has been done by few researchers (e.g. [7–15]) but no conclusive system has been standardized yet. The ASTM D5321 [16] test methods state that work is still in progress to develop the best gripping surface. This paper attempts to bring out the effects of poor geosynthetic gripping systems on a large direct shear device.

2 Large Direct Shear Device

Pullout, ring shear and large direct shear devices have been used in geosynthetic shear strengths testing. However, the large direct shear device remains the most popular method used for determining shear strength parameters. The ASTM-D5321 [16] standards state that a square or rectangular direct shear device with minimum dimensions of 300 mm and 50 mm in depth must be used during the test. This device is typically divided into two shearing blocks, the lower moving part, and the upper static part as shown in Fig. 1. The tested geosynthetics are gripped on the lower and/or the upper shearing block, translating laterally relative to the other during shearing. Although performing the shearing test is simple and straightforward, the quality of the results is highly affected by a large number of factors that include: the shear rate, normal stress, poor gripping mechanism and type of the geosynthetic specimens [17]. Among these factors, the poor gripping mechanism is considered to be the major source of errors [6], hence this paper focuses on the effects of poor gripping systems in shear strength testing.

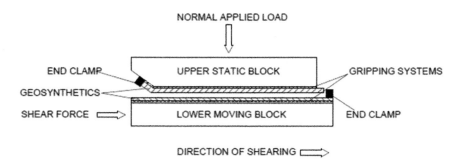

Fig. 1. Large direct shear device.

2.1 Specimen Gripping Systems

Generally, an inaccurately performed test produces unreliable results. Reviewing the relevant literature, one can recognize insufficient geosynthetic gripping systems as a major source of discrepancies in shear stress-displacement and strength behaviour. According to [18], the gripping system should provide adequate shear resistance against slippage between the tested geosynthetics and shearing blocks to avoid non-uniform displacement of the tested geosynthetics. The gripping system should also be able to transfer the entire applied shear load to the tested geosynthetic and allow free drainage of the specimen when required. End clamps are permitted in case extra strength is required to hold the tested specimen in position and force failure to occur at the desired interface. However, these end clamps and gripping systems should not interfere with the measuring of shear strength parameters of the tested specimen. Currently, the ASTM D5321 standard in use, states that the flat textured jaw-like clamping devices are normally sufficient but work is still in progress to define the best-textured surfaces that can be used to minimise or eliminate errors involved with the gripping surfaces.

2.2 Different Gripping Systems Used

Over the years, researchers have come up with different gripping systems that can be used to determine internal and interfacial shear strength of geosynthetics. The most common gripping systems used include textured steel plates, nail plates, glue and sandpapers. However, these gripping systems are sometimes reported to be not adequate to sufficiently secure the tested specimens to the shearing device. This has led to large variability in test results and difficulties in interpretation of results.

Textured Steel Grip. Good success with using a textured steel grip plate that consists of parallel woodworking rasps attached to the shearing blocks has been reported by [7–10].

During testing, geosynthetic clay liner (GCL) specimens were secured to a textured steel gripping surface to prevent slippage between the tested GCL specimen and the shearing block [7–9]. The textured steel grip was attached to both upper and lower half of the box and further clamped on both ends of the wooden substrate as shown in Fig. 2a. A similar, textured steel gripping system was also used by [10], to secure two carrier geotextiles to a porous rigid substrate as shown in Fig. 2b. This type of gripping provided a relatively uniform transfer of shear loads into the geosynthetic specimen which allowed minimal slippage to take place and for the shear failure to occur at the intended interface, [7, 8, 10].

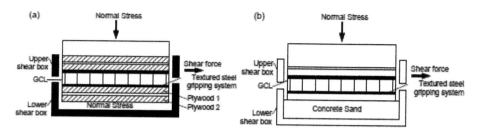

Fig. 2. Specimen gripping in direct shear.

Adhesive Bonding. The use of adhesive bonding is also another geosynthetic gripping method that has been reported, [11–13, 15].

[11, 13], conducted a study on interface strength testing where polyvinyl chloride (PVC) and water-resistant epoxy solvent was used to bond the geosynthetics to the direct shearing blocks, respectively. In addition, [11] used end clamps to improve confinement to the bottom platen. Similarly, [12, 15] applied glue (e.g. [15] used water-resistant spray-on adhesive and a two-part epoxy) as a gripping surface on the modified Bromhead torsional ring shear apparatus and pullout shear machine to measure the interfacial and internal shear strength parameter for the geosynthetics, respectively.

According to [11–13, 15], glue produced uniform shear displacement for each tested specimen indicating that slippage was minimal or prevented. However, [15] reported that this method was limited to lower normal stresses of less than 280 kPa (approximately). Additionally, [19] recommended not to consider glue for GCL specimens bonding as there are possible interference with the failure systems if careful steps are not taken (i.e. elimination of possible source errors).

Nail Plates. Special nail plates moulded in epoxy resin have been found to be another method of avoiding slippage during shearing tests on a direct shear box. [16] Investigated 2 mm short and finely distributed (1 nail per cm^2) nail plates placed on the upper and lower side of the GCL such that the two nail plates did not touch each other at any time during shearing. This method of specimen gripping provides adequate resistance against slippage and transfers the applied load across the tested specimen such that it led [20] to an improved stainless-steel nail plate gripping system with a better drainage.

3   Effects of Different Gripping Systems

Fox et al. [21] conducted a study where the effects of three different gripping systems (truss plate, coarse and medium sandpaper) were studied. The author sheared a hydrated woven (W)/nonwoven needle-punched (NW NP) GCL at a constant normal shearing stress of 37.8 kPa. Figure 3 shows the comparison of three results obtained during the shear tests. For the first test, the truss plate without end clamping was used and Fig. 3, shows a measured uniform stress-displacement curve of the tested GCL specimen. A second test was performed investigating coarse wet/dry sandpaper gripping surface with an end clamp system clamping the geotextile. The third gripping system used in their study was a medium coarse wet/dry, industrial strength sandpaper clamped at the edge of the shear box. Comparing the results obtained using the three gripping systems, the truss plates without end clamping produced a uniform shear failure with a higher peak of 75.9 kPa at a smaller displacement and marginally lower residual strength of 5.7 kPa as compared to the coarse and medium sandpaper. The coarse sandpaper produced a well-defined shear failure peak of 65.7 kPa and 6.9 kPa residual strength [21]. However, the medium coarse sandpaper did not mobilize sufficient friction to prevent tensile failure hence, slippage occurred rendering the test invalid [22]. With the difference of 15.5% in peak strength between the truss plate and coarse sandpaper, it shows that the type of geosynthetics gripping system in shear strength testing has a large effect on the accuracy of the results obtained.

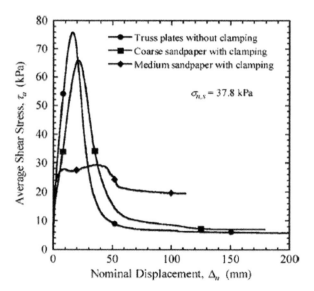

Fig. 3. Stress-displacement for GCL and three gripping systems [21].

In a separate study, [19] presented stress-displacement relationship curves for internal shear of hydrated needle-punched (NP) GCLs obtained using textured gripping systems and compared them to the results obtained by [23] using a modified metal connector plates without end clamps. Figure 4b–d displays results corresponding to different GCL lots, rolls, and products and may not be suitable for direct quantitative comparison. Instead, the shapes and similarities of the curves are important for the current discussion [19]. Generally, a stress-displacement curve representing the true performance of the material shear behaviour has a similar shape to one in Fig. 4a, [24]. The curve exhibits smooth transitions from the start of loading to peak shear strength and then to large displacement/residual shear strength, [24]. Relationships curves obtained for replicate specimens should show good similarity as normal stress is increased or decreased.

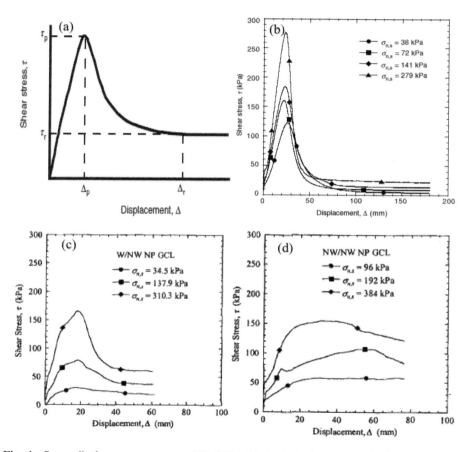

Fig. 4. Stress-displacement curves - NP GCL; (a) typical shear stress-displacement curve, (b) modified metal plate with no clamps, (c) modified metal plate with clamps, and (d) modified metal plate with clamps but the problem had occurred, [19].

The stress-displacement curves in Fig. 4b were the results obtained using the modified metal connector plate gripping system without end clamps. The relationship curves displaced sharp narrow similar peaks at lower displacements and according to [19], these curves are most likely the accurate representation of actual material shear behaviour. This can be confirmed by comparing high-quality stress-displacement relationships provided by [10] for torsional ring shear and by [15, 23] for direct shear.

In Fig. 4b, the stress-displacement relationships presented were investigated on a textured steel gripping system with end clamps. These relationships indicated slightly wider peaks with small stress peak but overall, the curves were in good similarity. Figure 4c, on the other hand, displaced double peaks, poor similarity, an absence of post-peak strength reduction ($\sigma_{n,s} = 96\,\mathrm{kPa}$) and undulations that were non-physical indicating that a problem occurred during shear testing. These inaccurate relationships in Fig. 4c were likely due to poor gripping systems that had caused slippage during shearing even though, machine friction problems are another possible cause of erroneous shear stress-displacement relationships, [19].

Fox and Kim [6] conducted eight direct shear tests on hydrated textured geomembrane (GMX)/GCL specimens at four different normal stresses (30.9, 41.2, 51.6, and 61.9 kPa) and the results obtained were compared to a simulated numerical model. For the first four shear tests, glue was used to bond the tested GMX specimen to the upper shearing block of the direct shear device whereas the last four tests, the GMX specimens were clamped at the end and allowed to slip on the shearing block (smooth gripping surface). Comparing the results obtained in Fig. 5a to the high-quality stress-displacement curves presented by [12, 15, 23], it can be observed that the relationships investigated when GMX was glued provided a close range to the true behaviour of the GMX/GCL interface.

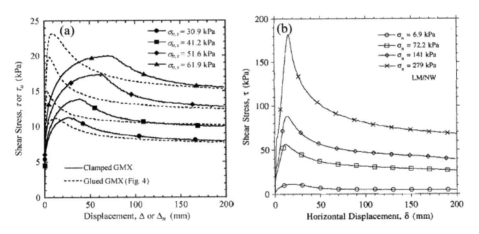

Fig. 5. Effects of gripping systems on GMX/GCL interface; (a) clamped vs glued specimen, [6] and (b) laminated texture HDPE/Nonwoven GCL, [15].

Fox and Kim [6] also compared the stress-displacement relationships curves for glued GMX/GCL interface tests to the simulated model as shown in Fig. 6a. These relationships (Fig. 6a) displayed excellent agreement at over certain ranges. [6] reported that each of these interfaces failed gradually with different points on the failure surface carrying shear stresses that range from pre-peak to post peak. Figure 6b shows the comparisons of failure envelopes for peak and residual shear strength of the glued GMX specimen with the clamped (smooth gripping surface) GMX specimen. The residual strengths were taken at displacements of 100 mm and 200 mm. The results indicated generally the same peak strength at the lowest normal stress, for both glued and clamped GMX interfaces. As normal stresses were increased the glued GMX specimen strength envelopes diverged displacing higher peak shear strength values. The largest error experienced at the highest normal stress for the clamped GMX tests was 14% [6].

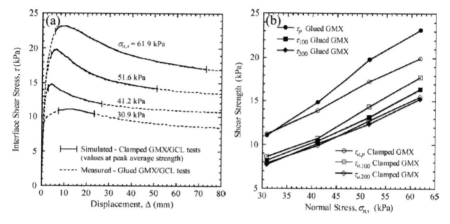

Fig. 6. Effects of the gripping systems in GM/GCL interface testing; (a) glued and simulated specimen, and (b) failure envelopes - glued and clamped specimen, [6].

4 Discussion

At present, it is not possible to know whether the discrepancies of the results caused by gripping systems are by a margin of 10%, 20% or even 50%. Researchers mentioned in the above sections, conducted their studies under different environments, (soil type, geosynthetic material, a gripping mechanism, applied load) which may have contributed to their differences in results. Hence, there is an urgent need for research to be carried out to quantitatively determine the effects of gripping systems in geosynthetic shear strength testing. In addition, a reflection on which gripping system would work best for various types of geosynthetic shear testing would be required. This system should avoid tensioning, transfer the applied shear stress across all failure surfaces, allow peak shear strength to occur anywhere and replicate the true behaviour of the stress-displacement of the tested specimen.

5 Conclusion

Reviewing the relevant literature, one can recognize slippage between the tested geosynthetic specimen and the gripping system as a major source of results discrepancy during shear strength testing. If the specimen is inadequately secured to the shearing blocks, the specimen experiences shear strength that deviates from the true material performance hence, providing inaccurate sets of design values in geosynthetic application projects. This may cause under or over-designing structures which may result in structural failure or cost ineffective designs, respectively. Hence, understanding or standardizing the geosynthetic gripping systems on a large direct shear device, pullout or ring shear will minimize the inaccuracies and increase reproducibility of test results. This will lead to accurate, safer and cost-effective design of geosynthetic structures.

References

1. ASTM D4439: ASTM D4439 (2006). https://compass.astm.org/EDIT/html_annot.cgi? D4439+17. Accessed 04 Sept 2017
2. Müller, W.W., Saathoff, F.: Geosynthetics in geoenvironmental engineering. Sci. Technol. Adv. Mater. **16**(3), 34605 (2015)
3. The Freedonia Group: World Geosynthetics - Market Size, Market Share, Market Leaders, Demand Forecast, Sales, Company Profiles, Market Research, Industry Trends and Companies. MarketResearch.com, Inc. (2017). https://www.freedoniagroup.com/industry-study/world-geosynthetics-3347.htm. Accessed 17 June 2017
4. ASTM International: ASTM 5321. https://compass.astm.org/EDIT/html_annot.cgi?D5321 +17. Accessed 04 Sept 2017
5. ASTM International: ASTM D6243. https://compass.astm.org/EDIT/html_annot.cgi?D6243 +16. Accessed 04 Sept 2017
6. Fox, P.J., Kim, R.H.: Effect of progressive failure on measured shear strength of geomembrane/GCL interface. Geotech. Geoenviron. Eng. **134**(4), 459–469 (2008)
7. Siebken, J.R., Swan, J., Yuan, Z.: Short-term and creep shear characteristics of a needlepunched thermally locked geosynthetic clay liner. Am. Soc. Test. Mater. Int. **STP11795S**, 89–102 (1997)
8. Trauger, R.J., Swan, J.R.H., Yuan, Z.: Long-term shear strength behavior of a needlepunched geosynthetic clay liner. Am. Soc. Test. Mater. **STP11795S**, 103–120 (1997)
9. Merrill, K., O'Brien, A.: Strength and conformance testing of a GCL used in a solid waste landfill lining system. Am. Soc. Test. Mater. **STP1308** (1997)
10. Zornberg, J.G., McCartney, J.S., Swan, J., Robert, H.: Analysis of a large database of GCL internal shear strength results. J. Geotech. Geoenviron. Eng. **131**(3), 367–380 (2005)
11. David, J.S.L., Robert, T.T.: Geomembrane interface strength tests (1991)
12. Eid, H., Stark, T., Doerfler, C.K.: Effect of shear displacement rate on internal shear strength of a reinforced geosynthetic clay liner. Geosynth. Int. **6**(3), 219–239 (1999)
13. Fox, P.J., Ross, J.D.: Relationship between NP GCL internal and HDPE GMX/NP GCL interface shear strengths. J. Geotech. Geoenviron. Eng. **137**(8), 743–753 (2011)
14. Lin, H., Shi, J., Qian, X.: An improved simple shear apparatus for GCL internal and interface stress-displacement measurements. Environ. Earth Sci. **71**(8), 3761–3771 (2014)
15. Triplett, J.E., Fox, J.P.: Shear strength of HDPE geomembrane/geosynthetic clay. Geotech. Geoenviron. Eng. **127**(6), 543–552 (2001)

16. ASTM International: ASTM 5321
17. Fox, P.J., Stark, T.D.: State-of-the-art report: GCL shear strength and its measurement – ten-year update. Geosynth. Int. **22**(1), 3–47 (2015)
18. ASTM D5321: Standard test method for determining the shear strength of soil-geosynthetic and geosynthetic-geosynthetic interfaces by direct shear. ASTM Int. **STP11795S**, 1–11 (2013)
19. Fox, P.J., Stark, T.D., Swan, J.R.H.: Laboratory measurement of GCL shear strength, January 2004
20. Zanzinger, H., Saathoff, F.: Shear creep rupture behaviour of a stitch-bonded clay geosynthetic barrier. In: 3rd International Symposium on Geosynthetic Clay Liners (2010)
21. Fox, P.J., et al.: Design and evaluation of a large direct shear machine for geosynthetic clay liners (1997)
22. Fox, P.J., Stark, T.D.: State-of-the-art report: GCL shear strength and its measurement. Geosynth. Int. **11**(3), 141–175 (2004)
23. Fox, P.J., Scheithe, J.R., Rowland, M.: Internal shear strength of three geosynthetic clay liners. Geotech. Geoenviron. Eng. **124**(10), 933–944 (1998)
24. Fox, P.J.: Internal and interface shear strengths of geosynthetic clay liners. In: 3rd International Symposium on Geosynthetic Clay Liners, pp. 1–17 (2010)

Hydraulic Performance of Geosynthetic Clay Liners with Mining Solutions

Yang Liu[1](✉), Li Zhen Wang[2], and Wei Jiang[1]

[1] Hunan Provincial Key Laboratory of Shale Gas Resource Exploitation,
Hunan University of Science and Technology, Xiangtan, Hunan, China
liuyang2585899@163.com
[2] Changsha Research Institute of Mining and Metallurgy Co., Ltd.,
Changsha, Hunan, China

Abstract. Geosynthetic clay liners (GCL) have been widely used as impermeable materials in landfill and other projects due primarily to their low hydraulic conductivity (k). Recently, GCLs are increasingly considered to contain mining leachates which may result in the degradation of their hydraulic performance. This study reports the hydraulic performance of a GCL under the mining solution from a mine in Hunan Province. Flexible-wall hydraulic conductivity test and free swell index test were conducted on GCL and bentonite specimens. The results indicate that the swell index decreased from 31 mL/2 g (in water) to 22 mL/2 g (under the mining solution). The hydraulic conductivity unexpectedly increased from 3×10^{-11} m/s to 4×10^{-9} m/s when changing the permeable solution from water to the mining solution. While if the GCL specimens were prehydrated with water for two days before contacting mining solution, the k value was decreased to an acceptable level (8×10^{-11} m/s). The greater effective stress also imparts a low hydraulic conductivity, the k value dropped from 4×10^{-9} m/s to 1×10^{-10} m/s when the effective stress elevated from 35 kPa to 200 kPa. The results of this study could provide a helpful reference for the application of GCLs under mining solutions.

Keywords: Geosynthetic clay liners · Mining solution
Hydraulic performance

1 Introduction

Geosynthetic clay liners (GCLs) are industry manufactured hydraulic barriers which generally consist of a layer of powdered or granular bentonite encased between two geotextiles and mainly combined together by stitching or needle punching [1]. GCLs have been widely used as hydraulic barriers in landfill liners/covers, surface impoundments and waste containment projects [2–7] due primarily to their low hydraulic conductivity to water ($k < 10^{-10}$ m/s). The low hydraulic conductivity mainly comes from the bentonite component which swells when in contact with water.

Studies have already been conducted to evaluate the hydraulic performance of GCLs with non-standard liquids. Higher hydraulic conductivity values were obtained when permeation with chemical solutions [8] and acid mine drainage [9], but potential

© Springer Nature Singapore Pte Ltd. 2018
L. Li et al. (Eds.): GSIC 2018, *Proceedings of GeoShanghai 2018 International Conference:
Ground Improvement and Geosynthetics*, pp. 430–436, 2018.
https://doi.org/10.1007/978-981-13-0122-3_47

pore filling resulted in decreased hydraulic conductivity of a GCL when permeation with alkaline solutions [10, 11]. Recently, GCLs are increasingly considered to contain mining leachates which may result in the degradation of their hydraulic performance. In mining industry, GCLs have a high probability to be in contact with acidic solutions, e.g., in heap leach containment systems [12] or the acidic leachate resulting from oxidation of impounded tailings [13]. Also heavy metals existed in mining solutions would result in the micro-structure change of the bentonite component and decreased swelling thus causing increased permeability of the whole GCL [14]. Consequently, the main purpose of this paper was to evaluate the compatibility of one GCL with mining solutions under different conditions. The result of this study would provide helpful reference to the use of GCLs in mining applications.

2   Methodology

2.1   Geosynthetic Clay Liner

All the tests (swell index and hydraulic conductivity) in this study were conducted on a commercial Chinese GCL. The GCL consists of polypropylene geotextiles (woven and non-woven on each side) and powdered sodium bentonite. The GCL is fiber-reinforced by needle-punching across the entire surface of the product then thermally locked to ensure a high shear strength. The photo and the properties of the GCL are shown in Fig. 1 and Table 1.

Table 1. Properties of the GCL specimen used in the present investigation

GCL properties			
Type of bentonite	Sodium	Initial unhydrated thickness (mm)	6 ± 0.4
Type of bonding	Needle-punched	Initial moisture content as received (%)	14 ± 1.2
Geotextiles (upper/lower)	Non-woven/woven	Mass per unit of area-Geotextiles upper/lower (kg/m^2)	0.25/0.15
Mass per unit of area-Bentonite (kg/m^2)	4.06 ± 0.45	Mass per unit of area-GCLs (kg/m^2)	4.54 ± 0.33

Fig. 1. Photo of the studied geosynthetic clay liner

2.2   Mining Solution

A mining solution collected form the heap leach was used in this study to test the compatibility of the GCL to the real solutions. The pH of the mining solution was approximately 5 measured using pH5+ meter from EUTECH Instruments Pty. Ltd.

2.3   Swell Index Test

Swell index tests were performed according to ASTM D 5890-Swell index of clay mineral component of geosynthetic clay liners.

2 g of powdered bentonite, previously oven dried at 105 ± 5 °C to a constant weight, was added in 0.1 g increment to a 100 mL graduated cylinder containing 90 mL of the test solution to undergo free hydration. The increments of bentonite were spread over the surface of the water every 10 min until 2.0 g of sample were added. Then the cylinder was topped off to 100 mL with the test solution. After a minimum of 16 h hydration, the volume of swollen bentonite was read in mL, and was reported as the swell index in millilitre per 2 g (mL/2 g) of dry bentonite (as shown in Fig. 2). All tests were duplicated.

Fig. 2.   Swell index test

2.4   Hydraulic Conductivity Test

The hydraulic conductivity tests were conducted generally according to ASTM D 6766-Standard Test Method for Evaluation of Hydraulic Properties of Geosynthetic Clay Liners Permeant with Potential Incompatible Liquids.

The hydraulic conductivity test equipment is shown in Fig. 3. Three flow pumps were used to provide the pressure for inflow, outflow and cell. Two interface chambers were installed in both inflow and outflow lines between the flow pumps and the cell to prevent potential damage to the pumps. Two pH and EC sensors were also inserted in-line between the interface chambers and the GCL sample to measure the pH/EC values of the influent and effluent. The pH and EC data were collected periodically.

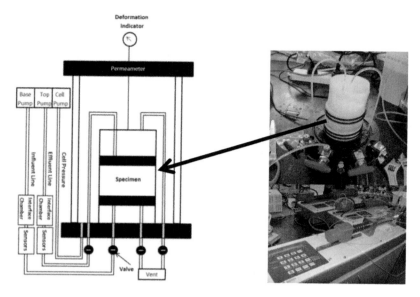

Fig. 3. Hydraulic conductivity test

The prepared GCL specimen was placed into the permeameter. One sheet of filter paper was placed between the top and bottom porous stones and the specimen to prevent intrusion of the material into the porous stones. Flexible membrane is used to encase the specimen against the leakage and rubber O-rings were also used to provide adequate seal at the base and cap. After assembling, the specimen was directly exposed to the mining solution for 48 h to mimic the worst scenario in the field.

Hydraulic conductivity tests were conducted with one mining solution at two effective stresses levels (35 kPa and 200 kPa) and also under the prehydration condition which means the GCL was prehydrated in deionized water for two days before contacting with the mining solution.

3 Results and Discussion

3.1 Swell Index

Data from swell index tests are shown in Fig. 4. As expected, the swell index values decreased from 31 mL/2 g to 22 mL/2 g when using water and the mining solution. Almost one third drop (9 mL/2 g) occurred for the studied bentonite.

As known, the industrial threshold value of swell index for a qualified GCL is considered at 24 mL/2 g. The initial swell index is much higher than the threshold (31 mL/2 g), however the value under mining solution is a little bit lower than the threshold (22 mL/2 g). The decrease of swell index is mainly due to the increasing ionic strength of the mining solution, the metals in the mining solutions replaced the sodium in the bentonite which should take the responsibility for the decreased swell

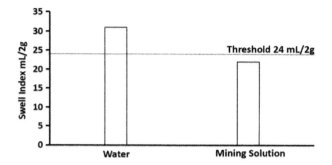

Fig. 4. Swell index test results

index value [15]. According to the previous research, the swelling properties of bentonite were closely related with the hydraulic conductivity of a GCL [6, 14]. The dropping swell index value would result in the increase in the hydraulic conductivity value which will be discussed in the following section.

3.2   Hydraulic Conductivity

Hydraulic conductivity test results under different conditions are shown in Fig. 5. The results indicate that the hydraulic performance of GCLs with mining solutions degraded compared to water as permeant, the k value was much higher than the industrial threshold (4×10^{-9} m/s vs 3×10^{-11} m/s), the ratio of the hydraulic conductivity values based on permeation with water to the value permeation with mining solutions was more than 100.

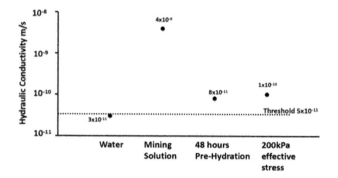

Fig. 5. Hydraulic conductivity test results

Low pH mining solution may result in the dissolution of smectite [16], and the release of multivalent cations (Fe^{3+}, Al^{3+}, and Mg^{2+}) could have resulted in a reduction in clay swelling, as shown in Fig. 4. Accordingly, little constriction of the pore space existed, and corresponding hydraulic conductivity values were greater.

The results of hydraulic conductivity for the specimens prehydrated with DI water for two days before contacting with the mining solution are shown in Fig. 5. Lower k values of the specimens prehydrated with DI water were observed compared with that exposed to mining solutions directly. The ratio of prehydrated k values to the non-prehydrated k values was ~ 40. Presumably these results were due to its better swelling properties in deionized water (swell index = 31 mL/2 g). As indicated in Fig. 5, prehydration was almost able counter the adverse effect caused by mining solution (8×10^{-11} m/s vs 3×10^{-11} m/s), which is acceptable as a waterproof material.

As expected, greater effective stress resulted in the decreased hydraulic conductivity. At 200 kPa, although the k value was still higher than the threshold, the ratio of k_w/k was lower than 4 (1×10^{-10} m/s vs 3×10^{-11} m/s). At high effective stress, the specimen was compressed compared to the lower effective stress indicating fewer pore space and resulting lower hydraulic conductivity.

4   Conclusions

The swell index results indicate that the SI value considerably decreased in the mining solution, which could be mostly attributed to the increased ionic strength, the constriction of the bentonite swelling would influence the hydraulic conductivity of GCL.

The hydraulic conductivity tests suggest the mining solutions exert adverse impact on the hydraulic performance of GCLs. The k value increased by 100 times over that of water. Mining solution lead to limited swelling due either to ionic strength or acid attack (or both) and resulting greater k. The elevated effective stress caused the closure of pore space and therefore restricted flow through the GCL specimens and the resulting lower hydraulic conductivity. The measured k at 200 kPa effective stress was less than 4 times to the k permeation with water. Prehydration almost counter the effect of the mining solutions, the k value dropped down to an acceptable level after 48 h prehydration before contacting with the mining solution. Further tests with different mining solutions are still needed to completely assess the compatibility/incompatibility of GCLs in mining applications.

Acknowledgement. The authors were funded by the National Natural Science Foundation of China (Project No. 51608192), the Natural Science Foundation of Hunan Province (Grant No. 2016JJ4031, 2017JJ3070) and the Scientific Research Foundation for the Returned Overseas Chinese Scholars, State Education Ministry (Year 2015, No. 1098).

References

1. Estornell, P., Daniel, D.E.: Hydraulic conductivity of three geosynthetic clay liners. J. Geotech. Geoenviron. Eng. **118**, 1592–1606 (1992)
2. Mazzieri, F., Di Emidio, G., Fratalocchi, E., Di Sante, M., Pasqualini, E.: Permeation of two GCLs with an acidic metal-rich synthetic leachate. Geotext. Geomembr. **40**, 1–11 (2013)
3. Bouazza, A., Gates, W.P.: Overview of performance compatibility issues of GCLs with respect to leachates of extreme chemistry. Geosynth. Int. **21**(2), 151–167 (2014)

4. Bouazza, A., Singh, R.M., Rowe, R.K., Gassner, F.: Heat and moisture migration in a geomembrane-GCL composite liner subjected to high temperatures and low vertical stresses. Geotext. Geomembr. **42**(5), 555–563 (2014)
5. Rowe, R.K.: Performance of GCLs in liners for landfill and mining applications. Environ. Geotech. **1**(1), 3–21 (2014)
6. Liu, Y., Bouazza, A., Gates, W.P., Rowe, R.K.: Hydraulic performance of geosynthetic clay liners to sulfuric acid. Geotext. Geomembr. **43**(2), 14–23 (2015)
7. Touze-Foltz, N., Bannour, H., Barral, C., Stoltz, G.: A review of the performance of geosynthetics for environmental protection. Geotext. Geomembr. **44**(5), 656–672 (2016)
8. Shackelford, C.D., Benson, C.H., Katsumi, T., Edil, T.B., Lin, L.: Evaluating the hydraulic conductivity of GCLs permeated with non-standard liquids. Geotext. Geomembr. **18**, 133–161 (2000)
9. Kashir, M., Yanful, E.K.: Hydraulic conductivity of bentonite permeated with acid mine drainage. Can. Geotech. J. **38**, 1034–1048 (2001)
10. Benson, C.H., Ören, A.H., Gates, W.P.: Hydraulic conductivity of two geosynthetic clay liners permeated with a hyperalkaline solution. Geotext. Geomembr. **28**, 206–218 (2010)
11. Gates, W.P., Bouazza, A.: Bentonite transformations in strongly alkaline solutions. Geotext. Geomembr. **28**(2), 219–225 (2010)
12. Hornsey, W.P., Scheirs, J., Gates, W.P., Bouazza, A.: The impact of mining solutions/liquors on geosynthetics. Geotext. Geomembr. **28**, 191–198 (2010)
13. Shackelford, C.D., Sevick, G.W., Eykholt, G.R.: Hydraulic conductivity of geosynthetic clay liners to tailings impoundment solutions. Geotext. Geomembr. **28**, 149–162 (2010)
14. Fehervari, A., Gates, W.P., Patti, A.F., Turney, T.W., Bouazza, A., Rowe, R.K.: Potential hydraulic barrier performance of cyclic organic carbonate modified bentonite complexes against hyper-salinity. Geotext. Geomembr. **44**(5), 748–760 (2016)
15. Ruhl, J.L., Daniel, D.E.: Geosynthetic clay liners permeated with chemical solutions and leachates. J. Geotech. Geoenviron. Eng. **123**, 369–381 (1997)
16. Liu, Y., Gates, W.P., Bouazza, A.: Acid induced degradation of the bentonite component used in geosynthetic clay liners. Geotext. Geomembr. **36**(1), 71–80 (2013)

Numerical Analysis of Different Effects of GCL and Horizontal Drainage Material on Moisture Field of Highway Subgrade

Feng Liu[1], Zhibin Liu[1(✉)], Shujian Zhang[2], Jian Zheng[3],
Songlin Lei[2], and Congyi Xu[1]

[1] School of Transportation, Southeast University, Nanjing 210096, China
seulzb@seu.edu.cn
[2] Jilin Provincial High Class Highway Construction Bureau,
Changchun 130062, China
[3] Inner Mongolia Xing'anmeng Transportation Bureau, Ulan Hot 137400, China

Abstract. The moisture field of subgrade soil has great influence on the engineering performance of road embankment as well as above pavement. Rainfall or capillary elevation of shallow groundwater can increase the moisture content of subgrade soil, which may lead to instability of the embankment, frost heave and thaw settlement of road subgrade. Geosynthetic clay liner (GCL) is mainly used in engineering practice to work as an effective material of sealing, isolation and leakage prevention. Horizontal drainage material (HDM) with high directional permeability is able to laterally drain water out of soils. Both of them can be introduced into highway subgrade to provide possible moisture filed adjustment. In this research, the moisture field variations of highway subgrade due to introduction of a layer of GCL or HDM are studied respectively based on numerical simulation. Calculation results indicate that both materials can effectively reduce the infiltration of rain water and keep the stability of the road embankment. Furthermore, GCL works better than HDM in stopping rain water infiltration, but a drainage layer is suggested to be placed above the GCL.

Keywords: Subgrade · Rainfall · Moisture field · GCL · HDM
Numerical simulation

1 Introduction

In the construction of highway projects, how to prevent water from damaging the subgrade is an important topic. The subgrade can be influenced by rainfall, evaporation and other natural factors, which may lead to changes of subgrade soil, and sometimes result in subgrade disease and even destruction. Water is one of the most important factors that cause subgrade diseases.

In recent years, many scholars have studied this problem. Salour and Erlingsson (2012) found that when the moisture content in unbound layers increased, the pavement bearing decreased significantly during spring thawing. Taamneh and Liang (2010) presented a six-year field monitoring of six fully repaired pavements and advised that the prolonged free water existing in the pavement structure was

© Springer Nature Singapore Pte Ltd. 2018
L. Li et al. (Eds.): GSIC 2018, *Proceedings of GeoShanghai 2018 International Conference: Ground Improvement and Geosynthetics*, pp. 437–445, 2018.
https://doi.org/10.1007/978-981-13-0122-3_48

detrimental because it not only adversely affected the carrying capacity of the pavement, but also led to premature failure of the road. Rainwater et al. (1999) used TDR probes to collect the meteorological data, subgrade water content, infiltration and temperature data in road. The results showed that the meteorological conditions could directly affect the water content of the stone base.

In order to solve the problem of water damage to the subgrade, some measures should be taken to adjust the subgrade moisture field to improve its stability. It is found that two kinds of materials can reduce the impact of water changes on the subgrade: Geosynthetic clay liner (GCL) is mainly used in engineering practice to work as an effective material of sealing, isolation and leakage prevention with a permanent waterproof capacity. In addition, a new type of horizontal drainage material (HDM) with high directional permeability is able to laterally drain water out from soils even under unsaturated conditions. Zhu and Feng (2016) used the finite element analysis software autobank6.0 to simulate the laying of the GCL, and the simulation results indicated that GCL's anti-seepage effect was significant. Wang et al. (2017) developed a physical model test to assess the effectiveness of a horizontal drainage material named wicked geotextiles in soil moisture reduction, and the results showed that the wicking geotextile could discharge water from the soil even if the moisture content was close to the optimum moisture content.

In this paper, the numerical simulation method is used to analyze the characteristics of moisture field of a highway subgrade with the existence of a layer of GCL or HDM, as few scholars have made comparison of such two types of materials. It is necessary to study the difference of these two materials. Three different subgrade models were created: without geosynthetics, with GCL, and with HDM. The SEEP/W module of GeoStudio is used to analyze the change of moisture field of subgrade soil under rainfall condition. The calculation results of three subgrade models were compared to draw a conclusion.

2   Basic Differential Equations and Boundary Conditions for Saturated - Unsaturated Seepage Flow

To solve the saturated-unsaturated seepage problem, it's necessary to establish the governing equation (assuming that the soil skeleton is not deformed and the water is an incompressible fluid) and the boundary conditions (Wu and Gao 1999):

$$
\begin{cases}
\frac{\partial}{\partial x}\left(k_x(n,s)\frac{\partial h}{\partial x}\right) + \frac{\partial}{\partial y}\left(k_y(n,s)\frac{\partial h}{\partial y}\right) = \rho_w g m_2^w \frac{\partial h}{\partial t} \\
h(x,y,t) = h_1(x,y,t), \text{ on the boundary of } s_1 \\
k_x\frac{\partial h}{\partial x}\cos(\overline{n},x) + k_y\frac{\partial h}{\partial x}\cos(\overline{n},y) = q(x,y,t), \text{ on the boundary of } s_2 \\
h(x,y,t) = z(x,y,t), \text{ on the boundary of } s_3 \\
h(x,y,t_0) = h_0(x,y,t_o)
\end{cases}
$$

In above equation: h - the head, $h = \frac{u}{\gamma_w} + z$; u - the pore pressure, γ_w - the density of water, z - the position head; s - the saturation; Cartesian axes x, y - the main directions of infiltration; k_x, k_y - the permeability coefficient along the main direction respectively

s_1 - the known head boundary, s_2 - the known flow boundary, s_3 - the exit section boundary, h_0 - the known head, q - the boundary flow, $\cos(\bar{n}, x)$ - cosine in the normal direction outside the boundary; t - the time.

3 Subgrade Model Establishment and Material Parameters

Due to the symmetry of the subgrade structure, half of the pavement is adopted for analysis. The subgrade configuration is shown in Fig. 1. It includes two layers: 0–8 m for silty clay, and 8–13 m for clay. The height of the model from the top of the embankment to the bottom of the calculated depth is 13 m. The initial groundwater level is 6 m from the bottom of the model. The physical and mechanical parameters of each layer of soil are shown in Table 1.

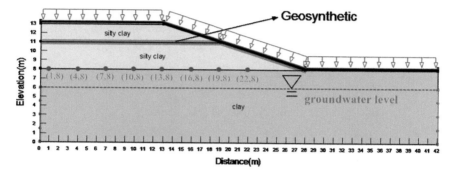

Fig. 1. The numerical model of subgrade section

Table 1. The physical and mechanical parameters of two layers of subgrade soil

Soil type	$\gamma/(\text{kN} \cdot \text{m}^{-3})$	c/kPa	$\varphi/(°)$	$k_{sat}/(\text{m} \cdot \text{h}^{-1})$	$\theta/(\text{m}^3 \cdot \text{m}^{-3})$
Silty clay	18	14	20	0.018	0.47
Clay	21	7	20	0.06	0.39

According to the provisions of the China Meteorological Administration, the amount of rainfall within 24 h is defined as the daily rainfall. Normally the daily precipitation is usually classified as light rain, moderate rain, heavy rain, rainstorm, etc. Light rain corresponds to the daily precipitation of 10 mm and below, moderate rain corresponds to the daily precipitation of 10–24.9 mm, heavy rain corresponds to 25–49.9 mm, rainstorm corresponds to 50–99.9 mm, and heavy rainstorm corresponds to 100–250 mm. In this research, the precipitation of 5 mm/h were chosen to simulate the impact. The rainfall lasted for 48 h, and the changes of subgrade moisture within 48 h after the end of the rain were analyzed too.

As shown in Fig. 1, GCL and HDM will be buried at a depth of 2 m below the surface of the subgrade. GCL is made of two layers (braided fabric and geotextile) and bentonite particles by suturing, acupuncture or bonding. Zhou and Feng (2016) studied the basic properties of braided fabric and geotextile and bentonite in GCL, finding that GCL has high strength and a good anti-erosion ability. GCL has low permeability coefficient and good waterproof effect. In this analysis, the permeability coefficient of GCL is 10^{-9} cm/s (Zhou et al. 2002). HDM can transfer water in the horizontal direction quickly. The horizontal permeability coefficient can reach 0.2 cm/s. Wang et al. (2017) studied the properties of HDM and found that the wide width tensile strength could reach 77 kN/m.

The soil water characteristic curve (SWCC) shows how much water is held in soil due to the corresponding matrix suction. SWCC reflects the water holding capacity of a soil (Li et al. 2007). The SEEP/W program provides a variety of estimated mathematical models. Normally the VG model is very similar to measured data, and the parameters are easy to determine. Therefore, VG model is used to estimate the parameters of unsaturated seepage property of subgrade soils here. The specific form of the model (van Genuchten 1980) is:

$$S_e = \frac{\theta - \theta_r}{\theta_s - \theta_r} = [1 + |\alpha h|^n]^{-m} \qquad h < 0$$

$$S_e = 1 \quad h \geq 0$$

In above equation: S_e - the saturation, θ - the volume of water content ($L^3 L^{-3}$), θ_r and θ_s - the residual moisture content and saturated moisture content ($L^3 L^{-3}$); h - the pressure head (L); α and n - the parameters of the shape of the curve, $m = 1 - \frac{1}{n}$.

Here, typical SWCC and infiltration functions of the two layers of soils suggested in Geostudio are adopted and shown in Fig. 2.

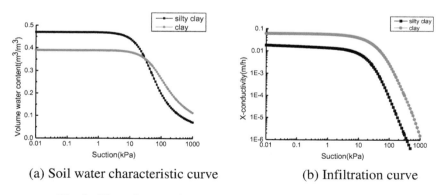

(a) Soil water characteristic curve   (b) Infiltration curve

Fig. 2. The soil water characteristic curve and infiltration function

4 Analysis of Subgrade Moisture Field

Figure 3 shows the pore water pressure of the subgrade in the initial state (0 h) in three subgrade models. Figure 4(a) and (b) show the pore water pressure of the subgrade without geosynthetics when the rain just stopped (48 h) and 2 days after the end of the rain (96 h). It can be seen from Fig. 4 that after the rainfall, the initial groundwater level in the subgrade is obviously increased. The pore water pressure above the groundwater obviously rises.

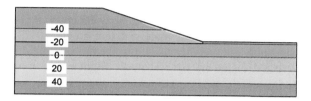

Fig. 3. Initial state (kPa)

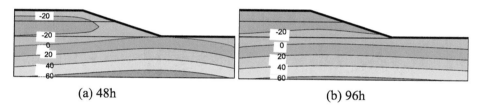

(a) 48h   (b) 96h

Fig. 4. Distribution of pore water pressure (kPa) in subgrade without geosynthetics

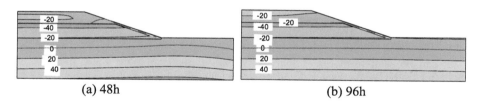

(a) 48h   (b) 96h

Fig. 5. Distribution of pore water pressure (kPa) in subgrade with GCL

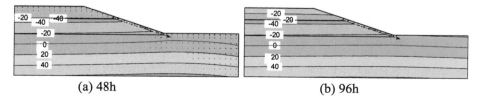

(a) 48h   (b) 96h

Fig. 6. Distribution of pore water pressure (kPa) on subgrade with HDM

Figure 5(a) and (b) show the changes of pore water pressure with GCL in subgrade. When GCL is laid in the subgrade, the rainfall has little effect on the pore water pressure in the area beneath GCL, and the groundwater level is slightly improved. But the pore water pressure above GCL becomes larger, which indicates that the water content above GCL increases. There may be water accumulation above GCL. The groundwater level dropped after the rainfall stopped.

Figure 6 shows the cloud map of pore water pressure over time in the case of HDM. As shown in Fig. 6, the presence of HDM, rainfall has little effect on the pore water pressure below HDM, and the groundwater level rises slightly near the slope. There may be some water accumulated above HDM, but the phenomenon has alleviated two days after the rain, which suggests that the water absorption capacity of HDM may be a feasible way to solve the subgrade moisture problem.

In order to better study the specific changes of the moisture field in the subgrade, a cross section is selected for quantitative analysis. Eight nodes are selected, and the coordinates are (1, 8) (4, 8) (7, 8) (10, 8) (13, 8) (16, 8) (19, 8) (22, 8) from left to right. As shown in Fig. 1, the section is 5 m from the top, 8 m from the bottom, between the groundwater level and the geotextile. The curve of pore water pressure over time in each node of the three models can be got, as shown in Fig. 7.

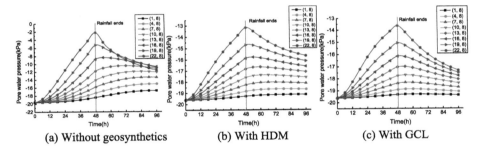

(a) Without geosynthetics   (b) With HDM   (c) With GCL

Fig. 7. Curves of pore water pressure

It can be seen from Fig. 7 that the variation trends of pore water pressure are basically similar in the three subgrade sections. During the period of raining (0–48 h), the pore water pressure continues to increase, and the pore water pressure reaches the maximum value at the end of raining. After the rainfall, the pore water pressure decreases, and the matric suction begins to increase. Finally the moisture field of subgrade soil tends to be stable. By comparing the pore water pressure curves of different nodes, it also can be found that the closer the nodes near the center, the pore water pressure of the nodes are smaller; the closer the nodes are near to the slope surface of the subgrade, the greater the pore water pressure will be. Obviously, it suggests that rainfall has more influence on the slope surface than inside. Two nodes are specially selected including (1, 8) and (22, 8). Then the effect of the two kinds of materials on the moisture field are compared.

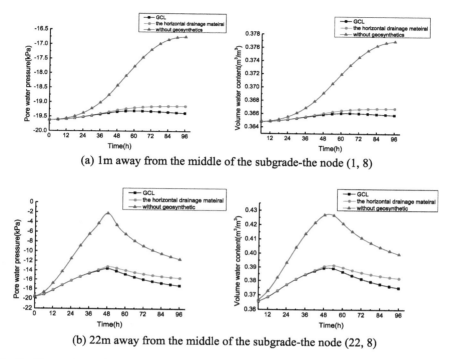

(a) 1m away from the middle of the subgrade-the node (1, 8)

(b) 22m away from the middle of the subgrade-the node (22, 8)

Fig. 8. Comparison of pore water pressure and volume water content in three subgrade models

As shown in Fig. 8, if GCL or HDM is installed, the pore water pressure and the volume water content below geosynthetics will be small. On the contrary, the pore water pressure and the volume water content in the subgrade without geosynthetics will increase quickly, much higher than that with geosynthetics. It can be concluded that GCL or HDM can effectively reduce the increase of pore water pressure and accumulation of water under the condition of rainfall. As the rain stops, the pore water pressure and the volume of water content begins to decrease. Through comparison between GCL and HDM, it can be found that GCL may be better in effectively preventing the increase of pore water pressure and volume water content. The change of water content in the area above the geosynthetics should be considered too. Therefore, a node above the geosynthetics is chosen to be discussed.

Figure 9 shows that during the rainfall, the pore water pressure and the volume of water content above the geosynthetics increases quickly. The geosynthetics will make the water above the geosynthetics unable to effectively infiltrate downwards, which results in the accumulation of water. In addition, HDM is more likely to cause the accumulation of water than GCL. However, after the rainfall stops, HDM can make the accumulated water in the subgrade horizontally flow out quickly leading to the decrease of pore water pressure. Thus, it is recommended to add a layer of drainage material above the GCL so that the water can be successfully discharged. In this study, it should be noticed that HDM is simply defined as a material with a high horizontal permeability coefficient. It may be the main reason for the accumulation of water above HDM.

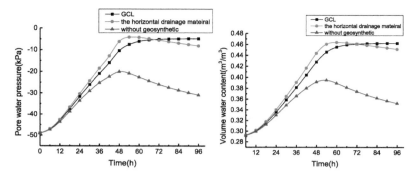

Fig. 9. Curves of pore water pressure and volume water content above the geosynthetics

5   Conclusions and Discussion

In order to learn the possible application of two different geosynthetic materials in highway subgrade to deal with the possible water-related destruction, the influence of GCL and HDM on the moisture field of the subgrade is studied by numerical simulation. Finally, the following conclusions are obtained:

(1) GCL and HDM can effectively prevent rainfall infiltration and maintain the stability of the road embankment.

(2) In general, GCL material is slightly better than HDM in preventing rainfall infiltration. In the area above the geosynthetics, water may be accumulated. Thus, it is recommended to add a layer of drainage material above the GCL material so that the water can be effectively discharged.

(3) In this research work, HDM is simply defined as a material with high horizontal permeability coefficient. It may be the reason for the accumulation of water in the area above HDM.

References

Li, Z.Q., Li, T., Hu, R.L., et al.: Methods for testing and predicting of SWCC in unsaturated soil mechanics. J. Eng. Geol. **15**(5), 700–707 (2007). (in Chinese)

Rainwater, N.R., Ronald, E.Y., Eric, C.D., et al.: Comprehensive monitoring systems for measuring subgrade moisture conditions. J. Transp. Eng. **125**(5), 439–448 (1999)

Salour, F., Erlingsson, S.: Pavement unbound materials stiffness-moisture relationship during spring thaw. In: Cold Regions Engineering, pp. 402–412 (2012)

Taamneh, M., Liang, R.Y.: Long-term field monitoring of moisture variations under asphalt pavement with different drainable base materials. In: Geoshanghai International Conference, vol. 42, pp. 453–459 (2010)

van Genuchten, M.T.: A closed-form equation for predicting the hydraulic conductivity of unsaturated soil. Soil Sci. Soc. Am. J. **44**(5), 892–898 (1980)

Wang, F., Han, J., Zhang, X., et al.: Laboratory tests to evaluate effectiveness of wicking geotextile in soil moisture reduction. Geotext. Geomembr. **45**(1), 8–13 (2017)

Wu, M.X., Gao, L.S.: Saturated-unsaturated unsteady seepage numerical analysis. J. Hydraul. Eng. **30**(12), 0038–0043 (1999). (in Chinese)

Zhou, J., Feng, J.G.: Analysis on the results of GCL (waterproof liner) anti-erosion test. Resour. Environ. Eng. **30**(3), 290–291 (2016). (in Chinese)

Zhou, Z.B., Wang, Z., Wang, J.Q.: An overview about properties and application of a new geocomposite-GCL. J. Yangtze River Sci. Res. Inst. **19**(1), 35–38 (2002). (in Chinese)

Zhu, Y.T., Zhao, W.X., Zhang, Z.H., et al.: The finite element analysis on seepage controlling effect of GCL. New Build. Mater. **2**, 68–73 (2016). (in Chinese)

DEM Simulation of Pullout Tests
of Geogrid-Reinforced Gravelly Sand

Chao Xu[1] and Cheng Liang[2(✉)]

[1] Key Laboratory of Geotechnical and Underground Engineering of Ministry
of Education, Tongji University, Shanghai, China
c_axu@tongji.edu.cn
[2] Department of Geotechnical Engineering, Tongji University, Shanghai, China
92chengliang@tongji.edu.cn

Abstract. Geogrid has been widely used to stabilize the earth structures and the interaction behavior between geogrid and its surrounding backfill soil is the key issues for this reinforcement to develop its reinforcing effects. In this study, a two-dimensional discrete element model was built by PFC^{2D} to simulate the pullout behavior of uniaxial geogrid-reinforced granvelly sand under tensile loads. The backfill soil was modelled as unbonded particles with linear contact stiffness model and the geogrid was modelled as bonded particles with piecewise linear model which was developed based on parallel bond model and was applied in every two adjacent two particles of geogrid. This numerical model was firstly calibrated and verified against the results obtained from laboratory experiments of geogrid tensile tests and direct shear tests of sand. Then, two different kinds of soil particle distribution of gravelly sand were simulated to study the effects of gradation on the pullout responses of displacement distribution along the geogrid, normal stress distribution in the geogrid plane and pullout force at different clamp displacements. The results showed that the pullout force of geogrid embedded in well graded gravelly sand increased with increasing the clamp displacements and was greater than that of geogrid embedded in poorly graded gravelly sand at larger clamp displacements.

Keywords: Geogrid · Pullout tests · Gravelly sand · Gradation
Discrete element method

1 Introduction

Geogrid, as an important reinforcing material, has been widely used to stabilize the earth structures, e.g. reinforced soil retaining walls, slopes, embankments, etc. The interaction behavior between geogrid and backfill soil is the key issues for this reinforcement to develop its reinforcing effects. And pullout tests have been regarded as a better way to study the load transfer mechanisms and to calculate the relevant parameters of geogrid-reinforced soil structures by conducting laboratory experiments and numerical simulations (Palmeira and Milligan 1989; McDowell et al. 2006; Sieira et al. 2009; Zhou et al. 2012; Tran et al. 2013; Wang et al. 2016). Though plenty of factors have been considered to study their influences on the responses of geogrid and backfill soil under pullout loads (Palmeira 2009; Moracin and Cardile 2012; Cardile et al. 2017),

© Springer Nature Singapore Pte Ltd. 2018
L. Li et al. (Eds.): GSIC 2018, *Proceedings of GeoShanghai 2018 International Conference: Ground Improvement and Geosynthetics*, pp. 446–454, 2018.
https://doi.org/10.1007/978-981-13-0122-3_49

the pullout behaviors of geogrid-reinforced gravelly sand have not been studied clearly, especially, by taking the effects of soil particle distributions into consideration.

The discrete element method (DEM) shows the obvious advantages in studying the properties of soil by considering its discontinuous nature (Gao and Meguid 2017; Miao et al. 2017). In this research, the Particle Flow Code (PFC^{2D}), a commercial DEM software, was employed to model the effects of particle distribution of gravelly sand on the pullout behavior. The backfill soil was modelled as unbonded particles with linear contact stiffness model and the geogrid was modelled as bonded particles with piecewise linear model which was developed based on parallel bond model. The results obtained in this research are expected to enhance the understandings of effects of soil particle distributions on geogrid-soil interaction behaviors.

2   DEM Modeling

2.1   General

The numerical model was built based on the experimental results reported by Moraci and Recalcati (2006) and Cardile et al. (2016). Uniaxial geogrid was used as reinforcement in their laboratory experiments, so this model test can be simplified as a two-dimensional problem and the PFC^{2D} software was suitable for this study. Figure 1 shows the details of the pullout box which had dimensions of 1700 mm long by 680 mm height and was modelled using wall element in PFC software. To account for the computational time, the soil particles were up-scaled by 20 times as well as the uniaxial geogrid for DEM simulations. Though the numerical results can be influenced by the up-scaling factor, this technology has been used in many DEM studies (Lin et al. 2013; Tran et al. 2013; Wang et al. 2016), and the results were satisfactory for investigated objects. More importantly, the soil particle distribution has not been changed after being up-scaled, as shown in Fig. 2. In order to reduce the effects of boundary conditions on the results of pullout tests, the friction coefficient between soil particles and walls was set to 0 during the whole process. A constant normal stress of 50 kPa was applied at the top of the specimen through servo control method, and a fixed rate of 1 mm/min was applied on the clamp which was simulated using clump

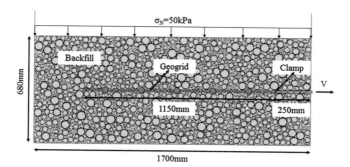

Fig. 1. Model of pullout test

element with a length of 250 mm. The diameters of the particles used to simulate the clamp were equal to that of the particles used to model the longitudinal members of geogrid. Furthermore, the friction coefficient between clamp and backfill was set to 0 to eliminate its pullout force induced during testing.

2.2   Backfill and Geogrid

Two different kinds of well graded (WG) and poorly graded (PG) gravelly sand were chosen as backfill soil, which have the same maximum and minimum particle diameters but different particle distributions, as shown in Fig. 2. In the numerical model, the sand was modelled as unbonded circular particles with the linear contact stiffness model. To make a dense specimen, a small friction coefficient (0.05) was set to the particles and the two-dimensional porosity of the specimen was set to 0.12 which was in a rational range according to the DEM studies conducted by Han et al. (2012). Multilayer compaction method was used during preparing the specimen and the it was divided into seven layers in this study. When the maximum contact force ratio equaled to 0.001, each layer will be considered to reach the equilibrium state and the calculation will be stopped. After finishing the preparation, the friction coefficient of particles was then increased to 1.6 to compensate the lack of angularity of circular soil particles used in the numerical model. The friction coefficient was obtained by calculating iteratively by comparing the DEM results with experimental ones, which was illustrated carefully in the later part of model verification.

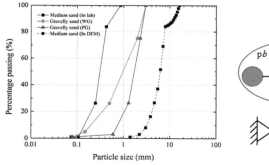

Fig. 2. Gradation curves of particle size distributions

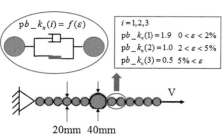

Fig. 3. The piecewise linear model

The geogrid was totally embedded in sand with a length of 1150 mm and was pulled out through a clamp device which was fixed to the geogrid without any relative displacement, as shown in Fig. 1. Their connections were simulated by applying the same tensile rate of 1 mm/min to the clump and the first particle of the geogrid in the DEM models. Meanwhile, the geogrid was modelled as bonded particles with piecewise linear model which was developed by Wang et al. (2014) based on parallel bond model. This model was applied in every two adjacent particles of geogrid, and the details were shown in Fig. 3. The parallel bond normal stiffness was the function of strain of geogrid, because the tensile results showed a nonlinear relationship between the force and strain and the pullout force was transferred progressively from pullout end to free end in pullout tests. However, a constant stiffness of parallel bond was always used in previous studies (Lin et al. 2013; Miao et al. 2017). Besides, to characterize the properties of transverse members of geogrid, circular particles of diameter of 40 mm (after being up-scaled) were used, as shown in Fig. 3.

2.3 Model Verification

To catch the mechanical characteristics of geogrid and medium sand used in laboratory experiments, numerical tensile tests and numerical direct shear tests were conducted using PFC, respectively. In the numerical tensile tests, the geogrid was modelled with a length of 200 mm as in the laboratory test. The last particle of geogrid specimen on the left was fixed and a constant tensile rate of 20 mm/min was applied to the first particle, as shown in Fig. 3. The parallel bond normal stiffness consisted of three different values which were computed based on the slope obtained from the relationship between force and strain in experimental tensile test. To simulate the nonlinear mechanical behavior of geogrid, this value was updated every 10 steps when calculating in PFC according to the strain (ε) of every adjacent two particles of geogrid along its length under tensile force. After being calculated iteratively, the final input parameters were shown in Table 1. The DEM results of tensile tests were coincident well with the experimental ones, as shown in Fig. 4.

Table 1. Parameters for DEM models

Density of particles (kg/m^3)	2650
Density of geogrid (kg/m^3)	1000
Contact normal stiffness of particles k_n (N/m)	2×10^7
Contact shear stiffness of particles k_s (N/m)	2×10^7
Contact normal stiffness of walls k_n (N/m)	3×10^7
Contact shear stiffness of walls k_s (N/m)	3×10^7
Parallel bond normal stiffness pb_k_n (GPa/m)	1.9, 1.0, 0.5
Parallel bond shear stiffness pb_k_s (GPa/m)	1.9
Friction coefficient of particles	1.6

The multilayer compaction method was also utilized to prepare the specimen in the numerical direct shear tests. And the tests were conducted at three different normal stresses including 25, 50 and 100 kPa, which were identical to those in the laboratory tests. The internal friction angle of sand was calculated based on the equation $\tau = \sigma \tan \varphi$, assuming that cohesion equaled to 0. It decreased with the increase of normal stress and the DEM results were just slightly larger than the experimental results, as shown in Fig. 5, which indicated that the model can capture the mechanical properties of sand. After being verified by these two tests, the authors believe that the numerical results are reliable and the DEM model can be used for further studies.

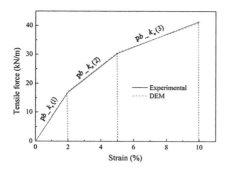

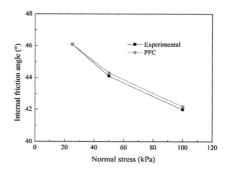

Fig. 4. Experimental and DEM results of geogrid tensile test

Fig. 5. Experimental and DEM results of direct shear tests

3   Analysis of Results

The calibrated numerical model was then used to study the effects of soil particle gradation on the pullout behaviors including displacement distributions of geogrid, normal stress distributions in the geogrid plane and pullout force at different clamp displacements (u). The input parameters were the same as those used in the numerical model of tensile tests and direct shear tests except the soil particle distributions.

Figure 6 shows the displacement distributions along the geogrid at different clamp displacements in WG and PG gravelly sand. It is obvious that the displacement increased with increasing clamp displacements and decreased nonlinearly from the clamp to the end. This pattern has been understood clearly by laboratory experiments and numerical simulations (Wang et al. 2014; Cardile et al. 2016). But there are some differences of displacement development of geogrid by considering the effects of soil gradation in this study. The entire geogrid almost has been activated in WG gravelly sand when the clamp displacement was beyond 70 mm, while the fourth knot of the geogrid embedded in PG gravelly sand has not been activated. This indicated that the pullout force has not been transferred to the fourth knot of the geogrid embedded in PG gravelly sand when the clamp displacement was less than 90 mm, which might be caused by the lack of some soil particles with poor gradation.

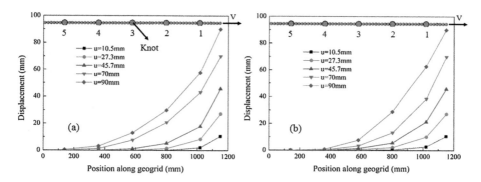

Fig. 6. Displacement distributions along the geogrid. (a) WG. (b) PG.

The normal stress distribution at the interface between geogrid and backfill soil (σ_{NG}) has always been regarded to be constant when calculating the relevant interfacial parameters in geosynthetic-reinforced earth structures. To verify this assumption, the normal stress distributions in the geogrid plane were monitored by using measurable element in PFC at different clamp displacements and the results were shown in Fig. 7. It was quite clear that the normal stress did not always equal to the target value of 50 kPa which was applied at top of the specimen. With increasing the clamp displacements, the normal stress closed to the pullout direction increased rapidly, while it decreased slightly at the latter part of the geogrid. Similar patterns have been found by Tran et al. (2013) and Wang et al. (2016), but they hold different explanations in terms of the stress increase. Tran et al. (2013) thought it was caused by using a rigid loading plate at the top of the specimen, while Wang et al. (2016) preferred to the horizontal forces induced by geogrid when pulling out and made comparative calculations showed that there was no obvious difference when using rigid plate or flexible plate. In this study, the distribution of normal stress can be divided into three sections approximately along the geogrid plane based on its calculated average value, as shown in Fig. 7. The normal stress of first section was larger than 50 kPa, the second one equaled to 50 kPa and the third one was smaller than 50 kPa. The width of the first section in WG gravelly sand was almost the same with that in PG gravelly sand, however, which possessed a wider second section. This can be attributed to the interaction behavior between geogrid and its surrounding backfill soil, because the latter part of the geogrid away from the pullout direction in PG gravelly sand has not been activated at the same clamp displacement compared with that in WG gravelly sand, and so was the surrounding soil particles at the end of the geogrid. In general, the nonuniform distribution of normal stress can be attributed to progressive development of geogrid displacement under tensile loads. Because the geogrid-backfill soil interaction can only be triggered by the movement of geogrid after all in the pullout test.

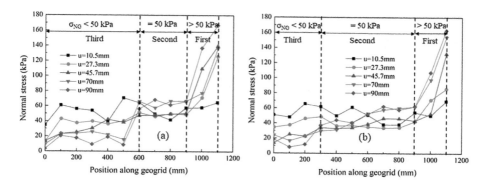

Fig. 7. Normal stress distributions along the geogrid. (a) WG. (b) PG.

The pullout failure needs to be checked in the calculation of internal stability of geosynthetic-reinforced soil retaining walls, but the effects of relative displacements between geosynthetic and soil on the pullout force are not considered in the current

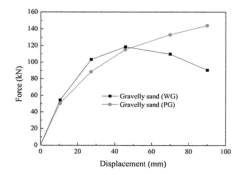

Fig. 8. Pullout force at different clamp displacements

design guidelines. Figure 8 shows the pullout force of geogrid embedded in WG and PG gravelly sand at different clamp displacements. The peak pullout force appeared in WG one when the clamp displacement was 45.7 mm. However, the pullout force in PG one increased with the increase of clamp displacements within 90 mm though it was a little bit smaller than that in WG one when the clamp displacements were less than 45.7 mm. The results of the pullout force can be ascribed to the distributions of displacements and normal stress shown in Figs. 6 and 7, respectively. Because the geogrid which was embedded in PG gravelly sand has not been entirely activated when the clamp displacements were less than 90 mm and its normal stress in the geogrid plain was greater than that in WG one overall.

4   Conclusions

In this study, a two-dimensional numerical model was built by PFC^{2D} to simulate the pullout behaviors of geogrid-reinforced gravelly sand. This model was calibrated firstly by the results obtained from laboratory experiments including the tensile tests of geogrid and the direct shear tests of medium sand. Then, two types of gravelly sand, WG and PG, were modelled to investigate their effects on the pullout responses under tensile loads while keeping other factors the same. Though there existed some limitations in this study by simplifying the pullout tests into 2D problems, the results can still provide some meaningful information about the interaction mechanisms between reinforcement and different particle distributions of backfill soil for uniaxial geogrid by utilizing the piecewise linear model in every two adjacent particles of geogrid and modeling its transverse members. The following conclusions can be drawn:

(1)   The displacements of the geogrid increased with increasing clamp displacements and decreased nonlinearly from the clamp to the end. The entire geogrid almost has been activated in WG gravelly sand when the clamp displacement was beyond 70 mm, while the fourth knot of the geogrid embedded in PG gravelly sand has not been activated.

(2) The normal stress in the geogrid plane did not always equal to the target value applied at the top of the specimen, especially under large clamp displacements. And the normal stress induced in the WG gravelly sand was less than that induced in the PG gravelly sand overall at the end of pullout tests.

(3) The peak pullout force of the geogrid embedded in WG gravelly sand appeared when the clamp displacement was 45.7 mm, while the pullout force of the geogrid embedded in PG one increased with increasing the clamp displacements within 90 mm and was larger than that of geogrid embedded in WG one when the clamp displacement was beyond 45.7 mm.

Acknowledgement. The support from the Key Research and Development Project of Chinese Ministry of Science and Technology under grants 2016YFE0105800 is gratefully acknowledged.

References

Cardile, G., Gioffre, D., Moraci, N., Calvarano, L.S.: Modelling interference between the geogrid bearing members under pullout loading conditions. Geotext. Geomembr. **45**(3), 169–177 (2017)

Cardile, G., Moraci, N., Calvarano, L.S.: Geogrid pullout behaviour according to the experimental evaluation of the active length. Geosynth. Int. **23**(2), 194–205 (2016)

Gao, G., Meguid M.A.: On the role of sphericity of falling rock clusters-insights from experimental and numerical investigations. Landslides (2017). https://doi.org/10.1007/s10346-017-0874-z

Han, J., Bhandari, A., Wang, F.: DEM analysis of stresses and deformations of geogrid-reinforced embankments over piles. Int. J. Geomech. **12**(4), 340–350 (2012)

Lin, Y.L., Zhang, M.X., Javadi, A.A., Lu, Y., Zhang, S.L.: Experimental and DEM simulation of sandy soil reinforced with H-V inclusions in plane strain tests. Geosynth. Int. **20**(3), 162–173 (2013)

McDowell, G.R., Harireche, O., Konietzky, H., Brown, S.F., Thom, N.H.: Discrete element modelling of geogrid-reinforced aggregates. Proc. Inst. Civ. Eng. Geotech. Eng. **159**(1), 35–48 (2006)

Miao, C.X., Zheng, J.J., Zhang, R.J., Cui, L.: DEM modeling of pullout behavior of geogrid reinforced ballast: The effect of particle shape. Comput. Geotech. **81**, 249–261 (2017)

Moraci, N., Cardile, G.: Deformative behaviour of different geogrids embedded in a granular soil under monotonic and cyclic pullout loads. Geotext. Geomembr. **32**, 104–110 (2012)

Moraci, N., Recalcati, P.: Factors affecting the pullout behaviour of extruded geogrids embedded in a compacted granular soil. Geotext. Geomembr. **24**(4), 220–242 (2006)

Palmeira, E.M.: Soil-geosynthetic interaction: modelling and analysis. Geotext. Geomembr. **27**(5), 368–390 (2009)

Palmeira, E.M., Milligan, G.W.E.: Scale and other factors affecting the results of pull-out tests of grids buried in sand. Geotechnique **39**(3), 511–524 (1989)

Sieira, A.C.C.F., Gerscovich, D.M.S., Sayao, A.S.F.J.: Displacement and load transfer mechanisms of geogrids under pullout condition. Geotext. Geomembr. **27**(4), 241–253 (2009)

Tran, V.D.H., Meguid, M.A., Chouinard, L.E.: A finite-discrete element framework for the 3D modeling of geogrid-soil interaction under pullout loading conditions. Geotext. Geomembr. **37**, 1–9 (2013)

Wang, Z., Jacobs, F., Ziegler, M.: Visualization of load transfer behavior between geogrid and sand using PFC2D. Geotext. Geomembr. **42**(2), 83–90 (2014)

Wang, Z., Jacobs, F., Ziegler, M.: Experimental and DEM investigation of geogrid-soil interaction under pullout loads. Geotext. Geomembr. **44**(3), 230–246 (2016)

Zhou, J., Chen, J.F., Xue, J.F., Wang, J.Q.: Micro-mechanism of the interaction between sand and geogrid transverse ribs. Geosynth. Int. **19**(6), 426–437 (2012)

Large-Scale Model Analysis on Bearing Characteristics of Geocell-Reinforced Earth Retaining Wall Under Cyclic Dynamic Load

Jia-Quan Wang$^{(\boxtimes)}$, Bin Ye, Liang-Liang Zhang, and Liang Li

College of Civil and Architectural Engineering, Guangxi University of Science and Technology, 268 Donghuan Road, Liuzhou 545006, China
wjquan1999@163.com, 602909193@qq.com,
804936022@qq.com, 791261540@qq.com

Abstract. In order to study the bearing mechanism of the geocell-reinforced earth retaining wall under the traffic load, the large-scale model test under dynamic load was designed. The laws of earth pressure distribution and acceleration response were also analyzed. The results showed that the vertical soil pressure at the same height under dynamic loading was the largest at 0.39H from the panel (i.e., vibration source). Moreover, the earth pressure near the panel was the second, and the earth pressure far away from the panel was the smallest. According to the stress diffusion rate in the soil calculated by corner method, the addition of geocells in the soil can enhance the diffusion rate of soil stress. In the early stage of loading, the stress diffusion rate under the same loading was significantly improved with the increase of frequency. The acceleration response at the same horizontal position decreased from the top of the retaining wall to the bottom of the retaining wall, and the acceleration response decreased with the increase of dynamic load. The horizontal acceleration response at the same distance from the vibration source was obviously larger than the vertical direction.

Keywords: Geocell · Reinforced soil retaining wall · Cyclic dynamic load
Bearing characteristics

1 Introduction

Reinforced earth retaining wall is a significant form of reinforced soil used in engineering. The reinforced earth retaining wall has not only the characteristics of structural flexibility, saving land and economical but also the advantages which many gravitational retaining walls can not match. More and more researchers have paid attention to this new structure [1–3].

In the experimental study, Han et al. [4] studied the failure mode and reinforcement mechanism of the reinforcement for geocell with different height, strength and welded spacing. Song et al. [5, 6] through the centrifugal model test and Plaxis numerical simulation of the geocell retaining wall showed that the aspect ratio and the slope of the retaining wall and the surface load of fillings have a significant effect on the deformation of the geocell retaining wall. Ling et al. [7] conducted seismic tests on five reinforced

© Springer Nature Singapore Pte Ltd. 2018
L. Li et al. (Eds.): GSIC 2018, *Proceedings of GeoShanghai 2018 International Conference: Ground Improvement and Geosynthetics*, pp. 455–462, 2018.
https://doi.org/10.1007/978-981-13-0122-3_50

earth retaining wall models and demonstrated that the geocell-reinforced retaining wall had strong resistance to seismic load and deformation. Chen and Chiu [8] performed model tests on the model geocell retaining walls with the different height and inclination to study the effect of the geocell and the failure mechanism. This showed that the displacement of retaining wall increases with the height and inclination. In the numerical simulation, Qu et al. [9] applied the finite element method to study the influence of the design parameters of geocell flexible retaining wall on earth pressure. Xie and Yang [10] conducted the numerical simulation using MARC and field test on the geocell flexible retaining wall. This result showed that the deformation of the retaining wall is related to the width of the wall, the spacing of the reinforcement and the foundation model. Chen et al. [11] used the finite difference program (FLAC) to examine the gravity wall models with three different angles. The results showed that increasing the geocell length can reduce the structural deformation and potential sliding surface.

The existing literatures mainly focus on the bearing behavior of geocell under static loading, but the performance of geocell under dynamic loading is less involved. This paper takes the highway reinforced earth retaining wall as the research background. The large-scale model box (3.0 m length × 1.6 m width × 2.0 m height) was designed. The MTS electro-hydraulic servo system can apply loads dynamically to simulate the traffic load. The model test of geocell-reinforced retaining wall under dynamic load was conducted, and then the earth pressure distribution and acceleration response was analyzed.

2 Materials and Methods

Referring to the relevant regulations of axle load in Technical Standard of Highway Engineering (JTG B01—2014), this test utilized a loading plate connected to the MTS to simulate the standard vehicle load acting on the top of the geocell reinforced soil retaining wall. Though the relevant test data obtained from the measurement components embedded in the model, the earth pressure distribution and acceleration response were analyzed.

2.1 Test Device

To simulate the real boundary conditions, the test of reinforced retaining wall utilized a large-scale model box whose main framework was welded. The side of the box is made of a double layer of 2 cm-thick tempered glass and the opposite side 6 mm-thick steel plate. The front of the model box is reserved for the panel, and the back was four 2 cm-thick steel plates, as shown in Fig. 1. The layout of the test components and the paving of geocells is shown in Fig. 2.

Fig. 1. Photograph of the test devices

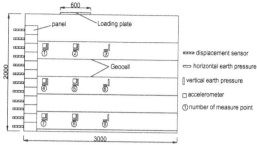

Fig. 2. Layout of components

2.2 Test Material

The granular soil used in the test was obtained from the banks of a local river in Liuzhou City. According to the Unified Soil Classification System, the soil was classified as well-graded sand. From the results of a sieve analysis, a uniformity coefficient (C_u) of 8.89, and a curvature coefficient (C_c) of 1.33. The detailed grain-size distribution parameters are shown in Table 1.

Table 1. Grain-size distribution of sand

Particle range (mm)	≤0.15	0.15–0.30	0.30–0.60	0.60–1.18	1.18–2.36	2.36–4.75	≥4.75
Particle composition (%)	15.08	19.73	17.95	21.67	18.64	6.87	0.05

The geocell used in all reinforced test was produced by Shandong Feicheng Friendship Engineering Plastics Co., Ltd. The size of geocell-reinforement was 280 cm × 160 cm (length × width), and welded spacing is 0.4 m. Table 2 summarizes the essential properties of the geocell.

Table 2. Geocell technical indicators

Cell height (mm)	Solder joint distance (mm)	Solder joint peel strength (N/mm)	Yield strength (N/MPa)	Tensile strength (MPa)	Elongation (%)
≥ 50	≥ 400	≥ 10	≥ 18	≥ 150	≤ 15

2.3 Test Loading Scheme

The sine waveform simulated the relationship between the size of the vehicle load and the time. When the vehicle load is applied, the vehicle load loading function is fitted as follows:

$$F = F_o + F_A \sin(2\pi ft)$$

In which F_o = center value (kN); F_A = amplitude (kN); f = frequency (Hz); t = time (s).

The loading plate was made of steel plates with dimension of 60 cm (length) 20 cm (width) × 3 cm (thickness). The grading dynamic loading method with the equal amplitude and different frequency is utilized. Starting from 10 ± 10 kN load, each gradated loading is divided into four different frequency (2 Hz, 4 Hz, 6 Hz and 8 Hz), and each frequency sustained 10 min. Then the center value increased by 20 kN, forming the progressive load relationship such as 0–20 kN, 20–40 kN, 40–60 kN…. The specific loading method is shown in Fig. 3.

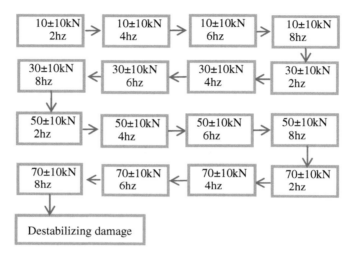

Fig. 3. The way of dynamic load

3   Results and Discussions

3.1   Analysis of Earth Pressure Under Dynamic Load

Figure 4 showed the relationship between the vertical additional soil pressure distribution under different wall height. The vertical soil pressure at the same height under dynamic loading was the largest at $0.39H$ from the panel (i.e., vibration source). Moreover, the earth pressure near the panel was the second, and the earth pressure far away from the panel was the smallest. The reasons are as follows: Position 1, 4, 7 are all the junction of retaining wall panel and the soil. As retaining wall panel, geocell and soil has different mechanical properties, they will form different interfacial interaction. Because the stiffness of panel is much larger than that of geocell and soil, the effect of panel on the delay of acceleration is more obvious. Based on the above-mentioned factors, the vertical additional soil stress near the panel side (position 1, 4, 7) is more difficult to diffuse than that at horizontal corresponding position (position 3, 6, 9), leading to the vertical additional soil stress at positions 1, 4 and 7 being greater than positions 3, 6 and 9.

In comparison with (a), (b) and (c) in Fig. 4, each point of each stage in the figure corresponds to the frequency values of 2, 4, 6 and 8 Hz. At the preliminary stage of loading, the frequency has a little effect on the increment of the earth pressure. However, with the loading progress, the effect of frequency on the increment of the earth pressure is obvious. In addition, at the instant of next loading is applied, the increment of earth pressure will rise sharply. It can be seen that the order of influence on earth pressure under dynamic loading is additional load → amplitude → frequency.

Figure 5 showed the stress diffusion rate from point 2 to the point 5 (the stress diffusion rate from the wall height of 1.5 m to 0.9 m below the load plate). According to the theory of stress diffusion in the soil (corner method) [12], the stress diffusion rate of unreinforced soil from 1.5 m to 0.9 m is 0.815 (the horizontal line in Fig. 5).

Figure 5 demonstrated that compared with the unreinforced retaining wall, the addition of geocells in the soil can enhance the diffusion rate of soil stress. This indicated that the soil reinforcement could effectively improve the earth pressure distribution. The difference of the bearing capacity between the reinforced soil and the unreinforced soil can be analyzed by comparing the theoretical stress diffusion rate of the soil without reinforcement. According to the stress diffusion curves at 2 Hz and

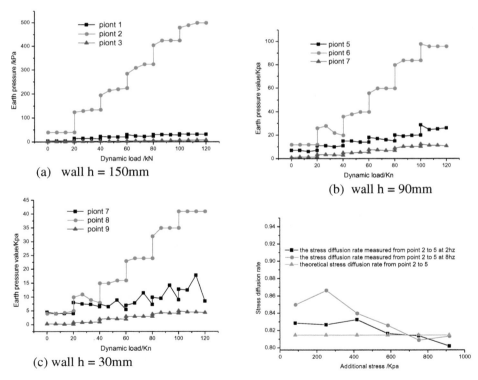

Fig. 4. Vertical additional earth pressure under dynamic loading

Fig. 5. Additional stress diffusion rate

8 Hz, the stress deformation of the whole retaining wall is divided into three stages: initial squeezing denseness, strengthening denseness and unstable failure. The first stage corresponds to 1st Level–2nd Level dynamic load (10 ± 10 kN, 30 ± 10 kN), the second stage corresponds to 3rd Level–5th Level dynamic load (50 ± 10 kN, 70 ± 10 kN, 90 ± 10 kN) and the third stage corresponds to the 6th Level dynamic load (110 ± 10 kN). The curves of the stress diffusivity at 2 Hz and 8 Hz showed that the stress diffusion rate at 8 Hz was significantly greater than that at 2 Hz in the early stage of loading. It is shown that the structure of reinforced soil is more compact under high-frequency dynamic loading, resulting in the rigidity of the soil being more easily increased. This condition is beneficial to the transmission of stress in the soil and increase the stress diffusion of the soil.

3.2   Acceleration Response of Reinforced Earth Retaining Wall

Figure 6 showed acceleration response of reinforced soil retaining walls. It can be seen from Fig. 6 that the acceleration response of the same horizontal position decreases with the top of retaining wall to the bottom of retaining wall at the same frequencies, and the vertical acceleration decreases with the increase of the vertical load value. In the process of soil compaction, the geocells and the soil form a "flexible raft foundation", the interaction between the soil and geocells makes the retaining wall to form a whole, and then exert effect of reinforced soil well to bear the external load. Moreover, due to the three-dimensional nature and fine flexibility of geocells, geocells being repeatedly pulled under dynamic loading, the acceleration dissipation was sped up. During the loading process of each level, this phenomenon will more obvious with the frequency increasing. The possible reasons are as follows: Due to the increase of frequency and cycle number of loading, the soil is compacted to make fillings and geocells close contact. The acceleration response can be stabilized, thereby the regularity of the curve become stronger.

When the soil is subjected to dynamic load, the whole wall will produce the corresponding acceleration response which spread from the vibration source to the surroundings through transmission between soil and soil. It can be seen from Fig. 6 that the acceleration peak of the position ① at the same frequency is larger than that at the position ⑤. These two places are the same distance from the vibration source. The possible reasons for the analysis are that geocell-reinforced retaining walls are different from other reinforced retaining walls on acceleration attenuation mechanism. On the one hand, the geocell is a three-dimensional reinforcement, and contact surface between geocells and soil is larger. It can not only play the interface effect similar to the geogrid, but also form the effect of "flexible raft foundation", making the geocell has a large lateral constraint on fillings. On another hand, under the excitation of the dynamic frequency and amplitude, the whole geocell-soil under the loading plate is fully compacted, and the adjacent cell is deformed. The contact between adjacent cell and internal soil is looser than that under the vibration source, resulting in hindering attenuating of acceleration at position ① in the horizontal direction.

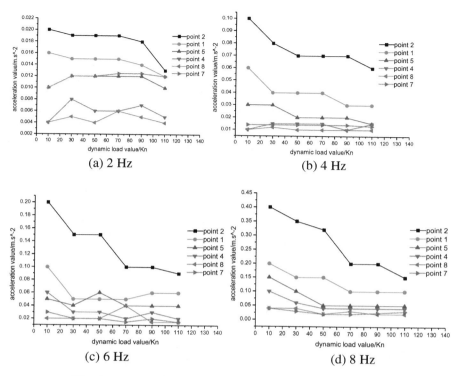

Fig. 6. Acceleration response of reinforced soil retaining walls

4   Conclusions

1. The results showed that the vertical soil pressure at the same height under dynamic loading was the largest at $0.39H$ from the panel (i.e., vibration source). Moreover, the earth pressure near the panel was the second, and the earth pressure far away from the panel was the smallest.

2. According to the stress diffusion rate in the soil calculated by corner method, geocell can enhance the diffusion of stress. This indicates that the soil reinforcement can effectively improve the earth pressure distribution. The stress deformation of the whole retaining wall can be divided into three stages: initial squeezing denseness, strengthening denseness and unstable failure. In the early stage of loading, the stress diffusion rate under the same level load is significantly improved with the increasing of frequency.

3. The acceleration response at the same horizontal position decreases from the top of the retaining wall to the bottom of the retaining wall, and the acceleration response decreases with the increase of dynamic load. The acceleration response at the equidistant distance from the excitation source is that acceleration response in the horizontal direction is significantly larger than the vertical direction.

References

1. Costa, C.M.L., Zornberg, J.G., de Souza Bueno, B., et al.: Centrifuge evaluation of the time-dependent behavior of geotextile-reinforced soil walls. Geotext. Geomembr. **44**(2), 188–200 (2016)
2. Yang, G.Q., Liu, H., Zhou, Y.T., et al.: Post-construction performance of a two-tiered geogrid reinforced soil wall backfilled with soil-rock mixture. Geotext. Geomembr. **42**(2), 91–97 (2014)
3. Latha, G.M., Krishna, A.M.: Seismic response of reinforced soil retaining wall models: influence of backfill relative density. Geotext. Geomembr. **26**(4), 335–349 (2008)
4. Barani, O.R., Bahrami, M., Sadrnejad, S.A.: A new finite element for back analysis of a geogrid reinforced soil retaining wall failure. Int. J. Civil Eng. **1**, 1–7 (2017)
5. Song, F., Xie, Y.L., Yang, X.H., et al.: Failure mode of geocell flexible retaining wall with surcharge acting on backfill surface. Chin. J. Geotech. Eng. **35**(s1), 152–155 (2013)
6. Song, F., Xu, W.Q., Zhang, L.Y., et al.: Numerical analysis of deformation behavior of geocell flexible retaining wall. Rock Soil Mech. **32**(s1), 738–742 (2011)
7. Ling, H.I., Leshchinsky, D., et al.: Seismic response of geocell retaining walls: experimental studies. J. Geotech. Geoenviron. Eng. **135**(4), 515–524 (2009)
8. Chen, R.H., Chiu, Y.M.: Model tests of geocell retaining structures. Geotext. Geomembr. **26**(1), 56–70 (2008)
9. Qu, Z.H., Xie, Y.L., Yang, X.H.: Influence of design parameters of flexible wall on earth pressure by numerical analysis. J. Chang'an Univ. (Nat. Sci. Ed.) **29**(6), 6–9 (2009)
10. Xie, Y., Yang, X.: Characteristics of a new-type geocell flexible retaining wall. J. Mater. Civil Eng. **21**(4), 171–175 (2009)
11. Chen, R.H., Wu, C.P., Huang, F.C., et al.: Numerical analysis of geocell-reinforced retaining structures. Geotext. Geomembr. **39**(8), 51–62 (2013)
12. Lu, T.H.: Soil Mechanics. Higher Education Press, Beijing (2010)

Shear Performance of Waste Tires, Geogrid and Geocell Reinforced Soils

Lihua Li[1(✉)], Feilong Cui[1,2], Henglin Xiao[1], Qiang Ma[1], and Langling Qin[1]

[1] School of Civil Engineering, Architecture and Environment,
Hubei University of Technology, Wuhan 430068, China
researchmailbox@163.com
[2] School of Civil and Transportation Engineering,
Hebei University of Technology, Tianjin, China

Abstract. The direct shear tests of the entire annular waste tires, half annular waste tires, triaxial geogrid, biaxial geogrid and geocell reinforced soils have been carried out respectively, through large direct shear apparatus, and the reinforcement mechanism and the shear performance have been analyzed. The test results show that reinforcement effect of reinforcement materials is obvious, and compared to unreinforced soils, the shear strength of reinforced soils increased greatly. The cohesion force of reinforced soils increased significantly, but the friction angle of reinforced soils has a small change, which illustrates that the shear strength of reinforced soils mainly come from the increase of cohesion force. The shear strength of reinforced soils is closely related to the formation of reinforcement materials. The shear strength of three-dimensional structure reinforced soils, such as entire annular tire, half annular tire and geocell reinforced soils, is evidently larger than that of two-dimensional structure reinforced soils, such as triaxial geogrid and biaxial geogrid reinforced soils.

Keywords: Waste tires · Triaxial geogrid · Biaxial geogrid · Geocell
Reinforced soils · Shear strength

1 Introduction

Geosynthetic reinforcement has been widely used in the various earthen structures such as roadways, slopes, embankments, and retaining walls. General geosynthetics include geogrid, geomembrane, geocell, and geotextile that are primarily made from polyethylene, polypropylene, and polyester (Koerner 2012). Waste tire is a new kind of reinforcement material, which has a great importance in dams, retaining walls and revetment engineering. The interface shear strength between reinforcement materials and soils is affected by various factors, including properties of the reinforcement materials, shear rate, etc. The shear test is one of the most important method to reveal the interaction between reinforcement materials and soils.

Many scholars adopt direct shear test method to study the properties of interface interaction between geosynthetics and soils. For example, Athanasopoulos (1996) adopt direct shear test method to study the properties of interface interaction between

L. Li et al. (Eds.): GSIC 2018, *Proceedings of GeoShanghai 2018 International Conference: Ground Improvement and Geosynthetics*, pp. 463–472, 2018.
https://doi.org/10.1007/978-981-13-0122-3_51

geotextile and clay. Goodhue et al. (2001) have carried the direct shear test on geosynthetics-sands. Farsakh et al. (2007) studied the effect of moisture content for the interface properties between geosynthetics and clay. Wu (2006) investigated the interface properties between different geosynthetics and sands, lime and fly ash. The interface friction performance between geogrid and soils has been carefully researched (Zhang et al. 2007, Bao et al. 2013, Shi et al. 2009). Liu et al. (2008) studied the shear properties of geocell reinforced soils through large direct shear tests, finding geocell reinforced soils could effectively improve the shear strength of soils. In addition, other scholars (Naein et al. 2013, Ding and Zhang 2005, Wei et al. 2005) have studied interface characteristics between different geosynthetics and soils by direct shear tests, and researched many useful conclusions.

In recent years, a number of researchers attempted to use waste tires as reinforcement materials in some applications (Foose et al. 1996; Zornberg et al. 2004; Li et al. 2011, Li et al. 2009, Zhang et al. 2011). The studies have shown that the waste tires used as reinforcement materials may improve the mechanical properties of the soils. In this study, the shear test on waste tires, geogrid and geocell reinforced sands have been carried out respectively and the results are compared.

2   Test Equipment, Materials and Method

2.1   Test Equipment

The shear tests were carried out on a direct shear apparatus shown in Fig. 1. Test system includes two shear boxed, the vertical loading device, horizontal shear loading device, hydraulic system, automated control system, and a data acquisition system combined with computer. The data acquisition system can automatically collect and process relevant data during the test. The dimensions of the upper and lower shear boxes are $\Phi500$ mm $\times$ 200 mm (diameter $\times$ height). The maximum normal and shear forces are 700 kN, and the maximum vertical and horizontal displacement are 50 mm and 100 mm, respectively.

Fig. 1. The large-scale direct shear test.

2.2   Test Materials

The test materials include the annular tires, half annular tires, geocell, biaxial geogrid, and triaxial geogrid are shown in Fig. 2, respectively. The grain size distribution of the sand is shown in Fig. 3, and its properties such as specific gravity, coefficient of uniformity (C_u), coefficient of curvature (C_c), maximum (γ_{dmax}) and minimum dry unit weight were determined as 2.67, 5.4, 1.4, 1.89 (g/cm^3) and 1.65 (g/cm^3), respectively. The parameters of waste tires, geocell, triaxial geogrid and biaxial geogrid are shown in Tables 1, 2, 3 and 4 respectively.

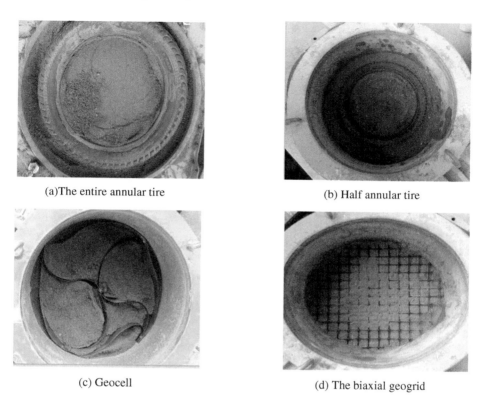

(a)The entire annular tire   (b) Half annular tire

(c) Geocell   (d) The biaxial geogrid

(e) The triaxial geogrid

Fig. 2. Test materials

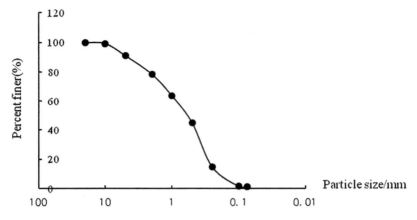

Fig. 3. Particle size distribution curve

Table 1. Physical properties of waste tires.

Diameter (cm)	40
Thickness (cm)	1
Width of tread (cm)	5
Width of tires sidewalls (cm)	4
Poisson's ratio	0.33
Elasticity modulus (kPa)	2.0×10^5
Tensile yield strength (MPa)	20

Table 2. Physical properties of geocell.

Type	TGGS-50-400
Thickness (mm)	1
Length of an aperture side (cm)	40
Tensile yield strength (MPa)	24
Tensile modulus (MPa)	6.5
An aperture size	40 cm × 40 cm
The height of geocell (cm)	5

Table 3. Physical properties of triaxial geogrid.

Type	TX160
Rib spacing	35 mm
Thickness of the node	4 mm
Effectiveness of the node	95%
The tensile modulus of low strain along neck (0.5%)	520 kN/m
The tensile modulus of quality control (2%)	315 kN/m

Table 4. Physical properties of biaxial geogrid.

Type	TSGS-30-30
Unit mass (g·m^{-2})	400 ± 40
Longitudinal tensile strength limit	30 kN/mm
Transverse tensile strength limit	30 kN/mm
Longitudinal ultimate elongation	$\leq 16\%$
Transverse ultimate elongation	$\leq 13\%$
The tensile strength of 5% longitudinal elongation	≥ 13 kN/mm
The tensile strength of 5% transverse elongation	≥ 15 kN/mm

2.3   Test Method

The direct shear test with strain control modes were conducted for unreinforced soil as well as reinforced soil with entire annular tire, half annular tire, triaxial geogrid, geocell and biaxial geogrid, respectively. The applied normal pressures were 100 kPa, 200 kPa, 300 kPa, and 400 kPa, respectively. The geocell and waste tire were placed on the surface between lower shear box and upper shear box, shown in Fig. 4 (a), while the triaxial geogrid and biaxial geogrid were placed on the interface between upper and lower shear boxes, shown in Fig. 4(b). The lower direct shear box was firstly filled with

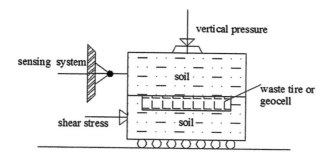

(a)The shear test diagram of waste tires or geocell reinforced soil.

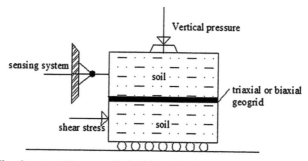

(b)The shear test diagram of triaxial or biaxial geogrid reinforced soil.

Fig. 4. The shear test models.

sands, which was compacted to a density of 90%. Then the reinforcement materials are placed on sands and backfilling sands was compacted into the cells of geocells or tires. At last, the upper direct shear box was filled with sands, which was compacted to the same density. The shear rate is 1 mm/min. Each test was not terminated until the recorded shear force became stable.

3 Test Results and Analysis

3.1 Test Results

The shear stress-displacement curves of unreinforced sands and reinforced sands under different vertical pressure are shown in Fig. 5. Figure 5 shows that the trend of shear stress-displacements curves of unreinforced sands are same to that of reinforced sands under different normal pressures. The shear strength of unreinforced and reinforced sands increases with the normal pressure increase. The shear stress-displacement curves of unreinforced and reinforced sands keep stable at the later stage of shear test.

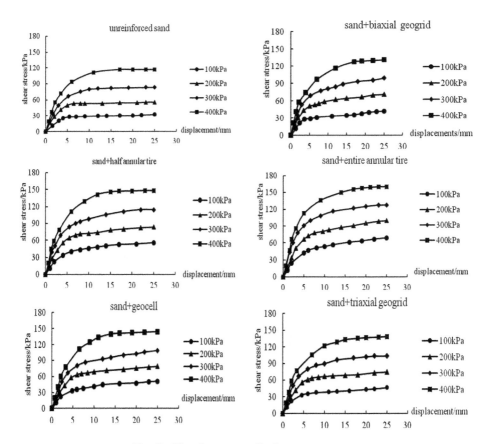

Fig. 5. The shear stress-displacements curves.

The curves of shear stress-normal stress of reinforced and unreinforced sands are shown in Fig. 6. Figure 6 indicates that the shear strength of the reinforced sands are significantly greater than that of unreinforced sands. The entire tires provided the most increase on the shear strength, while biaxial geogrid provided the least. At the normal stress of 100 kPa, the shear strength of the reinforced sands with the entire tire, half annular tire, geocell, triaxial and biaxial are almost greater 114%, 73%, 59%, 45% and 31% than that of the unreinforced sands, respectively. When the normal pressure increases to 400 kPa, the shear strength of entire tire, half annular tire, geocell, triaxial and biaxial reinforced sands are almost greater 37%, 27%, 24%, 19% and 12% than the that of unreinforced sands, respectively. Therefore, it is known that the reinforcement effect of reinforcement materials decreases with the normal pressure increase.

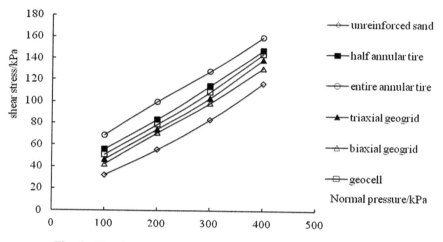

Fig. 6. The shear stress-normal pressure curves of reinforced sand.

The shear strength indexes of unreinforced sands and reinforced sands are shown in Table 5. According to Table 5, there is no significant increase in friction angle, but all the reinforcement apparently provides some cohesion force, which illustrates the reinforced sands mainly increase cohesion force to increase shear strength. And according to Table 5, the reinforcement with three-dimensional structure, i.e., the entire annular tires, half annular tires and geocell, provided higher cohesion force than planar reinforcement materials, i.e., biaxial geogrid and triaxial geogrid. This is an important reason that the reinforcement three-dimensional structure provided higher shear strength than planar reinforcement materials.

Table 5. The shear index of reinforced sands

Reinforcement	$\Phi/°$	C/kPa	$\Delta\Phi/°$	ΔC/kPa
Unreinforced sand	16	0	0	0
Entire annular tire	16.8	38.9	0.8	38.9
Half annular tire	17.1	23.4	1.1	23.4
Triaxial geogrid	17.0	14.5	1	14.5
Biaxial geogrid	16.4	12.2	0.4	12.2
Geocell	17.2	18.4	1.2	18.4

3.2 The Analysis of Reinforcement Mechanism

Compared to the internal friction angle of unreinforced sands, the internal friction angle of reinforced sands has a small change. But its shear strength is significantly greater than that of unreinforced sands, and reinforcement effect is obvious. The main reason is the resistant force between sands and reinforced materials. Although the reinforcement mechanism of several kinds of reinforcement materials are similar, but there are still some difference.

The shear surface of the unreinforced sands mainly depends on the friction force between the sands and sands. The shear stress of reinforced shear surface includes the friction force between the reinforcement materials and sands and the resistant force between sands and reinforcement materials. And the resistant force between sands and reinforcement materials increase with the increase of vertical pressure. In addition, the resistant force between sands and reinforced materials is also relevant to the shape of reinforcement materials. The resistant force between sands and entire annular tires, half annular tires, geocell mainly result from the lateral restraint force of tread and geocell. The resistant force between sands and triaxial or biaxial geogrid mainly result from the resistance of geogrid rib and aperture. The resistant force between sands and reinforcement materials increase with the increase of shear displacements, and reinforcement materials evidently show reinforcement effect. This is an important reason of that reinforcement materials can more better show reinforcement effect at large shear displacements than at small shear displacements.

In the three-dimensional structure reinforcement materials, the lateral confine effect of the entire annular tires is larger than that of half annular tires, and the lateral confine effect of the geocell is the smallest. The cohesion force of three-dimensional structure reinforced soils mainly come from the lateral confine effect of reinforcement materials for soils. The sidewall of the entire annular tires is greater than that of the half annular tires, which results in the lateral confine effect of the entire annular tires is greater than that of the half annular tires. Therefore, the cohesion force of entire annular tires reinforced sands is more evident than that of half annular tires reinforced sands, resulting in the reinforcement effect of the entire annular tires is superior to that of half annular tires. The deformation ability of the tire is extremely strong, which is far greater than that of geocell. The resistance of tire increase with the increase of shear displacements, and the resistance of tire still exists, even at the larger shear displacements. Therefore, though the width of entire annular tires sidewall is equal to the height of geocell sidewall, the shear strength of entire annular tires reinforced sands is greater than that of geocell reinforced sands.

The shear band mainly concentrate on the shear surface in the process of shear test. The sand particles far away from the shear surface have not been influenced obviously. This test phenomenon illustrates that the motion of waste tires and geocell reinforced sand particles is limited near the shear surface. Therefore, when the height of tires or geocell sidewalls is more than a certain value, the resistance effect of tires or geocell sidewalls in the process of shear test is not significantly increased. And tires and geocell mainly depend on the upper part of sidewalls to produce resistance for sands, and the confine effect of lower part of sidewalls is not significant. The deformation performance and confine effect of tires are stronger than that of geocell. Therefore, even if the height of geocell sidewalls is larger than that of half annular tires, the confine effect of half annular tires is stronger than that of geocell, which results in the reinforcement effect of half annular tires is better than that of geocell.

The triaxial geogrid and biaxial geogrid are two-dimensional structure reinforcement materials, which mainly depend on the friction between reinforcement materials surface and sands and resistant force between reinforcement materials apertures and sands to produce reinforcement effect. Obviously, the confine effect of two-dimensional structure reinforcement materials is smaller than that of three-dimensional structure reinforcement materials. Therefore, the reinforcement effect of two-dimensional structure reinforcement materials is also smaller than that of three-dimensional structure reinforcement materials. The triaxial geogrid has more geogrid ribs and apertures than biaxial geogrid, which results in the resistant force between triaxial geogrid and sands is larger than the resistant force between biaxial geogrid and sands. Therefore, the shear strength of triaxial geogrid reinforced sands is larger than that of biaxial geogrid reinforced sands.

4   Conclusions

On the basis of the presented study, the following conclusions can be drawn:

(1)   Compared to unreinforced soils, the shear strength of reinforced soils is saliently higher than unreinforced sand, especially the normal stress is low.

(2)   Among the reinforcement investigated, the entire tires provided the most significant reinforcement effect while the biaxial geogrid provided the least.

(3)   The internal friction angle of the reinforced soils had a small change. The cohesion force of reinforced soils was significantly improved, which implied that the reinforced soils mainly depended on the increase of the cohesion to increase shear strength.

(4)   The increased cohesion force of three-dimensional structure reinforced soils was larger than that of planar reinforcement reinforced soils.

(5)   The reinforcement mechanism of three-dimensional structure reinforcement materials is different from that of two-dimensional structure reinforcement materials. The reinforcement effect of three-dimensional structure mainly come from the lateral confine effect of reinforcement materials sidewalls, but the reinforcement effect of two-dimensional structure mainly come from the resistant force between reinforcement materials and soils.

Acknowledgements. The study is financially supported by the National Natural Science Foundation of China (No. 51678224, 51678223); National Key R&D Program of China (No. 2016YFC 0502208) and the Hubei Provincial Department of Education project (No. D20151402).

References

Abu-Farsakh, M., Coronel, J., Tao, M.: Effect of soil moisture content and dry density on cohesive soil- geosynthetic interactions using large direct shear tests. J. Mater. Civ. Eng. **19**(7), 540–549 (2007)

Athanasopoulos, G.A.: Results of direct shear tests on geotextile reinforced cohesive soil. Geotext. Geomembr. **14**, 619–644 (1996)

Bao, C., Wang, M., Ding, J.: Mechanism of Soil Reinforced with Geogrid. J. Yangtze River Sci. Res. Inst. **30**(1), 34–41 (2013)

Ding, J., Zhang, B.: Study of the interaction between geosynthetics and clay through direct-shear test. J. Hua Zhong Univ. Sci. Technol. **22**(2), 59–62 (2005)

Foose, G., Benson, C., Bosscher, P.: Sand reinforced with shredded waste tires. J. Geotech. Eng. **122**(9), 760–767 (1996). https://doi.org/10.1061/(asce)0733-9410(1996)122:9(760)

Goodhue, M.J., Edil, T.B., Benson, A.H.: Interaction of foundry sands with geosynthetics. J. Geotech. Geoenviron. Eng. **127**(4), 353–362 (2001)

Koerner, R.: Design with Geosynthetics, vol. 1, 6th edn. Xlibris Corporation (2012). 508 p.

Li, L., Tang, H., Xiao, B.: Discarded tire implications in reinforced slope. In: 4th International Conferenceon Technology of Architectureand Structure, vol. 9, pp. 1430–1433 (2011)

Li, Z., Zhang, H., Zhao, Y.: Engineering properties of tire shreds reinforced soils. Geotech. Invest. Surv. **6**, 19–22 (2009)

Liu, W., Wang, Y., Chen, Y., et al.: Research on large size direct shear test for geocell reinforced soil. Rock Soil Mech. **29**(11), 3133–3138 (2008)

Naein, K.I.: Interfcialshear strength of silty sand-geogrid composite. In: Proceeding of the Institution of Civil Engineers. Geotechnical Engineering 166 February Issue GE1, pp. 67–75 (2013)

Shi, D.D., Liu, W.-B., Shui, W.-H., et al.: Comparative experimental studies of interface characteristics between uniaxial/biaxial plastic geogrids sand different soils. Rock and Soil Mech. **30**(8), 2237–2244 (2009)

Wei, H., Yu, Z., Zou, Y.: Shear characteristics of soil reinforced with geosynthetic material. Shuili Xuebao **36**(5), 555–562 (2005)

Wu, J.-H: Study on interaction characteristics between geosynthetics and fill materials by pull out tests. Rock Soil Mechanics **27**(4), 581–585 (2006)

Zhang, W., Wang, B., Zhang, F., et al.: Test study on interaction characteristics between biaxial geogrid and clay. Rock Soil Mech. **28**(5), 1031–1034 (2007)

Zhang, D., Zhang, J.: Experimental research on pullout test of waste tire with and confinement. Rock Soil Mech. **32**(3), 733–737 (2011)

Zornberg, J.G., Cabral, A.R., Viratjandr, C.: Behaviour of tire shred sand mixtures. Can. Geotech. J. **41**, 227–241 (2004)

Performance of Multi-axial Geogrid-Stabilized Unpaved Shoulders Under Cyclic Loading

Xiaohui Sun[1(✉)], Jie Han[2], Steven D. Schrock[2], Robert L. Parsons[2], and Jun Guo[1]

[1] Shenzhen University, Shenzhen 518048, Guangdong, China
sunxhero@gmail.com, guoddx@live.com
[2] The University of Kansas, Lawrence, KS 66045, USA
{jiehan, schrock, rparsons}@ku.edu

Abstract. Low-volume roadways usually have unpaved aggregate shoulders for vehicles to stop during an emergency. To provide a good condition for vegetation, turf soil can be mixed with aggregate. However, the use of turf soil unavoidably results in the reduction of the load carrying capacity of the unpaved shoulders. To improve the performance of aggregate-turf shoulders, geogrid can be included to stabilize the base materials. In this study, four test sections of 0.15 m thick aggregate or aggregate-turf bases over subgrade with CBRs of 3% were constructed in a geotechnical box (2.2 m long × 2 m wide × 2 m high) to investigate their performance. The base courses of these test sections included one non-stabilized aggregate base, one non-stabilized aggregate-turf base, one geogrid-stabilized aggregate base, and one geogrid-stabilized aggregate-turf base. A cyclic load with a magnitude of 40-kN was applied on the surfaces of these test sections through a circular plate (0.305 m in diameter). Earth pressure cells and/or displacement transducers were installed at varying locations away from the loading plate to measure the vertical stresses, permanent deformations, and resilient deformations. The results show that the existence of geogrids reduced vertical stresses and permanent deformations but increased resilient deformations. The geogrid stabilized section with the aggregate-turf base outperformed that with the non-stabilized aggregate base.

Keywords: Multi-axial geogrid · Unpaved shoulders · Cyclic load

1 Introduction

Unpaved shoulders, including aggregate shoulders and turf shoulders, are commonly constructed next to traffic lanes of a low-volume road. Aggregate shoulders have a higher load carrying capacity as compared with turf shoulders while turf shoulders can provide a better condition for vegetation growth, which can help to reduce dust emission and control erosion [1].

To construct an unpaved shoulder with a good condition for vegetation and a relatively high load carrying capacity for traffic, Gantenbein [2] chose an aggregate-turf mixture as a base course for the shoulder. Since the use of turf soil weakened the base unavoidably, geosynthetics were applied to stabilize the aggregate-turf base and

© Springer Nature Singapore Pte Ltd. 2018
L. Li et al. (Eds.): GSIC 2018, *Proceedings of GeoShanghai 2018 International Conference: Ground Improvement and Geosynthetics*, pp. 473–482, 2018.
https://doi.org/10.1007/978-981-13-0122-3_52

improve its performance. Guo et al. [3] investigated the influence of geosynthetics on the growth of vegetation through a one-year-long vegetation test. It was concluded that there was no obvious evidence that geosynthetics could limit the growth of vegetation. In addition, it was found out that the aggregate-turf mixture at a ratio of 1:1 by weight had no significant difference from the native soil in terms of the growth of vegetation.

Geogrid is one type of widely used geosynthetic for aggregate base stabilization and subgrade improvement. Through the interlocking between its apertures and aggregates, geogrid can provide confinement to bases and resist the lateral movement of aggregates [4–6]. The performance of geogrid reinforced unpaved roadways has been investigated in many studies [7–12]. In general, the results revealed that the inclusion of geogrids reduced vertical stresses and permanent deformations but increased resilient deformations. However, the performance of sections with geogrid-stabilized aggregate-turf bases has not been well investigated.

This study aimed to evaluate the performance of multi-axial geogrids in stabilizing aggregate-turf bases. Four test sections with 0.15 m thick aggregate-turf bases over subgrade were constructed in a box with the dimension of 2.2 m long × 2 m wide × 2 m high and their performances were investigated. These test sections included one non-stabilized aggregate base, one non-stabilized aggregate-turf base, one geogrid stabilized aggregate base, and one geogrid stabilized aggregate-turf base. A 40-kN cyclic plate load was applied through a circular plate (0.305 m in diameter) to simulate traffic load. Vertical stresses, permanent deformations, and resilient deformations were recorded by preinstalled pressure cells and displacement transducers.

2   Test Materials and Test Setup

2.1   Materials

Base Course. Four test sections with a base thickness of 0.15 m were prepared in this study. Aggregate Base Class 3 (AB3) aggregate was selected as the base course material for two of the test sections since the aggregate is usually used for the construction of low volume roads in Kansas. The aggregate-turf mixture, prepared by mixing the turf soil and the AB3 aggregate at a ratio of 1:1 by weight, was used for the construction of the base courses of the other two sections. Figure 1 shows the modified compaction curves of the AB3 aggregate and the aggregate-turf used in this study.

Subgrade. The materials used for subgrade were prepared by mixing 75% Kansas River sand with 25% kaolin by dry weight. Pokharel [13] conducted the compaction test and CBR test of the subgrade. According to the CBR curve provided by Pokharel [13], to obtain a 3% CBR subgrade, the subgrade material was compacted in lifts with a water content of 10.3% approximately.

Geogrid. In this study, a triangular aperture geogrid with a radial stiffness of 270 kN/m (at 0.5% strain) was used to stabilize the bases.

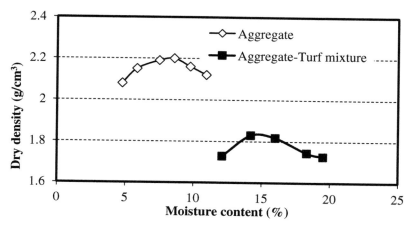

Fig. 1. Compaction curves of AB3 aggregate and aggregate-turf mixture based on modified Proctor compaction tests

2.2 Test Setup

Figure 2 shows the test setup of this study. At the interface of the base course and subgrade, four earth pressure cells were placed at locations with 0, 0.18, 0.25, and 0.38 m, respectively, from the center. At the surface of the test sections, displacement

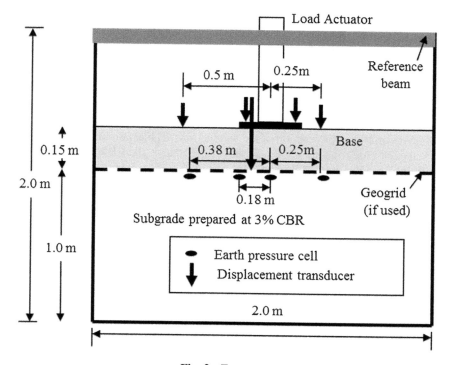

Fig. 2. Test setup

transducers were installed with distances from the center of 0, 0.25, and 0.5 m, as indicated in the figure. A telltale, with its bottom seated at the top of the subgrade, was installed to accommodate a displacement transducer so that the subgrade deformation could be measured, as indicated in Fig. 2.

A 40 kN cyclic load (frequency: 0.77 Hz) as shown in Fig. 3 was applied to the test sections by a steel loading plate (0.305 m in diameter). At the bottom of the steel plate, a rubber mat was attached to generate a relatively uniform applied pressure (552 kPa on average). During a single load cycle, the cyclic load wave was applied by following the details in Fig. 3.

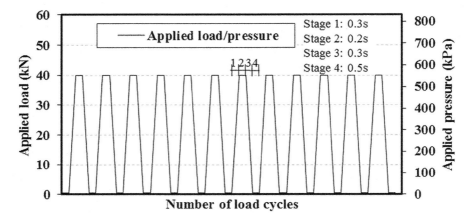

Fig. 3. Load wave applied in this study

3   Results and Discussion

3.1   Dynamic Cone Penetration Tests

Dynamic Cone Penetration (DCP) tests were conducted to investigate the CBR values of soil layers after the construction of the test sections. Table 1 summarizes the average CBR values of bases and subgrade. As shown in Table 1, the CBR values of the AB3 and AB3-turf mixture were 14% and 10% approximately, while those of the subgrade were around 3%.

Table 1. CBR values estimated by DCP tests

Loading	Stabilized condition	CBR (%)		
		Subgrade	Base course	
			AB3	Aggregate-turf mixture
40 kN cyclic load	Non-stabilized	3.3	14.3	
	Geogrid-stabilized	3.5	13.7	
	Non-stabilized	3.1		10.5
	Geogrid-stabilized	3.4		9.8

3.2   Vertical Stress

Figure 4 shows the vertical interfacial stresses of all the stabilized and non-stabilized test sections. In general, the existence of geogrids caused the decrease of vertical stresses at the top of the subgrade. Previous studies also showed similar results [9, 10]. In addition, the increasing load cycles leaded to the deterioration of the base courses and, as a result, the vertical stresses increased gradually.

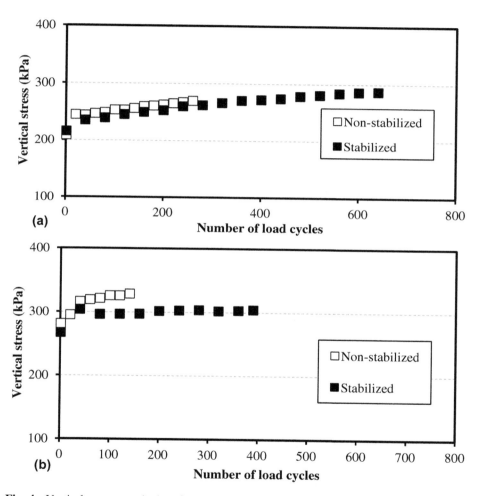

Fig. 4. Vertical stresses at the interface vs. number of load cycles of test sections with (a) AB3 bases and (b) aggregate-turf bases.

It is noted that the sections with aggregate-turf bases demonstrated higher vertical stresses than the sections with AB3 bases. The reason is that the CBRs (or resilient moduli) of the aggregate-turf bases were much lower than those of AB3 bases due to the mix of turf soils. In other words, the sections with aggregate-turf bases had a lower modulus ratio of base course to subgrade. According to Burmister's solution, a lower modulus ratio implies a higher vertical stress at the top of the subgrade. These results indicate that aggregate-turf bases were not as favorable as AB3 bases in terms of protecting subgrade.

Due to the nonlinear characteristic of soils, the ratio of the vertical stress to the bearing capacity of the subgrade influenced the accumulation of subgrade deformations significantly. The elastic limit of the subgrade can be estimated by the following equation:

$$q = N_c c_u \tag{1}$$

where q = the elastic limit, kPa; N_c = the bearing capacity factor, 3.14 (elastic limit); and c_u = the undrained shear strength of subgrade, which can be estimated as 20 *CBR* [13], kPa.

For a subgrade with a CBR of 3%, its elastic limit approximates 190 kPa according to Eq. (1). Clearly, the subgrade of the test sections in this study likely experienced a vertical stress higher than the elastic limit.

3.3   Permanent Deformation

Figure 5 shows the surface and subgrade permanent deformations of all the test sections. With the increase of the load cycles, permanent deformations were accumulated drastically at the beginning but the accumulation rate decreased gradually. As shown in the figure, the surface permanent deformations were mainly contributed by the subgrade. With the inclusion of geogrids, the surface and subgrade permanent deformations were reduced significantly.

As shown in Fig. 5, according to the number of load cycles at the same permanent deformation, the performance of the non-stabilized test section with an AB3 base was better than that with an aggregate-turf base. However, the geogrid-stabilized aggregate-turf base outperformed the non-stabilized AB3 aggregate base. A load improvement factor (LIF), which is the ratio of the number of load cycles for the geogrid-stabilized section to that for the non-stabilized section at the same permanent deformation, was utilized to quantify the benefit of the geogrid. According to Fig. 5, it was found out that the LIFs ranged from 1.8 to 3.2 with the permanent deformation varying from 40 to 70 mm. Similarly, the LIF concept was used to compare the non-stabilized section with the AB3 base to the geogrid-stabilized section with the aggregate-turf base. Correspondingly, the LIFs ranged from 1.8 to 2.3 with the permanent deformation varying from 40 to 70 mm.

3.4   Resilient Deformation

Resilient deformations at the surface and at the top of the subgrade were also recorded in this study. Figure 6 shows the resilient deformations versus the surface permanent deformation. As shown in the figure, in general, the aggregate-turf sections and AB3 sections demonstrated similar trends.

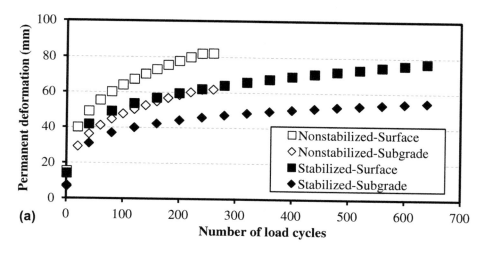

(a)

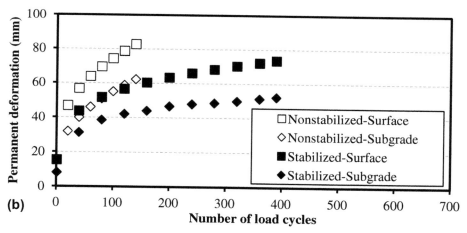

(b)

Fig. 5. Permanent deformation vs. number of load cycles of test sections with (a) AB3 base and (b) aggregate-turf base.

With the increase of surface permanent deformations, the surface resilient deformations did not reveal a significant change as compared with the increasing subgrade resilient deformations for non-stabilized sections. The possible reason for the increase of the subgrade resilient deformations is that the vertical stress at the top of the subgrade was increased (as indicated in Fig. 4) with the gradual deterioration of base courses under cyclic loading. Additionally, the deterioration of base courses mainly occurred in the shear band around the edge of the loading plate and the base course materials right underneath the loading plate experienced a hardening process due to compression. Therefore, the resilient deformation of base courses in the center would decrease. As a result, the measured surface resilient deformation in the center did not vary significantly.

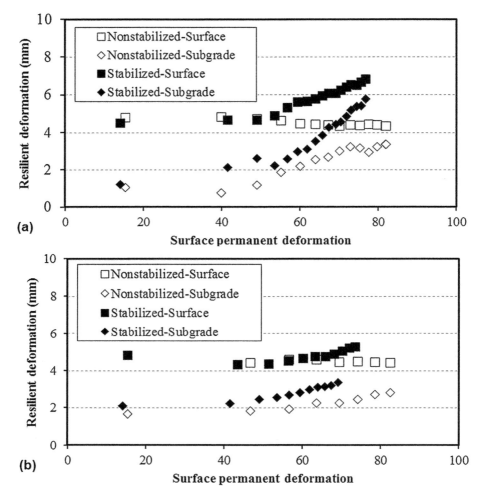

Fig. 6. Resilient deformation vs. surface permanent deformation of the test sections with (a) AB3 bases and (b) aggregate-turf bases.

As shown in Fig. 6, the geogrid-stabilized test sections had higher surface/subgrade resilient deformations than the non-stabilized test sections. This increase became more obvious at a higher surface permanent deformation. Previous studies also showed similar results [10]. One possible reason for this phenomenon is that the additional stresses in soils induced by geogrids increased the rebound of soils during the unloading stage. In addition, the increased resilient deformations due to the inclusion of geogrids might also be attributed to the discontinuity of the interface at a large permanent deformation, but this hypothesis needs further investigation.

4   Conclusions

In this study, sections with AB3 aggregate and aggregate-turf bases were tested under cyclic plate loading. Earth pressure cells and displacement transducers were used to measure vertical stresses and permanent/resilient deformations. The following conclusions can be drawn from this study:

(1) The vertical stresses increased at the top of the subgrade when using aggregate-turf bases due to the decrease of the base quality. The vertical stresses were also higher than the elastic limit of the subgrade (3% CBR) and the accumulation rate of the permanent deformations for aggregate-turf sections was more drastic. Generally, the aggregate-turf base had less capability in protecting subgrade as compared with the AB3 base.

(2) The geogrid-stabilized aggregate-turf base demonstrated a better performance than the non-stabilized AB3 aggregate base. The load improvement factor (LIF) obtained by the ratio of load cycles of the section with the geogrid-stabilized aggregate-turf base to those of the section with a non-stabilized AB3 base at the same permanent deformation ranged from 1.8 to 2.3.

(3) The geogrid-stabilized test sections had higher surface/subgrade resilient deformations than the non-stabilized test sections. This phenomenon might be due to that the additional stresses in soils induced by geogrids increased the rebound of soils during the unloading stage.

Acknowledgement. The Kansas Department of Transportation (KDOT) sponsored this study. Mr. Byron Whitted, an undergraduate research assistant, assisted in conducting the laboratory tests in this study.

References

1. Udo, K., Takewaka, S.: Experimental study of blown sand in a vegetated area. J. Coastal Res. **23**(5), 1175–1182 (2007)
2. Gantenbein, B.: Unique application: Greendale resurfacing job includes green shoulder. Western Builder, pp. 8–9 (2006)
3. Guo, J., Han, J., Schrock, S.D., Parsons, R.L.: Field evaluation of vegetation growth in geocell-reinforced unpaved shoulders. Geotext. Geomembr. **43**(5), 403–411 (2015)
4. Giroud, J.P., Han, J.: Design method for geogrid-reinforced unpaved roads. I: development of design method. J. Geotech. Geoenviron. Eng. **130**(8), 775–786 (2004)
5. Giroud, J.P., Han, J.: Design method for geogrid-reinforced unpaved roads. II: calibration and applications. J. Geotech. Geoenviron. Eng. ASCE **130**(8), 787–797 (2004)
6. Han, J.: Principles and Practice of Ground Improvement. John Wiley & Sons, Hoboken (2015). 432 p. ISBN: 978-1-118-25991-7
7. Das, B.M., Shin, E.C.: Strip foundation on geogrid-reinforced clay: behavior under cyclic loading. Geotext. Geomembr. **13**(10), 657–666 (1998)
8. Dong, Y.L., Han, J., Bai, X.H.: Numerical analysis of tensile behavior of geogrids with rectangular and triangular apertures. Geotext. Geomembr. **29**(1), 83–91 (2011)

9. Qian, Y., Han, J., Sanat, K.P., Parsons, R.L.: Performance of triangular aperture geogrid-reinforced base courses over weak subgrade under cyclic loading. J. Mater. Civ. Eng. ASCE **25**(8), 1013–1021 (2013)

10. Sun, X., Han, J., Kwon, J., Parsons, R.L., Wayne, M.H.: Radial stresses and resilient deformations of geogrid-stabilized unpaved roads under cyclic plate loading tests. Geotext. Geomembr. **43**(5), 440–449 (2015)

11. Sun, X., Han, J., Wayne, M.H., Parsons, R.L., Kwon J.: Determination of load equivalency for unpaved roads. In: Transportation Research Record: Journal of the Transportation Research Board, No. 2473, Transportation Research Board of the National Academies, Washington, D.C., pp. 233–241 (2015)

12. Sun, X., Han, J., Corey, R.: Equivalent Modulus of Geogrid-Stabilized Granular Base Back-Calculated Using Permanent Deformation. J. Geotech. Geoenviron. Eng. **143**(9), 06017012 (2017)

13. Pokharel, S.K.: Experimental study on geocell-reinforced bases under static and dynamic loading. Ph.D. dissertation, The University of Kansas (2010)

A Comparative Study on Shear Strength of Soil Using Geogrid and Geotextiles

Akash Chetty[✉], Akhil Jain, Devanshu Mishra,
and Kaustav Chatterjee

Department of Civil Engineering, IIT Roorkee, Roorkee 247667, India
chetty.akash1@gmail.com, akhil.jain46@gmail.com,
devanmishra1310@gmail.com, kchatfce@iitr.ac.in

Abstract. Failure of retaining walls, roads and collapse of soil have led geotechnical engineers to use reinforced soil to increase its strength. When the soil is unconfined and subjected to compressive stress, tensile stress are developed in the other direction. However if internal reinforcement layers are provided, frictional force is mobilized along the surfaces of the reinforcement layers and strength of the soil is considerably increased. It is a common practice among geotechnical engineers to use geosynthetics for soil reinforcement in various construction works. The major products of geosynthetics includes geotextiles, geogrid and geomembranes. Geogrids are used as reinforcement layers in retaining walls, embankments, road bases and below rail tracks. Geotextiles are used to separate different geomaterials and acts as filtration material to retain the soil particles. In the present study soil is collected from the bed of Solani River in Roorkee, India and soil classification is carried out on the unreinforced soil. Direct shear test is performed on unreinforced soil specimen and soil reinforced with geogrid and geotextiles (woven and non-woven) for different magnitudes of normal stress. It is observed that the shear strength parameters for geogrid reinforced soil are higher than that reinforced using geotextiles. Moreover, soil reinforced using non-woven geotextile is observed to have higher shear strength parameter than woven type. Hence the utility of geotextiles and geogrids as significant soil reinforcement materials to improve the shear strength parameters of unreinforced soil is highlighted in the present study.

Keywords: Sand · Geotextiles · Geogrid · Shear strength

1 Introduction

The stability of any structure is governed by the type of foundation on which it rests and the soil existing at that particular location. However under present circumstances due to space constraint, the design and construction of any structures within space domain is fixed and this forms a major economic importance in geotechnical engineering projects. For example in highway projects, a change in design might cause a significant change in elevation and the alternatives that the engineers are left with are a concrete retaining wall or unreinforced embankments with flat slope. Although concrete walls are simple to design but the material cost increases due to elevated

L. Li et al. (Eds.): GSIC 2018, *Proceedings of GeoShanghai 2018 International Conference: Ground Improvement and Geosynthetics*, pp. 483–490, 2018.
https://doi.org/10.1007/978-981-13-0122-3_53

construction. On the other hand, the design and construction of unreinforced embankments is controlled by overall stability and space limitations.

As the availability of suitable construction sites decreases, there is an increasing need to utilize poor soils for foundation support. Steel has been the most widely used reinforcement material and since poorly draining soils are usually saturated the possibility of corrosion of these reinforcements is high. Metallic reinforcements are not strong reinforcement candidates for poorly draining backfills because they do not provide lateral drainage to the cohesive fill. Although the different soil reinforcement systems have greatly extended the use of soil as construction material, their use has often been limited by the availability of good-quality granular material which has generally been specified for the backfill. With the introduction of polymer geotextiles and geogrids, non-corrosive reinforcement systems are now available. Geotextile reinforcements may be especially useful for reinforcing poorly draining soils because their drainage capabilities would help to increase the structure stability by dissipating excess pore water pressures. Geogrid reinforcements provide adequate tensile strength required for the design of permanent reinforced soil structures. Particularly, composite geotextiles, which combine the hydraulic properties of nonwovens with the mechanical characteristics geogrids or woven are probably the most appropriate reinforcement for marginal Soils.

Pasley (1822) gave the first explicit application of soil reinforcement for military construction. The performances of retaining walls, foundation and slopescan be significantly improved by using geosynthetics as a reinforcement materials in soils which has drawn considerable attention in recent years. The performance of large modeled foundations placed on reinforced soil was studied by Abu-Farsakh et al. (2007). Geosynthetics acts as an inclusive material within the soil by forming a reinforcement layer and starts developing tensile forces which contributes to the overall stability of reinforced soil. Henry (1969) used metal strips for improving the strength of soil and termed the composite material as Reinforced Earth. The major products of geosynthetics are nonwoven geotextiles, woven geotextiles, geogrid and geomembranes. Geogrids are used as reinforcement layers in retaining walls and embankments, road bases, below rail tracks and ground reinforcement. Lopes and Lopes (1999) proposed that shear resistance of soil-geogrid interface increased with soil particles having size greater than thickness of geogrid members but smaller than geogrids apertures. Liu et al. (2008) conducted large scale direct shear tests on soil-PET-yarn geogrid. Wu et al. (2008) investigated shear strength of soil-geosynthetics interface using tilt-table tests. Geotextiles are used to separate different geomaterials and act as filtration materials to retain the particles. The design of a reinforced soil structures is conducted using equilibrium method by equating external and internal stability of the structure. It also involves selecting the required tensile strength of reinforcement during design so as to achieve considerable margin of safety.

Direct shear test of soil-geotextile interface using different shear boxes was conducted by Ingold (1982) and concluded that friction angle obtained from shear box with lower cross sectional area was 2–3° higher when compared to a bigger shear box. Richards and Scott (1985) carried out direct shear test involving interfaces between soil and geotextiles while Bauer and Zhao (1993) conducted direct shear test involving interfaces between soil and geogrids. The behavior of plasticity index and normal stress on remolded clays using direct-shear tests was studied by Jesmani et al. (2010).

2 Experimental Investigations

2.1 Geosynthetics

In present study, two types of Geotextiles GSM 300 and Poly 250 are used. GSM 300 is a non-woven geotextile composed of polypropylene fiber. Poly 250 are made of artificial fibers with various petrochemical derivatives as their source. Geogrid used is Biaxial Monolithic SS 30 which is also composed of polypropylene fiber. Geosynthetics degrade when exposed to sunlight. So it's important to cover the geogrid with fill while testing them.

2.2 Soil Classification

In the present study soil classification as per IS 1498 (1992) is carried out to determine the various engineering properties of soil, as tabulated in Table 1. Grain size distribution as per IS 2720-4 (1985) is performed to classify the soil and different sieve sizes ranging from 4.75 mm to 75 microns are used. 500 grams of oven dried sample was taken and it was sieved manually. The soil retained on the sieves of various sizes were weighed. A curve is plotted between particle size and percentage fines of soil retained on each sieve and classification of the soil was carried out based on the uniformity coefficient and coefficient of curvature values. Specific gravity as per IS 2720-3 (1980) of the soil specimen was determined using pycnometer test and falling head permeability test as per IS 2720-17 is performed for finding out the co-efficient of permeability of soil, as tabulated in Table 1.

Table 1. Soil characteristics

Property	Value
Uniformity co-efficient	2.14
Coefficient of curvature	0.94
Soil type	Poorly Graded Sand (SP)
Specific gravity of soil	2.32
Dry density (gm/cc)	1.32
Co-efficient of permeability (cm/s)	0.0017

2.3 Tensile Strength Test

Tensile strength test is performed as per ASTM D4595-17 to find out the tensile strength, yield strength and ductility of geotextiles using geosynthetics creep testing machine and tabulated in Table 2. Because of the manufacturing process, Geogrids often have different tensile strength in different direction. So in order to calculate this test, we need to determine the tensile strength in main and minor directions. Tensile Strength per unit width of Geogrid came out to be 33.45 KN/m.

2.4 Apparent Opening Size Test

Apparent opening size test is performed as per ASTM D4751-16 to find out the apparent opening size (AOS) of geotextile by sieving glass beads through a geotextile. Apparent opening size is an indication of the approximate largest particle which would effectively pass through the geotextile (Table 3).

Table 2. Strength properties of geotextiles

	GSM 300	Poly 250
Test type	Tension – 10KN	Tension – 10KN
Sample type	Flat	Flat
Sample area (Sq. mm)	450	225
Peak load (KN)	0.20	0.15
Peak stress (N/Sqmm)	0.44	0.67
Displacement at peak load	90.73	73.59
Strain at peak stress (%)	90.73	73.59

Table 3. Apparent opening size of geotextile

	Sieve number	Bead size range (mm)
GSM 300	40–60	0.425–0.250
Poly 250	50–80	0.300–0.180

2.5 Large Scale Direct Shear Test

A large-scale direct shear test was performed as per IS 2720-13 (1986). A device having size 315 mm × 315 mm × 130 mm was used. The load applied by hydraulic jack moved through rigid frame and increases rigid load which is placed on top of soil. The normal load does not change throughout the test. The readings of proving ring, horizontal and vertical dial gauge had been recorded at regular time interval. Basically there are two types of corrections are applied.

Area Corrections: Shear stress (τ) and Normal stress (σ) are calculated as S/A and F/A Where S is shear force, F is normal force and A is the cross-sectional area of the sample. As the load applied by hydraulic jack increases, Area keeps on changing and where corrected area comes in action.

$$A_{corr} = A_0(1 - \delta) \tag{1}$$

Where δ is horizontal displacement on sample by shear force, A_0 is the initial area of the sample.

Dilatancy Corrections: The relationship between friction angle and normal stress is non-linear.

$$(\tau/\sigma) + (d_h/d_v) = \mu \tag{2}$$

$$tan\,\phi = \mu + tan\,\psi \tag{3}$$

Where σ is the normal stress; τ the shear stress; μ the friction coefficient of the particles sliding against each other; d_h the horizontal displacement; d_v the vertical displacement; ϕ the secant friction angle and ψ the Dilatancy angle.

3   Results and Discussions

1. The shear-normal stress relationship is linear and increase in shear strength is associated with decrease of column expansion and increase in normal stress illustrated in Fig. 1.
2. From Fig. 2 Maximum shear stress is obtained for a specific vertical confining stress.
3. From Fig. 3 at the initial stage, as the horizontal displacement is small, the reinforced soil undergoes vertical displacement.

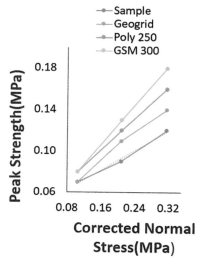

Fig. 1. Peak Strength (MPa) vs Corrected Normal Strength (MPa) for normal load of 1 ton

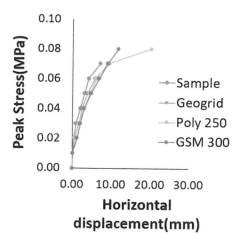

Fig. 2. Peak Strength (MPa) vs Horizontal Displacement (mm) for normal load of 1 ton

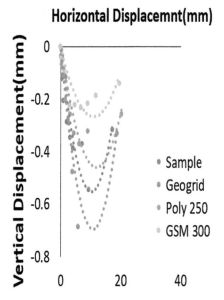

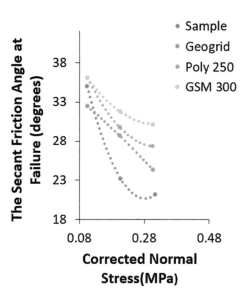

Fig. 3. Vertical Displacement (mm) vs Horizontal Displacement (mm) for normal load of 1 ton

Fig. 4. Secant Friction Angle (degrees) vs Normal Stress (MPa)

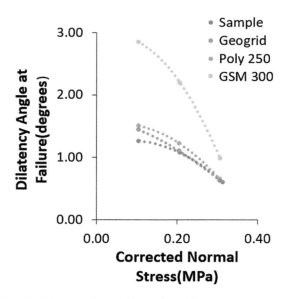

Fig. 5. Dilatancy Angle (degrees) vs Normal Stress (MPa)

4. The secant friction angle at failure decreases with an increase in normal stress. The relationship between friction angle and normal stress is non-linear with a quick decrease when normal stress is between 0.8–1.8 MPa and comparatively slower when normal stress is higher than 1.8 MPa as seen in Fig. 4.
5. The same as friction angle, the Dilatancy angle decreases as the normal stress increase from 0.1 to 0.3 MPa, however the rate of decrease keeps on increasing as shown in Fig. 5.
6. There are no possible reason from Figs. 3, 4 and 5 apart from, they follow same behavior in both reinforced and un-reinforced soil.
7. A lot of future research is needed to fully understand and describe the geosynthetics/ soil interaction as a composite material structure because of very high bearing capacity which cannot merely be explained by friction interaction of soil and reinforcing elements, but additional inter-locking effects with forces mobilized in front of cross bars.

4 Conclusion

1. The significant decrease in cohesion (c) of the soil was observed after using different geosynthetics in comparison with unreinforced soil. This decrement is due to the creation of shear resistance between the soil and surface, opening of geogrid and the bearing resistance provided by transverse ribs. The bearing resistance is induced when a relative moment occurs between transverse ribs and soil particles in apertures during processing of direct shearing.
2. There was not significant increase in angle of friction between the reinforced soils but there is a significant increase in angle of friction between reinforced and unreinforced soils.
3. Geogrid gives higher value of cohesion (c) and lower value of angle of friction (ϕ) as compared to geotextiles.
4. In the two geotextiles used GSM 300 gives lower value of cohesion (c) and higher value of angle of friction (ϕ) as compared to Poly250.
5. Secant Angle (Θ) and Dilatancy Angle (Ψ) keeps on decreasing as the Normal Load increases from 1 Tons to 3 Tons for a particular soil sample.
6. Secant Angle (Θ) and Dilatancy Angle (Ψ) are maximum for geogrid than Poly 250 and then GSM 300.
7. Failure plane is always horizontal and pre-determined which may not be the weakest. It starts at edge and progress towards the center replying non uniform stress distribution on failure plane.

References

Abu-Farsakh, F., Coronel, S., Tao, T.: Effect of soil moisture content and dry density on cohesive soil-geosynthetic interactions using large direct shear tests. J. Mater. Civ. Eng. **19**(7), 540–549 (2007)

Bauer, F., Zhao, S.: Evaluation of shear strength and dilatancy behavior of reinforced soil from direct shear tests. ASTM Spec. Tech. Publ. **1190**, 138–157 (1993)

Henry, F.: The principle of reinforced earth. Highway Res. Rec. **282**, 1–16 (1969)

Ingold, T.S.: Some observations on the laboratory measurement of soil–geotextile bond. Geotech. Test. J. **5**(3), 57–67 (1982)

Jesmani, F., Kashani, S., Kamalzare, T.: Effect of plasticity and normal stress on undrained shear modulus of clayey soils. Acta geotechnica slovenica **1**, 47–59 (2010)

Zomberg, J.G.: Performances of Geotextile-Reinforces Soil Structures. 1st edn. University of California, Berkeley (1994)

Liu, F., Ho, S., Huang, T.: Large scale direct shear tests of soil/PET-yarn geogrid interfaces. Geotext. Geomembr. **27**(1), 19–30 (2008)

Lopes, F., Lopes, S.: Soil and geosynthetic interaction influence of soil particle size on soil geosynthetic. Geosynthetics Int. **6**, 261–282 (1999)

Pasley, F.: Experiments on revetments, vol. 2. Murry, London (1822)

Richards, F., Scott, S.: Soil geotextile frictional properties. In: Second Canadian Symposium on Geotextiles and Geomenbranes, Edmonton, pp. 13–24 (1985)

Robert, M., Koerner, F.: Designing with Geosynthetics, 5th edn. Pearson Education, New Jersey (1933)

Wu, F., Wick, S., Ferstl, T., Aschauer, F.: A tilt table device for testing geosynthetic interfaces in centrifuge. Geotext. Geomembr. **26**(1), 31–38 (2008)

Sun, X.: Small and medium scale direct shear test of the Bremanger sandstone rockfill. Ph.D. theses. Delft University of Technology, Netherlands (2010)

Author Index

© Springer Nature Singapore Pte Ltd. 2018
L. Li et al. (Eds.): GSIC 2018, *Proceedings of GeoShanghai 2018 International Conference:
Ground Improvement and Geosynthetics*, pp. 491–493, 2019.
https://doi.org/10.1007/978-981-13-0122-3